EXPIRED:

COVID THE UNTOLD STORY

DR CLARE CRAIG

First published in 2023 by

Publishing Aloud Ltd

PublishingAloudLtd.com

A catalogue record for this book is available from the British Library

ISBN: (paperback) 978-1-7393447-0-2

ISBN: (ePub) 978-1-7393447-1-9

ISBN: (PDF) 978-1-7393447-3-3

Diagrams by Dr Clare Craig and Joel Smalley

Second Edition

ADVISORY

The information provided in this book is for educational and research purposes only. Every care has been taken in researching and presenting the medical information in the book (including that relating to mental health) but it is in no way intended to provide or supersede professional medical advice. Neither the author or publisher may be held responsible for any action or claim, loss, injury, damage or inconvenience caused or alleged to be caused, directly or indirectly as the result of information contained within it or its listed sources or any errors or omissions. Readers should obtain medical advice from a trusted and qualified professional pertaining to all matters relating to health including diagnosis and treatment.

Medical knowledge is constantly evolving and new information may become available and the information in this book may become outdated or obsolete over time. The author and publisher make no representation or warranties, either express or implied, with respect to the accuracy, completeness, currency, reliability or usefulness of the information contained in this book. The publishers and author have done their best to ensure the accuracy and currency of all information in *Expired* up to the date of publication. However, they can accept no responsibility for any loss or inconvenience sustained by a reader as the result of information contained within or any errors or omissions. If despite such efforts any inaccuracy should remain, they will correct any such typographical or other errors once notified on a subsequent reprint.

ACKNOWLEDGEMENTS

From the very beginning, I knew that writing non-fiction would be a monumental task, and my journey has proven me right. First and foremost, I am truly grateful to the courageous individuals who shared their personal stories, aiming to enlighten and educate others. The lively community on Twitter and the invaluable contributions from the HART group have enriched this work with their references, ideas, and engaging discussions.

I would like to express my heartfelt gratitude to Val Fraser, Martin Neil, Jemma Moran, my mother Jean Craig, my brother Robert Craig, and my husband Bryan Tookey for their invaluable feedback and support. To all those who have lent a helping hand along the way, your contributions have not gone unnoticed, and I sincerely thank you. You know who you are, and your assistance has truly made a difference.

Andrew Lockett has been an indispensable guide in both editing and navigating the intricate publication process. His expertise has been invaluable throughout this journey. I would also like to extend my thanks to Steff Cole for his exceptional work on the cover design, Algeron Waterschoot for his creative contributions to the AI artwork, and my typesetter for her meticulous attention to detail.

Finally, I must extend my warmest appreciation to my husband and four children. Their unwavering support, from taking on household chores to bringing a smile to my face each day (and graciously overlooking the occasional burnt sausage), has truly been the bedrock of my endeavours. Thank you, from the bottom of my heart.

CONTENTS

INTRODUCTION

Covid brought with it a new era of uncertainty. Assumptions had to be made about what might happen. Fear of a worst case scenario caused further exaggeration of these assumptions which were then presented as fact. After months and years of repetition these 'facts' became beliefs. The result has been two years (and counting) of most of our society living in a belief system based on myths. Welcome to Cloud-Covid-Land.

If we want to leave Cloud-Covid-Land, the first step is to understand how we arrived there. I have attempted to turn my diagnostic skills to focus on understanding the origins of each belief and where they conflict with reality. There are too many such beliefs to include in one book therefore this book focuses on those that relate to how the virus spreads.

It is possible to present counter arguments and claims, but rather than present alternative articles of faith, I have investigated the foundations for each belief, with a view to critiquing its evidence base. These foundations were often based on mistakes and careless regard for secure evidence. I attempt to show how these errors and the oversights that led to them were uncovered. I explore the major areas of conflict and how certain beliefs became not only widely embedded but placed on a pedestal such that they could no longer be questioned.

1

The book is called *Expired* for it encompasses all three meanings of that word. Firstly, covid and the response to it is a story about people taking their last breath. (I have decided not to capitalise covid just as we do not capitalise malaria, measles or flu. It does not deserve it). Certainly, there were deaths from the virus but there were also deaths from the disproportionate response to the virus, which itself arose partly from an overwhelming fear of death. Secondly, damaging mistakes were made because of people clinging onto untenable hypotheses past their shelf life and long after evidence had shown them to be misguided. Thirdly and fundamentally, this is a book about an airborne virus (*SARS-CoV-2*), the implications of its spread through expired air and the failure to acknowledge them.

Debates about covid have been polarised. Nuance died. The belief, on both sides of the debate, that lives were at stake, has meant that passions were raised and extreme positions were adopted. Covid has created immense division even among friends and within families. I explore how it was that simple narratives have taken hold and how we came to be so divided. The most polarising debates, such as those around, for example, abortion, happen when both sides believe they are arguing to prevent the most harm. One side of the covid debate believed every intervention was saving lives, while the other focused on the harms caused by interventions, and so irreconcilable differences were inevitable. It is important to take a step back and realise that everyone involved shares the common objective of wanting to minimise harm. That should be our starting point. From that shared understanding we can look together at the evidence to determine the best way to minimise harm overall. It is important to do so now to ensure mistakes are not repeated in the future. It will always be unhelpful to view others who disagree with us as agents of harm when they did not have that intent.

An understanding of Cloud-Covid-Land requires more than a rehashing of the scientific debate. First, it is important to understand the cultural, psychological and philosophical aspects to our own belief systems. Although everyone, especially scientists, like to think we dispassionately and logically weigh up the evidence before deciding what to believe, the reality is not so virtuous. Our prior beliefs determine what else we *can* believe. Even when there is conflicting evidence, rather than admit uncertainty, or adjust and replace our narratives, our default response is to pick a side. Changing our position means admitting mistakes (and perhaps gullibility) and most of us do not do that readily.

As well as this discussion of the scientific debate and how it evolved, I cover the factors that impact what we choose to believe, how we change our minds and what can influence those decisions. I have illustrated how beliefs can change with examples from my own changing opinions about covid. Fear played an important role in how our decisions were made and what we elected to believe at any one time. I present evidence for how that fear was generated as well as the impact it has had and was designed to have.

A scientist must be able to suspend their disbelief in order to rationally consider a hypothesis and judge it solely on the evidence available. Our ability to think rationally can be handicapped by our emotional responses and existing beliefs. When these are formed in an environment of fear and especially when they become part of our identity, our ability to challenge them rationally is strained. When authority figures constantly repeat false 'truths', myths are born.

Expired: Covid the untold story addresses twelve of these beliefs related to viral spread, extending to the lockdowns, asymptomatic infections, the efficacy of masks and more. Beliefs related to the spike protein covering the origin, treatments, vaccines and long covid will be covered in a second book, *Spiked: A Shot in the Covid Dark*. With each belief there is an area of conflict and evidence from both sides will be presented and critiqued. I will explore why people held differing assumptions based on the evidence available and the impact those beliefs had. Discussion around the issues raised is presented in chapters between the exploration of each belief.

From 2020, there was an inversion of ethical principles. The perceived needs of the elderly were prioritised over those of children. Governments enforced policies that were certain to result in harm with minimal opposition or even debate. Human Rights were disregarded. Laws were written that interfered with the minutiae of daily living, from the most intimate, like being able to say last goodbyes, to the most mundane, like sitting on a park bench. Huge amounts of money were spent on interventions that had no impact. A system of discrimination against a minority based on their health decisions was introduced and adopted across many nations. People responded in an unexpected manner: they outsourced their thinking to fact checkers, their morality to legal guidance and their bodily autonomy to the state. Leaving Cloud-Covid-Land means reinforcing ethical principles that have underpinned western societies for centuries and laying bare the damage that was done by overriding them.

The beliefs in Cloud-Covid-Land were sustained through the adoption of many attributes by which religions are maintained and there are references to these parallels throughout the book. We will meet the false prophets, high priests and puritans who drove the belief system uncovered in this book and go on to hear about blasphemers, witch-hunts and the worship of a saviour vaccine in *Spiked* but the themes and alignments to religions will be introduced here.

The book unavoidably presents a narrative of what happened. While every effort has been made to ensure its accuracy, it remains possible that it is imperfect. Over time more evidence may emerge that shows that the interpretation in this book was incorrect in places and new explanations will be constructed. I welcome that. That is how science progresses.

As a diagnostics expert my first concerns emerged around covid testing. I spoke out about them in September 2020. At the time I still believed many aspects of the official narrative. After that point I discovered other niche specialists had concerns around the handling of covid from their particular area of expertise and I started listening and investigating.

To fully understand covid requires an understanding of many disciplines including epidemiology, infectious diseases, virology, immunology, diagnostic testing, diagnosis of death and even physics. To understand all these aspects takes a large amount of work. Part of training in diagnostics is to be able to stop yourself from seeing what you expect to see and learn to observe dispassionately what the evidence says. Training also ensures that we remember that there is a real person behind every report and every number. After six months of working tirelessly, reading as widely as I could, I felt I started to understand what happened in spring 2020. Much more work was to follow to ensure I kept updated with subsequent developments.

I hope that this book summarises what I have learnt so far and provides a much more efficient way for others to understand what happened. I do not claim to cover all that has taken place but I trust the main issues regarding airborne long distance transmission and its implications have been addressed and that it may present a good, nuanced summary of *what* happened without it being a full chronological tale of *when* it all happened (see timeline at end of book). Rather than examining all the aspects of *why* events unfolded as they did, I focus instead on *why* some of the assumptions informing policy decisions proved in the end to be false.

Similarly, I have not attempted to include a roll call of all the people who played an important part in investigating or exposing the issues as that would be too long. I apologise for all those people who are not credited for their important work here. I have also not attempted to catalogue the conflicts of interest among officials and others responsible for bad decision making, which could in itself fill several books.

This book is not a self-help guide, rather it is a resource providing the evidence, explaining where it does not fit what has been the narrative presented by the authorities and mainstream media and proposing a clearer account that better fits the data we have. I don't claim – even remotely – to provide here all the answers nor do I profess great literary talent, but I do hope I present insights that many people have not encountered, and I hope too that they are considered to be worth sharing.

There will be a few statistics but the number crunching work can be found elsewhere whereas this book summarises the findings from that work. Although *Expired* is a book about science it is told through stories. There will be plenty of metaphors and we will touch on giants, bank robberies and Great Danes. Along the way we will find loneliness, poverty, errors and death but there will also be bravery, solidarity and successes.

Let us begin with fundamental aspects of covid itself that are accepted truths and can mostly be agreed upon. There was spread of a virus called *SARS-CoV-2*, globally first noticed in 2019. The virus causes a disease - covid which, like *SARS-1*, caused an acute respiratory syndrome. It could also cause the life threatening overreactive immune response known as a cytokine storm and a propensity to clot in both small vessels and larger ones leading to life threatening complications such as blockage of the vessels supplying the lungs. People died because of this. I realise there are people who dispute the evidence that there was a viral induced illness and I will address these points in *Belief Six: If you test positive you have covid.*

From these foundations we can begin to explore aspects of the disease that are more complex, including the science behind how it spreads, the human immune response to it, what a positive test result really means and more. If you are prepared to accept that some of what you've been led to believe was inaccurate and ready to hear an evidence-based appraisal of covid and its consequences, then read on and prepare to rethink some of what you thought you knew.

I'M A BELIEVER

Without a huge national effort to halt the growth of this virus, there will come a moment when no health service in the world could possibly cope; because there won't be enough ventilators, enough intensive care beds, enough doctors and nurses. **Boris Johnson, March 2020**[1]

In March 2020 I was scared. Every day I read BBC news. I was fully brought into the fear created and the story of a novel virus, originating in Wuhan, about which we knew nothing and which had the potential for a catastrophic outcome. Through January and February covid was just a news item but by March there was a real sense of dread emerging and it had become a frequent part of conversation with friends, colleagues and family.

Fear was certainly proving contagious. This started with images from China of young people dropping dead and others taken away by teams in Hazmat suits. Video footage illustrating the story of hospitals struggling to cope were shared on social media and some were even beamed into our houses via mainstream TV. In retrospect some of these look ridiculous. The Chinese video of the police trapping a man in a butterfly net had the words 'police' written on the vehicle and 'SWAT' written on the uniforms prominently in English rather than Chinese, making clear who the intended audience was. One of the men collapsing to the ground was unable to prevent his natural instinct to break his fall by stretching out his arms at the last minute.[2] When I first saw these videos I believed them.

Images were also shown, courtesy of the Chinese Communist Party, of doors to apartment blocks being supposedly welded shut to stop people leaving. People outside of China were aghast such events could happen and trusted that our governments would never condone such actions. Reports out of Wuhan claimed that thousands of patients had to queue for hours for treatment and hospital staff were too busy to have toilet breaks and had to wear nappies.[3] Images from Bergamo of full hospitals and the army moving bodies were really distressing. The fear induced was sufficiently overwhelming that within weeks, one by one, every country except Sweden and Belarus locked down, with little protest.

Despite the fear, from early on it was clear that even the worst case scenario would not be as bad as the feared and planned for pandemic flu and that the young and children were generally spared. That assuaged some of my fears. But fear doesn't disappear thanks just to reasoning. I was addicted. Fear results in the need for a next 'hit' such that the fearful will feel a compulsion to repeatedly assess the potential danger and would be consequently reinforcing the foundation of an addiction. Like all addictions it can never be satisfied. Every day I would tune in to hear the news hunting for information to reinforce this emotional belief. The warning that there would be an announcement from the Prime Minister on 23rd March 2020 ramped up the anxiety before the fear hit was delivered that evening when the first lockdown was announced.

LOCKDOWN

I first heard the term lockdown when I lived in New York City. My children were given lockdown drills intended to prepare for the school being attacked by a gunman. We had arrived midway through the school year so were new to the country as well as the school. The first time there was a drill, my eldest was in the bathroom and had no idea what she was meant to do. A teacher grabbed her and flung her into a nearby class of younger children who were cowering in the corner of the room, out of sight from the door. Lockdown drills in New York City schools were lessons in how to cower. Now, here we were back in the UK being instructed to cower from a virus.

But cower we must. Surely reducing contacts would indeed slow the spread and stop the NHS from being overwhelmed. I felt uneasy about schools being shut but, as it was only for three weeks, I did not voice that concern to friends or teachers. My youngest son, who was nine, was already sick with covid so neither of us waved goodbye to friends at the school gates. Just three weeks. We could get through that.

As we all cowered, the sound of the city hushed making the increase in the numbers of ambulance sirens more noticeable. Each passing siren was a further reminder of the fear and a momentary confirmation that our decision to cower was worthwhile. Our thoughts were with the people in those ambulances and all the staff who were not cowering but facing the dangers daily as they cared for the sick. The fear instalments from the news were closer to home now with headlines reporting daily the deaths of healthcare workers and bus drivers as well as occasional previously healthy young people.

DEATHS

Deaths began to rise in almost every town in Western Europe, then the number of fatalities surged. They rose from zero first gradually then rapidly and after four weeks of checking the numbers every night they finally began to slow. A full week went by in early April when the deaths were plateauing and each night we would check, hoping the deaths had fallen before finally the day arrived when they were lower than the previous day's. That was a critical week. In Wuhan the lockdown ended but elsewhere it was the week of peak fear. The prime minister left intensive care on Thursday 9th April, and spent three more days in hospital before being discharged. That was also the week that the *"three weeks to flatten the curve"* came to an end. Opening up at peak fear would take political bravery. There was none.

Deaths were higher in more densely populated areas and areas with economic deprivation. My elderly parents lived in central London and I worried about them being exposed. Common patterns emerged globally for risk factors for covid death with certain groups being at increased risk including the old, males, the obese, hypertensives, diabetics, the economically deprived and certain ethnicities. Our understanding of risk factors seemed to advance much faster than any advances in treatment.

NHS

I fully believed both the argument that intensive care might be overwhelmed and the strategy of trying to slow the spread to avoid a dangerous peak in pressure on resources. I even went as far as buying an oxygen concentration machine should anyone I know not be able to access medical care. It was made in China, bought through eBay, didn't arrive until May 2020 and has never worked.

When playing, *What was your most embarrassing covid behaviour?*, this is my winning entry.

In those first weeks there were some phenomenal achievements thanks to the immense efforts of so many people. A testing system was established from scratch and scaled sufficiently to deliver testing in hospitals. Capacity for covid testing was large enough that there was never a time when those being tested were more likely to test positive than negative. That means there was sufficient testing to include not just patients who could be diagnosed clinically but also plenty of patients where there was much more uncertainty as to the diagnosis. This is a good indication that, within hospitals, there was sufficient testing to find all the genuine cases and, given how quickly it was rolled out, it was a definite success.

Nightingale hospitals were established from scratch and staff recruited to fill them. Even if this was more hugely expensive political spin than practical solution, it was a phenomenal achievement. What was not anticipated was that patients sick enough to need hospital care for covid almost always had other medical issues. The Nightingales were designed to manage patients who needed respiratory support only. Only mainstream NHS hospitals had doctors from all major disciplines who could contribute to the care of these complex patients. Very few covid patients qualified as being pure respiratory failure patients who could use a Nightingale hospital. That, and a lack of staff, is why Nightingale hospitals were barely called upon.[4]

I believed then and I believe now that many NHS healthcare workers were being brave and working hard to save lives. I had little recent frontline experience but along with 20,000 other former NHS staff, I volunteered in case I could be of any use.5 On Thursday nights, I stood with my children on our balcony 'clapping for carers' and wondering if I would be called up to help. I never heard anything from the volunteering system and nor has anyone I know.

HOME

Aside from clapping on the balcony, and one son being infected, covid was not a part of day to day conversation for our four children. The prospect of home-schooling and being away from friends brought more fear than the fear of covid. Home-schooling was not new to our household. While in New York there was a six week period after we arrived when we had not yet secured a school place. Even once we had an address and could apply, my children had to prove they had chicken pox immunity and receive a full three dose course of Hepatitis B vaccine (which was not offered in the UK) before being allowed in the school. (The risk from Hepatitis B, which is bloodborne, comes from sharing needles and sex, so I am not sure what the school feared my children might be engaged in while unvaccinated).

Thanks mainly to my utterly distorted idea of how much a child can learn in a day, home-schooling had not been a happy experience for any of us at the time. In lockdown I aimed to have very limited ambition with only a few hours a day of school work, which made it less painful.

At the time I was working for a small diagnostics start-up and managed to set up a corner of the sitting room as a workspace with my computer and a large monitor for viewing cancer cases on a fold away table. To juggle work and teaching the children, I started at 6 a.m. and managed to squeeze in a couple of two-hour windows with my children, predominantly my youngest child, who did not manage to do much independent work. My husband used a desk in our bedroom and between us we just about managed to keep everyone sane and the teachers happy. We were hugely grateful our children were old enough to be relatively independent.

Although we had a paid subscription for supermarket delivery, we were aware of the pressures on delivery slots because of those trying to shield and we opted to join the socially distanced queues outside the supermarket that stretched across the car park, often taking an hour to reach the front. Shopping and walking the dog were a chance to escape the confines of the flat and the monotony of each day. Mealtimes and press conferences, which we tuned into religiously, were the other breaks from boredom.

So many neighbours volunteered to help those shielding that I was assigned to collect medicines from the pharmacy for one neighbour while someone else delivered her food. We messaged and spoke but I never saw her face as she was too scared to open her door. My brother helped in this way for my parents and ordered shopping delivery for them. From the outset my parents could see that lockdown was going to be longer than three weeks for them and they went into it with even more dread than we did. Their lives were already heavily restricted through disability and they lived for their brief contacts with the outside world.

After the three weeks passed and with a lack of any end in sight I started to be concerned about the children missing out on learning for so long. It also felt very wrong for them to be separated from their friends for such a long time. It would be another year before covid policy affected my daughter's exams and my son's last year of primary school. However, we were very aware that we'd had a relatively easy lockdown and very grateful that no loved one's life or major life events had been interrupted by it or covid.

There was no lack of frustration in our house at how slowly the lockdown ended. After over 100 days of lockdown, shops were allowed to open again in most of the country but schools remained shut to all but the most in need and key workers. My doctor friends did not have to juggle home-schooling the way their patients did. Even though the daily deaths dropped to low levels the atmosphere was still abnormal. There was much talk of carefully reopening to avoid a rebound in infections. Caution and responsibility were repeatedly mentioned. Come May, after (foolishly) rejecting an offer of furlough pay, I stopped work and the pressures became more manageable. By the end of July, masks were required in shops and I selected some fabrics I liked, lifted a sewing machine onto the fold away table and started sewing. I was still in Cloud-Covid-Land.

BEING RIGHT AND FEELING OMNISCIENT

On the whole, our indiscriminate enjoyment of being right is matched by an almost equally indiscriminate feeling that we are right. Occasionally, this feeling spills into the foreground, as when we argue or evangelize, make predictions or place bets. Most often, though, it is just psychological backdrop. A whole lot of us go through life assuming that we are basically right, basically all the time, about basically everything: about our political and intellectual convictions, our religious and moral beliefs, our assessment of other people, our memories, our grasp of facts. As absurd as it sounds when we stop to think about it, our steady state seems to be one of unconsciously assuming that we are very close to omniscient. **Kathryn Schulz, 2011**[6]

To understand belief we need to understand the brain. A simplified explanation of how the brain handles belief would sound like this. When we are presented with new information the brain metaphorically tags it before storing it. This can be simple, for example, is it good or bad, a truth or a lie? Once tagged it will fit neatly into our existing belief system and the satisfaction of having learnt something new will be felt. Information that does not fit our current beliefs falls into two categories. The first is the basis of humour and is associated with a surge of pleasure as we notice how it clashes with what we know. It is tagged as 'funny'

and we move on. The second leads to the uncomfortable feeling of holding conflicting beliefs – cognitive dissonance – where tagging and storing cannot happen.

An important factor that contributes to our feeling of omniscience is confirmation bias. We have a tendency to hunt out, focus on and recall information that confirms our pre-existing beliefs. This bias can further entrench our existing views, making it even more challenging to acknowledge and address conflicting information.

Unfortunately, holding and accepting conflicting ideas, while possible, is hard work and uncomfortable. It is like being handed a piece from the wrong jigsaw puzzle. A conscious choice must be made. Either the pain of cognitive dissonance must be accepted while working to a deeper, more complex understanding. In this case we would hunt for more jigsaw pieces from the other set and see if we can build a new puzzle to replace the existing one. Alternatively, we can take the hit of pleasure from labelling the contradiction as false and continuing to believe what fits into our current belief systems. In that case we just throw away the piece from the jigsaw that would not fit in. Tagging something that is correct as if it were false, without checking, is a shortcut to comfortably continuing on with our invalid beliefs.

By far the easiest solution to cognitive dissonance is to cling to our current beliefs. Even as conflicting information keeps coming it can be treated as false and dismissed. The alternative, of accepting a conflicting fact, requires hard work. The complex web of our belief system must be unpicked and then rebuilt to accommodate this new information. Take for example when someone has fallen in love with an unsuitable person and is presented with new evidence to that effect. Such evidence does not fit into their web of beliefs about their lover and is ignored. However, a perplexing tipping point comes where the evidence regarding their lover is overwhelming, the person admits they were wrong and the healing can begin. The process of recovering from love is, as we know, painful. Most of us are familiar with how that plays out. It takes time and hard work to rewire our thinking. Each mistaken belief must be corrected. Each piece of evidence that showed we were wrong must be taken on board and worked through, not least so that the same mistakes cannot happen again.

The emotional capture of being in love is a fertile ground for being wrong. Infatuation brings with it interruptions of intrusive thoughts over which we can obsess. There are aspects that are addictive and we look for affirmation that we are right to be so infatuated. It is claimed that infatuation lasts up to three years but for a few feelings something like infatuation can last for decades but without the anxiety. A similar emotional capture is seen with fear.

Fear can be as powerful as falling in love. When we are overwhelmed with fear, obsessive thinking is a healthy reaction, because constant assessment of danger when we are at risk protects us and so recognising when the danger has passed allows us to conserve our energy. Consequently, when in fear we actively hunt for information to assess the situation. Assessing the threat of an invisible enemy, however, leaves us open to manipulation. Like our blindness to flaws when we fall in love, when we are fearful we can be blind to reassuring news and focus only on information that inflates the fear. The constant pushing of fear narratives by the government and mainstream media and the incessant repetition of incorrect ideas, has massively exaggerated this problem.

Infatuation and fear are both so powerful that our usual systems for getting things right go awry. Open-mindedness, curiosity and checking our emotional response are all victims of our need to feed the addiction to our emotional state. The hunt for new information is not done with an intent to learn so much as an intent to justify our infatuation or our fear. Even when we are fully aware that this is what we are doing, the temptation is still to believe that everything we are feeling is right.

When we are not emotionally captured with love or fear then as individuals we can correct for our lack of omniscience through listening to multiple other voices and opinions. As a society we had multilayered systems designed to correct for errors. These ranged from formal rules of free speech which aim to ensure that a full range of opinions can be heard, through to science which tests ideas and democracy. All of the above require debate from opposing voices whether they be political parties and the media or just colleagues or neighbours. Opposition and challenge is our means of error correction. All of these mechanisms can appear noisy and messy to outsiders. They may make those in power feel weak. But together they have enabled the flourishing of democratic free nations because efficient error correction stops disastrous mistakes. They are not a weakness; they are our greatest strength.

Error correction ensures that the knowledge base from which decisions are made is sound. In democratic societies we also have defined ethical principles to guide those in power as to the right course of action. In the last two years these error correction mechanisms were broken, including through censorship of scientists and parliamentarians deciding they were not essential workers and making themselves redundant! In addition, ethical principles like the duty of adults to protect children were set aside. Even principles that had been firmly established in international agreements such as that of informed consent and bodily autonomy, were disregarded without thought to the consequences of such action. There seemed to be a belief that ethical principles only applied outside of emergencies. To fully leave Cloud-Covid-Land may take decades while we attempt to rebuild these error correction mechanisms and ethical principles are strengthened such that they cannot be blithely ignored in future.

As an individual leaving Cloud-Covid-Land the first step is to alter your relationship with the media and the fear narrative. People first need to be able to think rationally to take on board the evidence. Only then will it be possible to understand the misrepresentations and mistakes that form the basis of much covid belief. The first step is to quell the fear. The perceived danger, whether from the virus or the response to it, can be inflated by fear and understanding that is fundamental to being able to approach the problem rationally. Behavioural changes such as establishing a routine, eating healthily, exercising, immersion in nature, sleep, social engagement and relaxation are all seemingly trivial parts of life but together they can provide a powerful barrier to anxiety and fear. Reducing the hunt for negative news about the virus or the response to it and forming habits to correct intrusive thoughts are also needed. Finally, self-compassion is essential: be kind to yourself.

If you want to leave Cloud-Covid-Land then you must be prepared to work through multiple aspects of cognitive dissonance to enable a new web of beliefs to be built based on the accumulated evidence from real world data. It can be emotionally draining and tiring but there is no reason why it should not also be interesting and fulfilling.

Non-academics typically deal with cognitive dissonance less often. Therefore, trying to untangle the myths and rethink beliefs can be incredibly difficult. Unfortunately, the truth may present more questions than answers. It might not leave you feeling omniscient but may mean you are more likely to be right.

BELIEF ONE: COVID ONLY SPREADS THROUGH CLOSE CONTACT

> *When someone with the virus breathes, speaks, coughs or sneezes, they release small droplets containing the virus. You can catch COVID-19 if you breathe in these droplets or touch surfaces covered with droplets.* **NHS Website, December 2022**[7]

The changing official narrative about how covid spreads takes in the full breadth of scientific mistakes and illustrates why humans make them. Scientific errors arise when scientists *extrapolate* too far from what they can prove; make *excuses* for conflicting evidence and *exclude* certain facts from their thinking altogether. In the case of *SARS-CoV-2* transmission the whole theory was extrapolated from a mistake. Officials have subsequently conceded the mistake[8] but not accepted the implications of that nor done anything to educate the public about it as doing so would expose them as having been previously wrong. In this case the mistake was to think covid only spread through close contact. Everyone now accepts it can also travel through the air.[9]

BEFORE COVID

The slogan used in the 1918 flu pandemic was:

"Cover up each cough and sneeze. If you don't, you'll spread disease" [10]

One hundred years later the advice had not moved on. The government's plan for any emergency, the "UK National Risk

Register Of Civil Emergencies 2017" planned a campaign around the use of tissues:

"Catch it. Bin it. Kill it." [11]

This slogan was adopted in the government plans to manage covid in the UK in March 2020, within its self-described 'battle plan' to flatten the curve.[12]

In 1918, germ theory – the idea that infectious diseases are spread through micro-organisms – was several decades old and, rightly, accepted wisdom. If the infected were the source of the virus and the disease made them cough and sneeze then it would stand to reason that the cough and the sneeze would be the source of the infectious material, right? Keeping a distance and reducing the number of contacts people had would therefore stop the virus in its tracks.

By the 1930s and 1940s, our understanding of what happens when we cough or sneeze moved forward dramatically with the use of high resolution cameras. The smallest drops emitted while breathing still could not be seen and these new pictures were limited to only one or two metres away from the source but the results were nevertheless illuminating and showed that spread of respiratory infections was not due to droplets from coughs and sneezes alone but smaller aerosols that were breathed out and remained suspended in the air.

Somehow these findings were lost in history, as we shall see. Thanks to scientists being wedded to old ideas and a dash of human error, public health guidelines were based on the idea that the infected are constantly depositing almost all infectious material onto the ground in front of them rather than into the air. It is an odd belief to hold but it was firmly established for many years.[13]

For example, pandemic planning was based on influenza being spread predominantly by droplets which would *"fall rapidly to the ground."* Before getting on to the details of the mistake it is important first to understand a bit about spread.

MYSTERIES OF TRANSMISSION

The story starts in the nineteenth century with the battle of ideas between miasma theory – the idea that disease was spread in air as bad smells – and germ theory of microorganisms causing the spread of infectious disease. The miasma theory had existed for hundreds of years with people attaching metal spheres full of pungent scents called pomanders to their belts and carrying 'pockets full of posies' which were fragrant flowers to protect the wearer from bad smells. The word malaria itself comes from the Italian for bad - 'mala' and air – 'aria.' The work of Koch, Lister, Pasteur and others finally falsified the miasma theory proving that microorganisms were the source of infectious disease and germ theory was established. This debate from 150 years ago was still influencing decision makers in the covid era.

It is naive to assume we fully understand transmission of any respiratory virus let alone one that has only recently been studied. But before coming on to covid, it is important to first understand more about what is known about influenza transmission. Despite hundreds of years of trying to understand influenza there remain many mysteries in its transmission. These numerous paradoxes mean the model of close contact transmission cannot be the only way spread occurs.

Dr Edgar Hope-Simpson first started working as a GP in 1933, and carried out extensive research into how influenza spread. He worked from an 18th century cottage on Cirencester High Street where he based his GP practice. After securing public health funding, he named the cottage the "Epidemiological Research Unit." He dedicated his life to the study of influenza and highlighted a number of well-established phenomena that did not fit the conventional model of disease spread seen for other respiratory viruses. For example, he noted how for hundreds of years there were reports of influenza[14] outbreaks in boats that had been at sea for weeks with no external human contact, making close contact transmission an impossibility in those particular cases.

While believing that the accepted model was over-simplified, he acknowledged that close contact transmission played an important role with epidemics beginning in towns and spreading later to more rural areas. Unlike others, he emphasised the importance of the susceptibility of the host and the role of immunity.

Influenza waves he noted, would begin simultaneously, with the same genetic variant, in locations at the same latitude but continents apart which could not be explained by close contact transmission. Hope-Simpson focused particularly on how influenza outbreaks would occur that were genetically identical to outbreaks from years earlier. The expected mutations that should have been evident if there had been an undetected chain of transmission were not there. It was as if these viruses had been frozen in time and then brought back to life.

Certain influenza mysteries have marked parallels with *SARS-CoV-2* and other viruses. An outbreak of the common cold was reported in 1973 in the British Antarctic Survey base with six out of twelve men affected a full seventeen weeks after their last contact with others.[15] In December 2021, an outbreak of covid occurred on the Belgian Antarctic base.[16] The outbreak occurred one week after the researchers arrived despite them being vaccinated and having no contact with others at that time. The researchers tested negative before flying to South Africa, isolated in South Africa for ten days, with two more sets of negative tests and tested negative again five days after arrival, yet still the outbreak occurred.

Other covid mysteries include an outbreak of a thousand cases diagnosed within two days of each other in a garment factory in Sri Lanka, without an identified super-spreader, at a time when there was minimal community covid.[17] Finally, an Argentinian fishing vessel had an outbreak after a full five weeks at sea, despite everyone testing negative before setting sail.[18]

Public Health scientists failed to show any humility and preferred to ignore these phenomena entirely than admit that we cannot explain them via a close contact model. The tragedy is that these are precisely the types of mysteries that should pique the interest of curious scientists allowing us to expand our knowledge through investigation.

One international group of biologists did game changing research in this area which was published at the beginning of 2018.[19] By sampling air one to two miles above sea level in the mountains of Spain, they demonstrated that billions of viruses (and bacteria) could be collected each day. These viruses were contained in smaller aerosols than bacteria and lingered longer in the atmosphere. Separately, viruses collected from the air above the sea, travelling miles in the wind, have been shown still to be capable of causing infections.[20] The biologists from the Spanish mountain study concluded that viruses are capable of long persistence and dispersal sufficient to explain outbreaks of identical genetic sequences occurring simultaneously in distant corners of the globe. It would also explain outbreaks seen at remote locations. Exposure through the air is therefore a possibility that has an evidence base and its role needs to be taken seriously.

SMALL SIZES

The key to understanding transmission lies entirely in getting to grips with tiny sizes and large numbers. How *SARS-CoV-2* spreads is critical to how we might respond to it so it is worth spending some time understanding it better. People are not good at visualising large numbers, or small sizes particularly those beyond our day-to-day experiences of the world. Together these two challenges make understanding the spread of a virus hard for people to grasp.

In order to understand size better, and seeing as this is a book about myths, I want to take you into a land of giants. Everything will be doubled in size ten times over, making it a thousand times bigger than it is now. Let us start with the size of the virus. It would just reach a size big enough for us to see – like the tiniest grain of salt. Imagine the virus was floating in the air in your nostril. Your nostril (front to back) would be the size of a bus. A three storey building would be taller than Everest and a one year old would be the size of the tallest building in the world, the Burj Khalifa. Viruses are tiny, even among things too small to see. On this scale pollen would be somewhere between the size of a grape and a regular tomato.

Why does size matter? When someone expels virus in a sneeze, cough or just through breathing, it leaves them in fluid droplets of various sizes. The size of the droplets determines what happens next and most importantly how far they can travel from the infectious person. The failure of public health authorities to understand the tiny sizes involved has led to an embarrassing mistake with big implications for their recommended covid policies. It's a mistake that has barely been talked about.

What was known? The public health and infectious disease scientific community had a belief that there were two ways in which respiratory viruses transmit – in most cases via large droplets but specifically for measles and tuberculosis via fine droplets called aerosols. (An aide memoir would be that droplets drop rapidly to the ground while aerosols remain longer in the air. However, the mistake as to where the cut-off between the two lies means this labelling does not work as intended). The cut-off between what was described as a droplet and what was fine enough to be counted as an aerosol was five microns (the size of a lentil in the giant analogy). The mistaken belief was that when someone coughed, sneezed or talked, everything emitted that was larger than five microns (a lentil) would fall to the ground within two metres (hence calling them 'droplets'). Therefore, spread of most respiratory diseases could be managed by ensuring this two metre or six foot danger zone was dealt with. The belief was firmly held, not least because it was what the textbooks and public health guidance said, but it was wrong.

HISTORICAL DETECTIVES

Physicists who specialised in aerosols and disease transmission and others struggled to get their voices heard by the public health authorities. They knew that aerosols much larger than five microns could remain suspended in the air, travel significant distances and be inhaled.

Physicist Linsey Marr, originally worked on air pollution. It was only when her eldest child started attending childcare that her interest shifted to respiratory disease spread. She noticed how hygiene measures seemed to have little impact on the spread of coughs and colds between the children and started reading about infectious disease transmission, wondering if the source of infection was in the air. As she read the textbooks she kept coming across the same, often unsourced, claim that all material larger than 5 microns would drop rapidly to the ground. She said, *"I'd see the wrong number over and over again, and I just found that disturbing."* [21]

To test her hypothesis she (and her colleagues) set up air samplers in childcare venues, aeroplanes and a health centre.[22] She demonstrated that each cubic meter of air contained thousands to tens of thousands of viral particles, the majority in tiny aerosols that could remain airborne for hours. Her manuscript describing these findings was rejected for publication by the major medical journals. She, like other physicists and engineers who were also experts in the mechanics of aerosols, found themselves caught between disciplines. Conferences on infectious disease transmission would be dominated by epidemiologists and physicists were not invited to share their knowledge on aerosol spread. According to Trisha Greenhalgh, Professor of Primary Care at Oxford University, aerosol scientists were *"systematically excluded from key decision-making networks and committees."* [23] Perhaps it was Marr's first-hand experience of the difficulties sharing her work that made her so determined to find out how long the infectious disease community had been incorrect about the evidence on aerosols plummeting to the ground.

A breakthrough came when Marr teamed up with other physicists including Jose Jiminez and Lydia Bouroubia and historians Katherine Randall and Thomas Ewing to set about trying to understand the foundation of the myth.[24] Randall, between juggling a graduate dissertation and a six-year-old daughter in 'remote schooling' because of covid, discovered the work of William and Mildred Wells, a Harvard engineer and his physician wife.[25] Together the Wellses showed that the ability for different sized droplets to remain suspended in the air was dependent on the balance between gravity and evaporation. Their experiments predicted the duration of suspension. Any droplet larger than 100 microns (the size of a grapefruit in the giant analogy) would sink to the ground within a second and therefore could only travel a limited distance.[26] Droplets smaller than that would be affected primarily by evaporation rather than gravity. As the liquid evaporated, the aerosols would become lighter still.[27] Gravity no longer pulled them rapidly to the ground; they remained suspended in the air for extended periods subject to air currents.[28] The Wellses had discovered evidence of a size of droplet for which spread beyond a short distance would not be possible – a grapefruit for the giants and 100 microns in real life.

The theory of using social distancing to stop such spread therefore had an evidence base. But the size of the particles that would behave in this way (grapefruits) were in a different league to the ones in which most *SARS-CoV-2* particles are emitted (lentils and smaller).[29]

When the Wellses carried out this work in the 1930s and 40s they struggled to find acceptance of their findings, not least with Alexander Langmuir the first Chief of Epidemiologic Services at the CDC (the United States Centers for Disease Control and Prevention). As with many errors, this was the result of humans being wedded to an idea in an emotional way. Germ theory was the bedrock of infectious disease knowledge and therefore any theories around transmission through the air that resembled miasma theory were dismissed out of hand. The work of the Wellses did not contradict germ theory but it was dismissed as if it did. It was not until the 1980s, in a retrospective of his work at the CDC, that Langmuir admitted the Wellses had been right – aerosol spread was real.[30] The difference between droplets and aerosols was incorporated into public health guidance and textbooks – but the wrong number was used. The definition of a droplet should have been anything larger than 100 microns (grapefruits) but a 5 micron (lentil) cutoff was used instead. The claim that anything larger than a lentil would plummet to the ground was born.

How did this mistake happen? Further detective work led Randall finally to the origin of the five micron (lentil) size cut-off. After their work had been dismissed by the CDC the Wellses carried on with their experiments. Experiments on tuberculosis had shown that aerosols larger than five microns would be filtered in the upper airways or caught in mucus and only those of five microns or less would penetrate the deeper lung and be able to cause infection.[31] The same cut-off was not thought to apply to other respiratory viruses, like *SARS-CoV-2*, which can infect cells in the upper respiratory tract. So whether a droplet was larger or smaller than five microns was meaningful in terms of infections in tuberculosis (and measles) but it was not meaningful in terms of droplets falling obediently to the ground.

Publications by William Wells included the importance of this five micron (lentil) size cut-off for successful infection in tuberculosis. Somehow, this number was conflated with his earlier work such that textbooks on the subject were wrong, claiming that droplets over five microns would plummet to the ground. Two totally unrelated thresholds had been transposed: the size at which droplets could not spread through the air beyond two metres or six feet and the size at which they could not penetrate the lung and cause tuberculosis infection.

Part of the reason for the error was because of how the use of the term 'airborne' changed over time. Although since 2000 it has been used fairly consistently to mean any *"particles suspended in the air,"* before then it was sometimes used to mean *"particles which could be inhaled,"* or *"particles which are infectious."* Katherine Randall carefully mapped out how references using the historical meaning were used to support claims using the modern meaning.[32] It was this change in use of language that led to the size of infectious particles in tuberculosis being used to support the idea that only particles larger than 5 microns could remain suspended in the air.

Another reason for the error was due to an oversimplified dogma. Public health doctors and scientists liked to remove nuance to give people simple narratives to persuade them to act and in doing so one in particular, Dr Charles Chapin, may have contributed to the error on droplet size.

DR CHARLES CHAPIN

Dr Charles Chapin was Health Officer of Providence, Rhode Island from 1884 and, in 1926, president of the American Public Health Association. Alexander Langmuir described him as *"the greatest American epidemiologist."* He made an important contribution to emphasising that the source of infection was another person rather than the air or objects, but took things too far. He firmly believed that almost all infectious disease is spread through bodily secretions. He particularly emphasised both the *"danger from fingers"* which have touched the mouth or nose, and the larger droplets from *"mouth spray."*[33]

In 1910 he published *The Sources and Modes of Infection* which became and remained a seminal public health textbook for decades.[34] In it he was evangelical about close contact transmission as the route of all infectious disease and derisive of the theories of transmission through the air. He thought any residual belief in the idea of infectious air would lead to excuses not to be rigorous about control measures to stop close contact spread.

His efforts to introduce safety precautions, such as separating hospital beds and hygiene measures were hampered by people who could not see the point given that the patients would still be sharing the same air. He wrote, *"If the sick-room is filled with floating contagium, of what use is it to make much of an effort to guard against contact infection? If it should prove, as I firmly believe, that contact infection is the chief way in which the contagious diseases spread, an exaggerated idea of the importance of air-borne infection is most mischievous. It is impossible, as I know from experience, to teach people to avoid contact infection while they are firmly convinced that the air is the chief vehicle of infection."* [35]

Although he addressed the evidence separately for each infectious disease, he applied his conclusions to them all. He made a *"distinction… between the larger droplets of mouth spray which contain the most bacteria and which settle out of the air in the space of a few feet from the mouth and the smaller droplets which float for a longer time and may pass to some distance from the speaker and which alone may be considered as properly constituting an infection of the air."*[36] He theorised that diseases were *"spray-borne only for two or three feet."*[37] He chose to ignore evidence of infectious agents being collected from the air some distance from the source. He admitted that *"droplets from speaking"* could *"float for from five to six hours"* and be *"carried fifty-five meters along a corridor, and up two flights of stairs, and also a considerable distance out of doors."*[38] However he added, it should not be assumed that *"because bacteria are observed to fall on agar plates from the air of a room, the air is infectious."*[39]

He dismissed these findings of airborne bacteria saying only sputum droplets were infectious not those from saliva. This is particularly odd given his concerns around direct spread through saliva which suggest he had a phobia of germs. *"The cook spreads his saliva on the muffins and rolls, the waitress infects the glasses and spoons, the moistened fingers of the peddler arrange his fruit, the thumb of the milk-man is in his measure, the reader moistens the pages of his book, the conductor his transfer tickets, the 'lady' the fingers of her glove. Everyone is busily engaged in this distribution of saliva, so that the end of each day finds this secretion freely distributed on the doors, window sills, furniture and playthings in the home, the straps of trolley cars, the rails and counter and desks of shops and public buildings, and indeed upon everything that the hands of man touch. What avails it if the pathogens do die quickly? A fresh supply is furnished each day."* Despite having no evidence, he tried to justify his contradiction, saying, *bacteria… found floating in the air… may not after all be dangerous, either because they have wholly or partially lost their virulence, or because they are too few in number, or for some other unknown reason."* [40]

This allowed him to present a neat narrative that all infection was from touch or large droplets that would banish *"the sewer-gas bogey."* [41] Over a hundred years later and we are still suffering the effects of his oversimplification and evangelism.

The only possible exception he allowed was that *"Tuberculosis is more likely to be airborne than is any other common disease."* [42] Was this dogma of all transmission being through large droplets, with the exception of tuberculosis, part of the reason for the error in droplet size in the textbooks? Could information on the size of aerosols that cause the spread of tuberculosis have been conflated with the size that could travel further than two to three feet because of this dogma?

It seems he recognised that he may have taken his conclusions too far. *"While the tendency is thus away from air infection we must be on our guard lest our generalization carry us too far. It may be a fact that most diseases are not air-borne, and yet further investigation may show that certain other diseases concerning which we are still in doubt may be usually transmitted in this way."* [43] However, this warning was lost amongst the key thrust of the book that all transmission is through close contact.

After Chapin, the deck was loaded in favour of the idea of close contact droplet transmission. The actual evidence for droplet transmission is weak. In March 2020 one analysis of droplet transmission concluded, *"reviewing the literature on large droplet transmission, one can find no direct evidence for large droplets as the route of transmission of any disease."* [44] In spring 2020, a physicist from Hong Kong, Professor Yuguo Li, carried out mathematical simulations which showed that combining the falling ballistic-like projectile of droplets and the tiny surface area of eyes, mouth and nostrils the chances of infection from droplets during close contact was miniscule. [45] In reality, aerosols (with or without touch) could have been responsible for all the close contact transmission that we see.

The default dogma after Chapin was that droplet transmission was key. Watertight evidence had to be produced demonstrating that transmission was in fact through aerosols before droplet transmission could be questioned, despite its lack of supporting evidence. [46] For measles and tuberculosis aerosol transmission was accepted thanks to such comprehensive incontrovertible evidence but other respiratory infections were still considered to spread through droplets despite strong evidence of aerosol transmission. It is noteworthy that much of the evidence of aerosol transmission comes not from human studies but from publications on animals contracting the same viruses.[47] Were vets able to remain more open minded about transmission?

THE SPREAD OF *SARS-COV-2*

Through 2020, there was a growing body of evidence that people could catch *SARS-CoV-2* while remaining a considerable distance away from the infected source. Outbreaks occurred in restaurants, fitness classes and in a room after the index case had left. Large outbreaks occurred in hospitals despite precautions against droplet transmission. People in quarantine hotels caught virus with the same genetic fingerprint as people down the corridor who they had never met.[48] Animal studies demonstrated that sharing air through ducts between the cages was sufficient to spread infection.[49]

Nevertheless, the WHO insisted that large droplet transmission was the significant source of spread which must be through close contact. This formed the basis for advice on social distancing, mask wearing, one way systems and perspex screens. Perspex screens prevent ventilation of an even kind and can create pockets of poor ventilation.[50] Even once evidence emerged that *SARS-CoV-2* could spread in aerosols much smaller than 100 microns (grapefruits) the WHO and public health bodies did not change their advice because of their mistaken belief that only aerosols less than five microns (lentils) could travel beyond one or two metres.

Linsey Marr and others started a campaign in July 2020 to recognise aerosol transmission, starting with an open letter to the CDC from over two hundred scientists.[51] It took until spring 2021 before information on aerosol transmission was included in guidance from the WHO and CDC. The scientists involved were very careful not to undermine the public health authorities, always making clear that they considered aerosol transmission to be an additional challenge and that the current advice on preventing droplet transmission with masks remained relevant.

WHO pronouncements on transmissions show the absolute certainty with which unevidenced claims were made and the very slow process of correction.[52] It was not until December 2021 that they first used the word *"airborne."*

Feb 2020: *"droplets land on objects and surfaces around the person. Other people then catch COVID-19 by touching these objects or surfaces, then touching their eyes, nose or mouth."*

Mar 2020: *"FACT: #COVID19 is NOT airborne..."* *"These droplets are too heavy to hang in the air. They quickly fall to the ground."*

Jul 2020: *"short-range aerosol transmission, particularly in specific indoor locations, such as crowded and inadequately ventilated spaces over a prolonged period of time with infected persons cannot be ruled out."*

Oct 2020: *"Aerosol transmission can occur in specific settings, particularly in indoor, crowded and inadequately ventilated spaces, where infected person(s) spend long periods of time with others, such as restaurants, choir practices, fitness classes, nightclubs, offices and/or places of worship."*

Apr 2021: *"The virus can also spread in poorly ventilated and/or crowded indoor settings, where people tend to spend longer periods of time. This is because aerosols remain suspended in the air or travel farther than 1 metre (long-range)."*

Dec 2021: *"aerosols can remain suspended in the air or travel farther than conversational distance (this is often called long-range aerosol or long-range airborne transmission)."*

The December 2021 admission of long distance aerosol transmission should have totally changed policy. Unfortunately, changing the words on the WHO website does very little to reverse the beliefs that are strongly held from the first pronouncements the WHO made when they had undivided global attention. The WHO and other authorities are unwilling to draw attention to their mistaken beliefs from years before.

The story of the aerosol transmission of *SARS-CoV-2* encapsulates much of what this book is about – scientific beliefs and the harms they can cause. The whole saga was based on a medical myth derived from a mistake. The mistake itself arose because of an inability to rationally examine the evidence and disconnect emotions related to the debate between germ theory and miasma theory from a hundred and fifty years earlier. Finally, it illustrates the difficulties faced by those trying to be heard when their expert view is in conflict with those in power.

SEEING IS BELIEVING

The 100 micron (grapefruit sized) cut off is an interesting one as it is the cut-off for what is visible to the naked eye. When we see dust caught in the light or spring seeds drifting past us, it can be hard to believe they are under the influence of gravity as they rise and fall in the warm currents of air, but they are and eventually they will descend to the ground. Droplets smaller than we can see, strange as it may seem, rise rather than fall because they are affected by the warm current of air from our breath. They can remain suspended in the air for hours and can drift miles upwards in thermal currents.

In a world of mile high giants, with bus-sized nostrils, grapefruits and larger droplets would occasionally come tumbling to the ground when the giants coughed or sneezed. For every grapefruit or larger droplet, thousands of smaller droplets would be produced just with breathing normally. Most of the time the giants would walk around breathing out a cloud of droplets smaller than grapefruits like the cloud of dirt permanently surrounding Pigpen in the Charlie Brown cartoons. These droplets would then shrink with evaporation and drift further and further from the original source of infection. If they hit more humid or colder air they may condense and become large enough for gravity to pull them towards the ground until they reach warm air again. However, in Cloud-Covid-Land the entire cloud of aerosols would plummet to the ground in seconds leaving nothing relevant behind in the air.

In the real world, the droplets that are large enough to have a limited trajectory are much more likely to be emitted on sneezing and coughing. Those emitted when just breathing or talking are the smaller aerosols that can spread much further. Knowing this does not tell us which of these transmission routes is of more importance to spread overall. There remain two potential sources of spread – close contact droplet transmission or close and long distance aerosol transmission. Examining other evidence is needed to determine the role of each of these mechanisms in spread.

HOW IMPORTANT IS CLOSE CONTACT TRANSMISSION?

The inability to think about alternatives to the close contact transmission model led to frustration. By spring 2021, at a time when the UK test and trace system had plentiful resources and there were few cases nationally, politicians called for explanations for the failure to identify those elusive chains of transmission.[53] Canadian data showed that, even at times of low prevalence, only around 60 percent of infections were from a known source.[54] The remaining infections had no identifiable source. Even in China, an authoritarian state where public health officials would be heavily reprimanded for seeming incompetent if they could not find a source, there were problems. In a study of over 10,000 cases, only 17 percent of cases had more than two other people who could be linked.[55] For more than half of the outbreaks there were just two other people linked, often including a spouse. A third either had no identifiable source or other missing data. There were publications on small outbreaks where chains of transmission could be determined but no-one wanted to highlight the bigger picture where tracing failed. They believed when contact could not be determined it was a failure of those trying to do the tracing. The lack of an identifiable source contributed to the idea of asymptomatic transmission. The fact that a source could not always be identified was combined with the belief that transmission occurred only through close contact to conclude that those spreading the disease must have been unaware they were infected. The government's repeated claim that *"1 in 3 people with covid-19 don't have any symptoms but can still pass it on"* should have been reversed. The reality was that more than one in three people were catching it without any close contact with an infected person.

There are plenty of anecdotes and case studies apparently demonstrating a clear chain of transmission from person to person. The question remains what proportion of an epidemic is due to spread occurring without close contact transmission. This question sits at the edge of our understanding of viral transmission because it is so hard to measure.

Long distance aerosol transmission does have an evidence base. Cambridge University demonstrated that the full range of genetic variation of the virus seen in the general community was present in the viruses sampled from the care home population.[56] Either someone from every outbreak was entering a care home and breathing right next to the residents or else the air from the community was being shared in the care homes such that exposure was inevitable. Chains of close contact transmission were also identified but the presence of close contact transmission is not evidence against transmission via long distances in the air.

Furthermore, there have been several occasions when Australian authorities have struggled to understand the source of Delta variant infections in the community at times of very low prevalence. The original source of Delta in Australia was identified through genetic tracking to a traveller returning from Sri Lanka who tested positive on arrival and quarantined in two hotels.[57] None of his fellow plane passengers or crew, nor the staff or residents of either hotel tested positive but other members of the public did despite having no contact with him. Within a week of him leaving quarantine there were 44 genetically linked cases in the community none of whom had contact with the source.[58] Rather than consider long distance aerosol transmission, the assumption was made that there must have been a chain of transmission that they had failed to identify. James Merlino, the acting premier of Victoria, said transmission had occurred due to people *"being in the same place, at the same time, for mere moments."*[59] Instead of attributing this to airborne transmission over long distances it was attributed to asymptomatic spread because of scientists being wedded to the close contact transmission model.

To take another example, in the UK, the Cheltenham Festival was considered to be an important seeding event. Two postcodes adjacent to the festival recorded the highest case rates in the country in the following week. It seems unlikely that the staff at the festival all lived on the doorstep. Part of this could have been due to focusing testing on this area. However, spread through the air would also explain the phenomenon.

At the outset much was made of the need to wipe surfaces or even every item from the supermarket because of the risk that the virus could be transmitted through touching them. I remember ensuring I did not touch anything I did not intend to buy to avoid the silent opprobrium of other shoppers. It is possible to detect the part of the virus which enables replication, its nucleic acid or RNA, on surfaces. However, this is like finding forensic evidence at a crime scene when the criminal has long since fled. What matters is whether intact whole virus can be identified.

For other respiratory viruses intact virus has been shown on every surface from tables and sofas to toys and cushions.[60] Four studies have demonstrated viable *SARS-CoV-2* virus on hospital surfaces and one from the packaging of frozen cod.[61] A further eleven studies failed to find viable virus on surfaces in the real world.[62]

Experiments in a laboratory setting showed how ubiquitous the virus might have been. In November 2022, the UK Food Standards Agency posted results of a detailed experiment where foodstuff and packaging had been deliberately contaminated with virus to measure how long it remained infectious.[63] They showed virus remained infectious on every food item. For refrigerated high protein, fat and water containing foods like ham and cheese, viable virus was still at significant levels at the end of the experiment, after seven days. The role of refrigeration may be important as was seen in outbreaks in meat processing plants. It is always hard to be sure if laboratory experiments translate to the real world. If virus really was that ubiquitous everyone would have been exposed regardless of attempts at avoidance. The authors failed to recognise this and claimed their results *"highlight[ed] the importance of proper food handling."* If only people had scrubbed their croissants and raspberries harder, lives could have been saved!

The assumption with surface contamination is that it is caused by an infectious person coughing or sneezing into their hand before touching the surface. However, given that aerosols contained the majority of viral particles,[64] there is no reason to exclude the possibility that the surfaces were contaminated by aerosols that themselves were already at a distance from the source. Virus was ubiquitous in the air and may have been ubiquitous on surfaces.

It has been suggested that spread can occur through transmission from faeces. Spread in a Hong Kong apartment block was attributed to modification to the venting pipe of the sewage system.[65] (Chapin's sewer-gas Bogey had returned!) As well as finding viral nucleic acid in faecal samples from infected patients, intact virus capable of infecting cells in culture was demonstrated. Any transmission that occurs through touch and the spread of faecal contaminants would probably be difficult to distinguish from close contact transmission in terms of who would be exposed. However, the possibility that viruses in sewage can become aerosolised has been mooted and such spread would be harder to trace as a chain of transmission, at least at the beginning of an outbreak.[66]

There was therefore evidence of two modes of transmission. Close contact person to person spread but also spread at a long distance from the source through aerosols or surface transmission. The question then becomes which was the dominant mode that drove the trajectory of each wave? Was close contact transmission the primary driver, with aerosol transmission over a long distance through the air just topping up the total number of cases? Alternatively, was long distance spread through airborne aerosols or surface deposits the driver of outbreaks with consequent close contact spread finishing the job? (We will come back to this question in *Belief 9: Lockdowns saved lives*).

HOW FAR COULD *SARS-COV-2* SPREAD?

To judge how far the virus could spread first we must address how long it can last in the air. Aerosols can be artificially kept suspended by producing them in a metal drum which is rotated at just the right speed for the centrifugal forces that pull the aerosols towards the top of the drum to cancel out the downward pull of gravity. One study, including four laboratories, demonstrated that virus capable of causing infection in cells, remained at similar levels for the whole 16-hour experiment in aerosols in the air.[67] That is long enough to survive an entire winter's night. The authors commented on the lack of reduction in viable virus over that time. It meant they could not estimate the time after which only half the viable virus would remain. Experiments where measurement was carried out after much longer time intervals will be needed to understand exactly how long virus can last in the air.

Having established that the virus could remain suspended in the air for a significant time the question arises as to how far the virus could spread from an infectious person and still cause an infection. Within that question lie many other questions. How far can the virus travel? Is a virus that has been exposed to air for a long time still able to cause an infection? How many virus particles are needed to infect someone?

Experiments of infectious disease spread were often carried out by placing culture plates at different distances away from the (bacterial) infectious source and seeing how far spread could occur. The extent of possible spread seemed to be limited by how far away people put the culture plates. One experiment in the House of Commons in 1906 was carried out by bacteriologist, Dr Mervyn Gordon.[68] He gargled with a solution containing an unusual relatively benign bacteria, *Chromo Prodigiosum* as it was then known (*Serratia marcescens*), which multiplies into bright red, easy to identify colonies when grown. He then loudly recited passages from Shakespeare for an hour. Culture plates were placed around the debating chamber to determine how far bacteria could be spread through talking. The plates nearest him had more numerous cultures but cultures were also present on plates at the far end of the room, 20 metres away. Could they have travelled further if the room had been larger?

A different type of coronavirus, causing infection in pigs at a farm, was shown to spread through the air over 10 miles away and still be capable of causing infection.[69] In cattle, foot and mouth disease was shown to spread 190 miles over the sea and 37 miles over land from France to the Isle of Wight in 1981.[70] Sand from the Saharan desert, which is definitely larger and heavier than the aerosols, nevertheless can be blown from Africa turning the UK sky red and covering cars in a film of dust. An understanding that viruses can spread considerable distances through the air should totally change our approach to interventions.

The majority of virus particles are emitted in aerosols less than 5 microns (lentil sized in the giant analogy).[71] The vast majority of droplets emitted when coughing or sneezing are greater than 100 microns (grapefruit sized) and would fall to the ground within one to two metres. However, particles emitted when talking are smaller and the majority would remain suspended. Keeping a distance of two metres when someone is sneezing or coughing directly into your face seems like generally good advice and, thankfully, advice that would not be needed very often as few people are that rude. The distance needed to sufficiently reduce the exposure dose when talking to someone would depend entirely on how susceptible you were to infection. Whether there is a totally safe distance given the 190 miles that foot and mouth virus travelled to the Isle of Wight is highly unlikely. Nevertheless, the restrictions imposed by the government were based on the idea that all spread was through droplets during close contact.

LARGE NUMBERS

How much virus is out there? An unimaginable amount. A sick person breathes out around 100,000 particles a minute.[72] That amounts to 8,000 virus particles with every breath and 16 particles in every millilitre of exhaled air. It would take over two hours to count each particle exhaled in a single breath.

Each night, every infected person breathes out about 72 million particles into the night air. If you had started counting each of those particles at a rate of one per second on the day a child is born, they would be 27 months old and learning to count themselves before you finished counting.

Over the course of an infection one person can produce somewhere between 3 billion and 3 trillion virus particles. In one pre-Omicron wave therefore, between 80,000 and 100 million virus particles were produced every hour for every uninfected person in the country.

The number of particles needed to cause an infection is debatable. For other infections, a single virus particle has been shown to be enough to cause infection in one laboratory study.[73] Other studies show that single digit numbers are enough such that a single virus laden aerosol would more than suffice.[74] According to the UK government, *"sampling of environments where people have influenza or Monkeypox show far more viral RNA than for SARS-CoV-2, yet the outbreak data indicate that both are much less transmissible. This suggests that a lower viral dose is needed to initiate a SARS-CoV-2 infection than for these other diseases."* [75] The answer will always be highly dependent on the immune status of the person being infected. A susceptible person would succumb at a much lower dose but when exposed to a large amount of virus, even someone who otherwise would have seen off an infection can become infected.

Assuming everyone was susceptible to one particle then hypothetically, a single person would emit enough virus particles in ten hours to infect every person in the UK. Within 55 hours there would be enough virus to infect the whole of the USA.

The virus was tiny but ubiquitous. It could be argued that breathing air alone was enough to become exposed. Indeed, it had been shown in previous years that public indoor environments, pubs, supermarkets, restaurants, libraries, contained sufficient influenza virus in winter for an infectious dose to be inhaled after an hour of breathing the air.[76] The air on covid wards contained enough virus in each litre to infect multiple susceptible people.[77] (The key to understanding this is to realise that not everyone is susceptible to a particular variant, see *Belief Two: Everyone was susceptible*).

Of course reality is more complex than the air being full of virus and everyone being exposed. Some people were more likely to be exposed but the proportion of people who would manage to totally avoid exposure to any particular variant by the end of its wave was likely very small. Was it possible to avoid exposure? Given it was airborne and ubiquitous this seems unlikely. Close contact with someone who was infected early in the wave may have caused them to be exposed earlier than otherwise. Comparing the proportion of household contacts who became infected (which represented the proportion who were susceptible) to the percentage of the population who developed antibodies in each wave shows a close match. This and the shape of the trajectory of a viral wave shows that eventually the majority of the susceptible who managed to avoid exposure are infected before the tail of the wave fades to nothing.

The government's response to evidence that the virus was airborne was to advise people to open a window. The idea that this would solve the problem and prevent exposure was fanciful. Public information videos were produced of unsuspecting infected people exhaling clouds of black couscous-sized particles which were at a totally different scale to reality. Opening a window may reduce the concentration of virus but does not make it disappear – it just moves it outside. It might be hoped that ultraviolet light would destroy the virus. If there is plenty of ultraviolet light around, viruses are thought to struggle to survive. In winter, in the UK, ultraviolet light is in short supply and there is none at all for fifteen hours on the shortest day. Could wind disperse the virus? Perhaps, but in the winter, fog can stay trapped in valleys for days. It is also worth remembering how far smoke travels thanks to the wind, with forest fires in Indonesia creating a smog every summer in Malaysia, more than a thousand miles away.

In cities, thermal currents would cause air outside a window to travel up to the top of the building where it would move with the wind. However, once the air reaches a colder region it would start to descend only to rise again in a warm current adjacent to another building. During the first SARS epidemic, spread between neighbouring blocks was shown to follow the distribution predicted from such air currents with those living in apartments below the infectious source not being exposed.[78]

The above evidence is all slightly hypothetical. What about real world evidence? There was an enduring belief that outdoor spread was very unlikely that originated during summer 2020. Ultimately this belief was born out of the belief in close contact transmission. If you believed that spread was only possible with close contact, then an increase in contacts outside that did not lead to a rebound in infection must surely be because spread outside was not possible. At that time there was also negligible spread indoors as most people had been exposed to that variant of the virus and were either immune at the outset or had since developed immunity from infection. The idea that being outside prevented covid was repeated so often and by so many people in authority that it has become very difficult to refute despite not having any evidence base. Once new variants started spreading indoors they could spread outdoors too. Real world examples include spread among football fans watching the Euros at Wembley and people who attended a large number of outdoor music festivals across the world.[79] (It is possible that transmission occurred inside indoor facilities at some of these events but the question of outdoor spread has not been excluded).

Trying to prove that viable virus was present in outside (or indoor) air is really hard. Attempts to measure its presence at the tail of a wave have failed in Italy[80] and Spain.[81] However, viral material (but not necessarily intact virus) was detected in the air in Italy[82] and in air from three out of five hospital gardens and three out of seven urban sites in Turkey.[83] It is not possible to definitively say that *SARS-CoV-2* has been demonstrated in aerosols in the outside air but neither can that be ruled out on the current evidence. Despite putting huge effort into trying to demonstrate aerosols in the air almost all the relevant papers start with a declaration of faith to the mantra that *SARS-CoV-2* is spread through close contact droplet transmission.

Even if there are infectious aerosols outside, the dosage that people are exposed to will be highest indoors especially over long periods. That is why the highest risk of spread occurs between household contacts. However, the fact that indoor spread is a higher risk does not mean that outdoor spread through the air has been ruled out as another possible route.

If spread through the air was unimportant, how was a measurably higher risk of catching covid seen in areas with higher air pollution?[84] (This risk of catching covid was distinct from the increased risk of worse illness having caught it). One possibility is that immunity is impacted either through lower ultraviolet light leading to lower vitamin D levels or directly as a result of pollution. The lower ultraviolet light could also lead to higher levels of virus in the air. However, it has been shown that other viruses can spread between people by piggybacking on particulate matter in the air.[85] It is hard to reconcile finding a measurable difference in covid transmission due to air quality and still concluding that long distant airborne spread is not significant.

The amount of virus that was produced by each infectious person combined with the way in which virus could remain suspended in the air with thermal currents and wafted by the wind meant that exposure must have been inevitable except for those in the most remote areas.

The misinformation from those in authority about close contact spread led to the cruel idea of locking children in their rooms to avoid spread within a family. If someone is infected in your household then you will be exposed to the virus. Thankfully, 80-90 percent of the time, household contacts did not succumb because they were not susceptible to that variant (see *Relief 2: Everyone was susceptible*).

IDENTIFYING THE CHAIN OF TRANSMISSION

A failure to acknowledge the low susceptibility rates led to real difficulty in understanding transmission as a whole. Measuring transmission is far from straightforward. Not only is it difficult to tease out the size of the aerosols responsible for spread but it can even be difficult to know who was the source of transmission within a household. The challenge comes from the variation in incubation period, which ranges from two to eleven days.[86] Say a woman, Layla, was infected at an event she attended with two flatmates, Sara and Isobel, and became symptomatic two days later. Sara became symptomatic five days later and Isobel after seven days. Maybe Sara and Isobel had longer incubation periods? Maybe they did not catch it at the event but from Layla afterwards, or even from each other? Who can tell? It is easy in a situation like this to create a mirage of person to person transmission, but is it real?

Close contact spread is real, but is not the exclusive method of spread. The literature is inevitably biased towards reporting chains of transmission that could be traced and therefore contains many examples of close contact transmission. The reality is that measuring spread is very difficult and so there will inevitably be debate about how much spread can be attributed to aerosols. This is one of many remaining mysteries about transmission.

GLOBAL PATTERNS

How do different methods of viral transmission explain global patterns of spread? In the close contact transmission model the fact that cases and deaths peak at much the same time is explained by international travel resulting in extensive seeding. Because chains of transmission take time, rural and remote areas should be spared once restrictions on close contacts are put in place. The UK's Scientific Advisory Group for Emergencies or SAGE published minutes from 10th March 2020, just two weeks before cases peaked, which say, *Modelling suggests the UK is 10-14 weeks from the epidemic peak if no mitigations are introduced."*[87] Similarly, pandemic flu modelling carried out in a joint project between a team of mathematicians and the BBC using mobile phone proximity to simulate asymptomatic close contact, concluded that it would take 14 weeks for influenza to reach the most remote regions of the country if people continued interacting as normal.[88] Only then could a wave begin in those areas. That evidence clearly contradicts what we know of influenza spread in the real world where every region is affected within 4 to 5 weeks and the whole epidemic phase each winter is over within about 16 weeks. What is more, the simultaneous peaks miles apart were evident through history long before significant international travel. These facts clearly demonstrate quite how inadequate the close contact transmission model is for explaining real world findings. Years of real world evidence from influenza and global evidence from the first covid wave falsified this model and it should have been adjusted long ago.

There have now been numerous covid waves globally with the timing of peaks predicted much more accurately by geographical location than by any restrictions on human behaviour. The influenza pandemic in 1918 appeared to travel from east to west. Although the peaks were at a similar time, excess deaths peaked first in Berlin, with Paris and London following shortly thereafter. The USA also showed regions affected sequentially from east to west in 1918. With covid, rather than uniform seeding through travel, the 2020 wave moved rapidly across Europe from east to west. First Italy saw peak deaths, before Spain, then France, UK and lastly Ireland. Eastern Europe escaped a notable spring 2020 wave altogether. In winter 2021 the peaks shifted from west to east with Ireland and the UK peaking before Italy and Spain and lastly Eastern Europe. Eastern Europe went on to have a significant spring 2021 wave not seen in most of Western Europe. Could this relatively slow spread in one direction be due to progress through the air rather than close contact transmission?

Despite travel being severely restricted, the same pattern of rapid global dissemination seen in spring 2020 was seen for variants. The Alpha (aka Kent) variant appeared first in Montenegro in July 2020, before Sweden and Italy, and was detected at a low level in almost every country by the last week of November or first week of December. Likewise the Delta (aka Indian) variant arrived in Australia and New Zealand at the same time as arriving in the UK despite the attempts of the antipodeans to hermetically seal their borders. These broad patterns of global spread do not falsify the model of close contact transmission but do add weight to the idea that there are multiple other factors at play.

There was room for uncertainty about how this particular coronavirus might behave but surely the prime assumption should have been that it would behave similarly to other coronaviruses or influenza. Public Health advice was given on a precautionary basis – what if it is spread through close contact, we fail to reduce close contact and people die needlessly? There is a defence to be made for some caution in this case but everything should have been done to gather evidence to better understand transmission. Instead authorities remained closed-minded. Did they fear the backlash that admitting their error on droplet size would bring?

When virus can spread through the air in aerosols, then being exposed to a small quantity of virus in each wave will be virtually inevitable. Our immune systems protect us and only a fraction succumb (see Belief Two: *Everyone was susceptible*). There is a critical question that remains unanswered by the mainstream narrative where all spread occurs through close contact. Why did every covid wave, in every country, have a similar trajectory in terms of the time to peak deaths? Blaming changes in human behaviour for causing the peak in deaths no longer holds water after multiple waves where people have not changed their behaviour at all. When assuming that everyone is exposed, the dynamics of a wave of infection reflect the spread through only the fraction of the population who are susceptible. If everyone will be exposed at some point during each wave then avoiding exposure through changes of behaviour would only delay the inevitable by a short time. It was this fatalistic approach that was taken with respect to influenza for hundreds of years. Yet, somehow, the aerosol transmission story was packaged as a reason to do more, not less to try and slow the spread.

Aerosol transmission can explain many mysteries around how covid and influenza spread but there may be other factors too. Anyone who claims to fully understand the transmission of influenza or *SARS-CoV-2* is lying. Trying to persuade the public, journalists and politicians that it is not as simple as close contact transmission is a huge challenge.

When I have written on *SARS-CoV-2* spreading through the air, I have been attacked as not believing in germ theory. The evidence of viral spread through tiny aerosols in the air is now widely accepted, yet saying so still causes some people to picture me as a plague doctor advocating filling beak shaped masks with lavender to filter the smells.

The proponents of germ theory had to fight to be believed. Conventional wisdom even drove Von Pettenkofer, a miasma evangelist, to give himself cholera by drinking cholera culture in the hope of proving his hypothesis.[89] Eventually the germ theorists overturned hundreds of years of beliefs about bad smells and miasma in the air being the route of disease. It is astounding that the expired arguments from the mid 1800s are still able to have such influence on today's thinking. But they do.

Could anything have been done to prevent aerosol transmission? Some argue that we should filter the air in all indoor environments. Unfortunately, we cannot provide a source of clean public air because, unlike water, a supply of air cannot be separated. Updating hospital ventilation systems to include UV-C light seems like a sensible way to protect our most vulnerable. Trying to clean the air elsewhere, say in schools or offices, misses the point that exposure will be inevitable when you inhale the infected air from your sick neighbour overnight. A national debate must be had about whether to also provide as sterile an environment as possible in care homes. Yes, people in care homes are more vulnerable to covid. So too are certain groups in private accommodation of varying quality. Should people in care homes be safer from disease than people in their own homes? Is it more important to invest money in extending the length of life for those near death, or should we prioritise investment in the quality of their life?

Overall, there is no evidence that has disproved the commonly held belief that transmission occurs through close contact with an infected person. However, there is substantial wide ranging evidence that long distance transmission with aerosols is a reality. *SARS-CoV-2* spreads through the air and there is very little we can do about that.

At the end of this chapter I have probably raised more questions than answers. How important is long distance transmission as a driver of each wave? How many virus particles are needed to start an infection? Are there any interventions that work? These questions will be tackled in forthcoming chapters but having the foundation of knowledge about the sheer quantity of virus present and the minute aerosols in which it is airborne is the first step to understanding the broader picture. Alarmingly, if airborne infection was enough to spread covid, it is possible that all interventions were futile or at best, their impact in their ability to prevent rather than delay infection, massively overestimated.

TOP THREE MYTHS
1. Droplets were the main source of transmission and infectious material would fall to the ground within 2m or 6ft
2. Infections with no close contact source must be due to asymptomatic spread
3. Long distance spread was not possible because such an occurrence sounds like miasma theory

HOW DO SCIENTISTS GET THINGS WRONG?

Epidemiological hypotheses must provide satisfactory explanations for all the known findings – not just for a convenient subset of them. **Fred Davenport, 1977**[90]

How can science be so muddled up with beliefs? Surely, at the heart of science there is neutrality with evidence leading decision making? Unfortunately, in reality, scientists are subject to all the flaws of human decision making. All scientists. No-one owns the truth. Given enough time, everyone will be proved wrong about some of their beliefs. Some people find that very easy to forget. Some scientists seem to believe that no future generation will look back at us and laugh about what we got wrong, despite that happening for every previous generation. Being open to accepting error requires adopting contrary views and handling the inevitable cognitive dissonance that comes with doing that, which is unpleasant and takes work.

Science progresses, not when more evidence is found that supports what is already believed, but when evidence proves the conventional narrative wrong. It is hard to prove certain things are right just through experience. If you believe that all swans are white then finding more and more white swans adds weight to that belief but it does not prove it. In contrast, finding a single black swan disproves your belief. Science progresses by proving theories wrong rather than collecting more and more evidence that fits the existing theory.

People hate being wrong (including me). Nevertheless, the way science is approached is to take a belief, a hypothesis, and set out to see if it can be proved wrong. First you think of an explanation of the world that would mean you can explain something, that is your hypothesis. Then you set out to prove that the interpretation is wrong. If you were wrong you have progressed. The foundation of science is the exploration of being wrong. Every scientific breakthrough has proven those in powerful positions to have been wrong. Each one has had to be fought for and many have taken years to be accepted.

The obvious counter to scientists who believe they are omniscient is that the complexity of the world far exceeds the ability of one person to understand. Even within a niche specialty, there will be plenty of areas which are poorly understood. Therefore, if we want to improve our collective understanding of the world, it must be done through listening to every voice and not letting those who believe they are omniscient drown out those with interesting questions.

Any debate about covid was heard only by select audiences willing to engage. There were very few scientific breakthroughs. Instead there was a strange inversion. Normally, old established assumptions believed by the majority are challenged by a minority with new evidence. With covid the majority beliefs were based on new assumptions and when they were challenged it was done so on the basis of old, established evidence. For example, the idea that natural immunity would not be protective, despite decades of experience around immune memory, was ignored. Why was this old, established evidence being ignored?

It was not simply that some scientists were being fooled into believing something outside of their discipline. The belief that catastrophe was around the corner meant that reassuring evidence was considered dangerous. The focus was on preventing catastrophe, so anything that distracted people from that goal must surely be, not only wrong, but immoral. From that foundation arose the ill-founded fears that everyone was susceptible to each variant, that the healthy or asymptomatic were spreading disease, that viral spread could only be stopped by intervention and that every intervention would work.

In order to maintain these incorrect assumptions there needed to be distortion of the truth. Aside from plain old errors, there are three key ways in which the evidence can be exaggerated, misrepresented and distorted to allow scientists to defend their preconceived ideas. I have called this the Evidence Manipulation Triad and both sides of the covid debates are guilty of using some of these techniques. The three elements are:

1. Extrapolating

2. Excusing

3. Excluding

EXTRAPOLATING

Extrapolating happens when weak evidence is given overly significant weight or when evidence is synthesised. Synthesised evidence comes in three flavours:

a.) Modelled results based on assumptions only

b.) Weak evidence that is adjusted to produce more impressive results

c.) Evidence that addresses only part of the question at hand

The close contact transmission model was based on a closed-minded understanding of the mechanisms through which germs can spread persisting as a hangover from debates taking place over a hundred years ago. Extrapolating from evidence that bad smells were not the source of disease led to an inability to conceive of the notion that viruses could transmit through the air. Worse still, the mistaken idea that infectious particles would almost all fall straight to the ground was based on a model of droplet dispersion with a mistake at its very heart, yet was extrapolated and used in public health guidance for many years. Additionally, some, myself included, extrapolated from the mathematics of the first wave, showing most the population were not susceptible, and assumed the epidemic was over in summer 2020.

EXCUSING

Excusing happens when evidence is presented that contradicts the scientist's current belief. A scientist should think openly about all implications of the new evidence and devise several new hypotheses to test. Instead, our reaction is often to find reasons why this evidence might be wrong. Perhaps it was measured incorrectly, or the sample was biased or there was some factor that was not controlled for that meant things did not progress as expected. For example, the public health authorities were so wedded to close contact being the only route for transmission that evidence of spread from unknown sources was excused as being due to asymptomatic spread. Similarly, Von Pettenkofer, the miasma proponent, who had seen the germ theory evidence discounted it because of years of expert knowledge about bad air and must have made excuses for that evidence in his mind before drinking the cholera culture and catching the disease.

EXCLUDING

Excluding happens when there are no excuses to be made and instead the evidence is totally ignored. The paradoxes of outbreaks in remote locations like Antarctica and on boats at sea or the genetically identical variants of influenza appearing simultaneously across the northern hemisphere even in eras before international travel are totally ignored. By excluding such evidence, scientists may overlook crucial information that could enhance their understanding of the phenomenon being studied.

The way science has historically handled these biases is through open debate. In theory, if all voices are heard and the evidence is presented from every angle then the conflicting evidence becomes apparent. Further experiments or measurements can then be carried out to clarify the point of contention. In practice, further work is only carried out if it can attract funding. Funding is almost all secured through powerful institutions which are often either linked to government or, directly or indirectly to industry. There was no funding for work that challenged the official covid narrative.

It was a struggle to be heard, especially for those working free of charge and without a job title or institutional backing and it was worrying to see preprint servers, where submissions can be made prior to peer review, rejecting papers from established professors. The editors of the journal *Science* commented on how they had discussed whether it was *"in the public interest to publish the findings"* before printing a peer reviewed paper.[91] The danger is clear. Though this is just one example – inspired by a superficially benign motive – the question arises as to whether other editors filter publications, not because they are unreliable, but because the findings are not politically helpful, creating a marked bias in the published literature.

Breadth and depth of scientific knowledge is immense. Too much for anyone to keep on top of. To keep up, people trust the experts and take things on faith. Because the majority of the population are taking things on faith, a consensus answer takes on immense power. It is hard to weigh up the evidence yourself and admit uncertainty. It is much easier to dismiss someone who is scientifically questioning the consensus by claiming they are mistaken. In many cases, the background knowledge required to be able to carry out an assessment is not within reach anyway, so the consensus will win by default. The shortcut of believing what authorities say and what the majority believed did not favour the truth around covid.

4

FALSE PROPHETS AND THE SCIENCE™

At a time when things are most uncertain, we turn to the most certain thing there is: science. Science can overcome diseases; create cures; and, yes, beat pandemics. It has before. It will again. Because when it's faced with a new opponent, it doesn't back down, it revs up, asking questions 'til it finds what it's looking for. That's the power of science. Science will win. **Pfizer campaign video, April 2020**[92]

Much trust was placed in *The Science*, a new belief system sold on the basis of the public understanding that science is omniscient. *The Science* was a belief system of fictional-science based on trusting experts over evidence. Science fiction is limited by having to be believable whereas fictional-science seems to have no such restrictions.

Real science measures the world around us. Measurements can be taken directly as natural events take their course or else experiments are run which are designed to answer specific questions and the measurements reveal whether our assumptions were wrong. It does not keep going *"til it finds what it's looking for."* An honest experiment cannot be carried out with that approach.

The whole point is that you do not know what you might find. The same cannot be said for 'scientific' modelling on which much of *The Science* around covid was based.

Science uses measurements to reduce uncertainty about the world. Guessing the size of, say, a door would give us a range of possible sizes but measuring it reduces the uncertainty and may expose totally erroneous guesses. Experiments are simply measurements carried out in a way that allows an answer to a particular question, or hypothesis. Using measurements allows data driven decisions for which there is a degree of justification.

In Cloud-Covid-Land a significant proportion of scientific literature published on covid involved no measurements. Instead assumptions were made to feed into a computer model. The word 'model' indicates that they are not based on reality. In fact, when you see the word model, replace it with hypothesis, for that is the word we should use in science for ideas that are based on assumptions. These models produced eye catching, beautifully smooth and very scientific looking graphs and then conclusions were drawn from these graphs and presented as results. Scientific measurements are invariably presented as graphs (although these are never as perfect looking) so this was a neat camouflage for what was otherwise just projections. These papers were then laundered through the scientific peer review process into the published literature and presented as *The Science*.

Neil Ferguson's infamous February 2020 predictions[93] were based on the following assumptions:

1. We were all susceptible to varying degree;

2. A third of people would never know they had it, yet would spread it at half the rate of the symptomatic;

3. One in twenty-two would need hospital care and for every ten patients in hospital three would need critical care;

4. Out of a thousand people infected, nine would die.

We now know that only a fraction of the population were susceptible to the original variant; only a small proportion of people were responsible for most of the spread; and the estimates from the model for hospitalisations was massively overblown with deaths also being exaggerated. The estimates included that over two thirds of hospitalised over 80 year olds would require critical care. This was farcically wrong. A stay in intensive care is akin to climbing a mountain. To survive and benefit from intensive care requires a certain amount of underlying strength.[94] Anyone with an understanding of elderly care or critical care could have set them straight in March 2020, but none were asked. Although they assumed nearly one in three[95] of those hospitalised would need critical care, in retrospect only one in eight did.[96] A combination of these overestimates led to the conclusion that demand would be 30 times greater than maximum capacity in both the UK and USA.[97]

The narrative was far removed from reality. The actual hospitalised proportion in later 2021 turned out to be so small it was hard to show any benefit in trials for antivirals. The modellers had assumed a 10 percent rate of admission in those at risk of severe disease[98] but the trials only had a 3 percent admission rate.[99]

Not only was the model fed with flawed assumptions but the Ferguson computer model itself was riddled with bugs and written in code over a decade old and in an obsolete computer language.[100] The model was designed to run basic building blocks as a foundation and then deviate into different possibilities. That way it could be re-run to provide different answers. Such a design would be legitimate as a way of predicting a range of possible outcomes. However, this particular code couldn't produce the same basic building blocks each time it was run.[101] Neil Ferguson's programmes had been used to devastating effect for decades providing advice on mad cow disease, foot and mouth disease, bird flu, swine flu and more. Each time the models massively overestimated the size of the problem.

Had the code worked smoothly there would still have been huge issues because of the assumptions being used. It is hard to predict the future and assumptions have to be made. The less that is known, the wider the range of assumptions that should be used. It was quite disingenuous to point at the algorithm and say the findings were based on 'science' when using different assumptions would have produced a very different answer. Assumptions could have been fed in that would have shown no significant impact at all from covid. The truth lay somewhere between this extreme and the assumptions that were used. A much more honest approach would have made clear the weakness of the assumptions and the massive range of possible outcomes.

Instead, the figure of 510,000 deaths in the UK and 2.2 million deaths in the USA were taken and pushed by the media until it was no longer a prediction, ripe with uncertainty and readily contradicted, but a belief meaning we must take action.[102] Attempts to caveat the belief and to widen the view of the range of possible outcomes were not forthcoming. Later corrections of the model based on refining the assumptions never appeared even as more and more real world evidence demonstrated its failures By the end of 2020 there had been 80,000 deaths attributed to covid and even after a full second year that figure had reached 150,000 but with only 114,000 deaths above expected levels. In

May 2021, the 510,000 figure was still being quoted by government advisors, including Dominic Cummings, a powerful government advisor, as an inevitability.[103] He seemed to believe that this prediction still held true rather than just using it to illustrate what previous decision making was based on.

The 510,000 was not plucked out of thin air but it was based on two assumptions. First, that 85 percent of the population would catch it in the first wave and then that 0.9 percent of those would die. Those were the two assumptions needed to reach a figure of 510,000 deaths in the UK in a single wave. Rather than presenting it in this more honest way, it was dressed up in scientific language with many accompanying figures and graphs. However, ultimately that is what Imperial were proposing. If there had been more openness about those two assumptions and the fact that those two alone are all that is required to reach the total death estimate, then everyone, including politicians, would have been able to join a debate as to whether those assumptions were reasonable or not.

In retrospect some members of Parliament expressed such opinions with Bob Seeley saying, *"Never before has so much harm been done to so many by so few based on so little, questionable, potentially flawed data. I believe the use of modelling is pretty much getting up there for a national scandal. Modelling and forecasts was the ammunition that drove lockdown and created the climate of manipulated fear and I believe that creation of fear was pretty despicable and pretty unforgivable."* [104]

WAS IT REASONABLE TO ESTIMATE THAT 85 PERCENT OF THE POPULATION WOULD CATCH IT?

To reach an estimate of 85 percent an assumption must have been made that no one had any immunity to it. While it is true that no one had immunity derived from having been infected by it, that is not the same as saying that no one had sufficient immunity, from other infections, to protect them from it.

Had the influenza model been considered then the range of possible people who would be susceptible in any one full winter season would be 5 to 15 percent.[105] A further comparison can be made with the 1918 Spanish Flu epidemic. Spanish flu is thought to have been caused by more than one pathogen but there is uncertainty about this. In total, it is estimated that one third of the population were infected over the course of four separate waves over three winters. That means that the average number infected in any one wave was 8 percent and 11 percent were infected each winter on average. A paper in 2012 estimating the fallout should there be a man-made pandemic as a result of a lab leak predicted 15 percent of the population would be infected.[106] As early as 5th March 2020, Chief Medical Officer for England, Chris Whitty admitted that under 20 percent had been infected in Wuhan.[107]

In retrospect, the antibody studies carried out in May 2020 demonstrated that covid had infected about 7 percent of the population. Similar figures were found in Western Europe and the USA. Was it ever reasonable to assume that more than about 15 percent of people could be infected in any one winter? Fears that this could have been a man-made bioweapon may have warranted an extra degree of caution. However, even then, an estimate that it would infect more than five times as many people as the worst nature had ever produced, should have been heavily caveated.

WAS IT REASONABLE TO ESTIMATE THAT 0.9 PERCENT OF THOSE INFECTED WOULD DIE?

It is a well-established predicament in epidemics that early measurements of mortality will be overestimates. The initial measurements of the number of deaths should be reliable but an accurate measure of the number of cases that led to those deaths depends on detecting mild cases that may go uncounted. Therefore any estimate of mortality at the outset needs to be scaled down when considering the likely total impact.

In an epidemic the initial estimates for how deadly it may be are always overestimates. It is an inevitability because the deaths are measured per case, the case fatality ratio. Testing cannot reach every case so the deaths appear to have occurred from a smaller number of cases than was in fact the case. In retrospect, it is possible to measure who had the disease by testing for antibodies to it. All the initially undiagnosed cases should be captured with this testing enabling a more realistic estimate of the chances of dying, the infection fatality ratio or IFR. A more useful measure from the point of view of someone assessing their own risk would be the proportion of people exposed to the virus who died, an exposure fatality ratio. Such a measure would always be hypothetical and shown in retrospect because exposure is not simple to measure. How can you differentiate those who were exposed but already immune from those who were lucky and not exposed? However, it would have been more meaningful for people trying to assess their individual risk.

Because the mortality rate is so variable by age, taking a single value for the population is overly simplistic. For the under-70s, the actual percentage who died was less than a third of the Imperial estimate.[108] Imperial's estimate was only ever a projection based on no measurements. Once antibody data became available more accurate measurement was possible. How close these measurements come to an accurate IFR would be dependent on both the accuracy of antibody testing and on which deaths are labelled as covid deaths (see *Belief Four: Death certificates are never wrong*).

After collating the sum of the evidence on IFR, Prof Ioannidis, Professor of Medicine and Epidemiology at Stanford University estimated a global IFR of 0.15 percent (i.e. 99.85 percent would survive overall).[109] He estimated that the overall IFR in America and Europe was 0.2 percent for people living outside of institutions in the community, and 0.3–0.4 percent overall. Based on these estimates, the Imperial estimate had exaggerated risk by two to three fold. In October 2022, Professor Ioannidis refined those estimates to only 0.07 percent for 0–69 year olds.[110] Unfortunately, even Professor Ioannidis' estimates rely heavily on antibody testing which has such large variation between studies and regions that the results need to be treated with care. Using the England data alone, there is no way you could reach such a conclusion. Prior to Omicron the IFR figure for England for the whole population under 69 years old was around 0.2 percent. That is 998 of every 1,000 infected would survive. If Professor Ioannidis' revised estimate is right, that would suggest two in three covid deaths in England in the young were over diagnosed. That could happen for example, where a young person, dying from another cause in, say, a hospital or hospice, tested positive for covid and the doctor included a mention of that on their death certificate.

CONTINUED MODELLING MISTAKES

After the initial estimates were proved catastrophically wrong when places that did not intervene saw similar outcomes to those that did, the modellers continued to predict doom. The same flawed assumptions led to models in autumn 2020 that led to tiers and then a second lockdown. Even as late as May 2021, the same assumptions were in use. Modellers argued that they struggled to include a seasonality effect in their model as they had no real world data to base that on.[111] In reality these same modellers had no real world data for many of their assumptions but, by then, had a year of data on seasonality from the whole world. Although there were confounding variables the seasonal effect is so profound, particularly across the Northern Hemisphere, that it is quite astonishing that it was dismissed. Seasonality is discussed further in the next chapter, *Belief Two: Everyone was susceptible*.

The result of the modelling was an upper estimate that in six weeks up to July there would be the same number of hospital admissions and deaths as had occurred in the preceding 14 months. In reality, there were fewer than 2 percent of the predicted admissions and fewer than 0.3 percent of the predicted deaths. (The modellers' upper bound prediction was 346,000 hospitalisations and 169,000 deaths. In reality there were 6,500 admissions and 450 deaths.)

The harms from modelled data were not restricted to predictions of hospitalisations and deaths. In Cloud-Covid-Land modelling was used to massively amplify assumptions about asymptomatic spread and masking. Papers were published that made assumptions about the proportion of the population who could spread covid asymptomatically and how effective they would be at spreading it. They then, produced graphs demonstrating how much such spread would contribute to a hypothetical epidemic and drew conclusions about what a huge impact asymptomatic spread was having. These conclusions were just the assumptions that had been fed in but now dressed-up as 'science'. These were works of fiction. No new measurements took place and there was nothing scientific about how they were carried out.

Other papers modelled the impact of mask wearing. Again assumptions were fed in about the size of particles that were important for transmission and the degree to which these could be reduced by mask wearing. Again multiple, beautiful, often colourful graphs could be drawn but these were not graphs of results based on experimental measurements. It was pure storytelling. The authors' logic was that because they believed certain things (the assumptions fed into their model), they therefore also believed other intangibles (the outputs of the model). It is hardly surprising that a paper founded on an assumption of reduced transmission would yield results showing that masks reduced transmission and deaths.[112]

Evidence from modelled data has been accepted as valid. Much of this would not reach the threshold for submission as evidence in a Court of Law, and yet many people are happy to use it when it suits their purposes.

In 1692, the court in Salem, Massachusetts, accepted 'spectral evidence' based on dreams and visions, as legally valid in the Salem witch trials. The trials led to the arrests of 150 men and women and the hanging of 19 of them, convicted as witches on the basis of such testimony. Throughout the trial most people thought such evidence invalid. It was still submitted and this was justified, on the basis of necessity, such as 'the terrors of war'. A claim that they were at war against the devil completed the justification. By October 1692, the court was dissolved and spectral evidence was no longer admissible in court.

Like spectral evidence, there is no real world proof from the covid models, only the imaginings of individuals. On that basis the models would be rejected as evidence by a court. With scientists presenting models as truth, dressing them up as scientific publications and getting the seal of peer review publication it proved to be a convincing illusion. Neil Ferguson's hypothesis was treated as evidence by governments; other hypotheses were not heard and conflicting submissions were ignored.

It was not just journalists and politicians that were taken in. Numerous scientists were swept along, readily presenting such fictional 'evidence' to defend a position they had already adopted on covid. The alternative for them was to start digging into the rapidly expanding literature with an open mind and weigh up the much less straightforward real world evidence. After the huge amount of work this would entail they would likely be left with more questions and large areas of uncertainty. These people did not have time for that! They had day jobs, families, and lives to be getting on with. How much easier to find a paper that supports your position and just stick to that.

Rather than be honest about complexity and uncertainty '*The Science™*' was presented as a series of simple narratives. The narratives could change over time and even contradict the previous position. For example, the switch from saying masks don't work to saying they were needed to prevent spread or from saying that the vaccines were 95 percent effective against infection to saying they were only ever designed to reduce severe disease and death. People accepted these narratives and defended them passionately. After all, lives were at stake and they had no reason to doubt the best intentions of the people giving the advice.

Those who dared to question the assumptions, predictions and advice given by the government and their scientific advisors were accused of wanting to put lives at risk. Such accusations reveal a lack of thoughtfulness and imagination in those making the accusation. The motivations of the supposed heretics were in fact the same motivations as their critics. They could see that each intervention came with harms and were concerned about the wider public health impact of government policy. Every intervention comes with a balance of risks and benefits and those who thought the benefits of interventions had been oversold and the risks of these were significant, were just as passionate about preventing harm and saving lives as those who believed every intervention was necessary.

BELIEF TWO: EVERYONE WAS SUSCEPTIBLE

> *While many people globally have built up immunity to seasonal flu strains, COVID-19 is a new virus to which no one has immunity. That means more people are susceptible to infection, and some will suffer severe disease.* **WHO, March 2020**[113]

WHAT DID TEAM IN-THEORY THINK?

In theory, the virus was novel and no-one had any immunity to it. The Imperial model was extrapolated from what was known about measles. A measles outbreak, in the Faroe Islands, in 1846[114], infected around 6,100 out of 7,864 people who lived there and killed just under 3 percent of those infected but spared those who had been infected 65 years previously in the last outbreak.[115] The measles model supposes that everyone is susceptible until infected. The Imperial team, led by Neil Ferguson, predicted 510,000 deaths in a year from Covid based on the assumption that 85 percent of the population would catch it and that only lockdowns and vaccination could protect people from the virus 'ripping' through the population.[116]

Based on that model, the peak in cases only occurred because of interventions and the slackening of restrictions should have resulted in a rebound of infections. People's immune systems would not be able to protect them. The Imperial modellers declared that every relaxation in restrictions would result in a rise in hospitalisations that would need to be managed through repeatedly re-introducing restrictions.[117] In this model only human behaviour and intervention could prevent hundreds of thousands of deaths. Neil Ferguson warned that the faster relaxation in the USA meant *"they are probably at a higher risk of a significant second wave than most European countries."* [118] The 85 percent figure was never changed and was the basis of future claims concerning merits of lockdown and vaccination. (See previous chapter).

WHAT DID TEAM REALITY THINK?

As the data accumulated it was increasingly clear that the covid waves globally were behaving in a mathematically predictable way, approximating to what is called a Gompertz curve. The virus growth slowed from the start as it found it increasingly difficult to find the next susceptible host. At the outset of any wave there are plenty of susceptible hosts and an infected person can easily pass on the virus to another susceptible person. By the end of a wave, the variant is struggling to find the last remaining susceptible people in the population and so the wave fades gradually to nothing. The Gompertz curve follows a set formula such that the deaths on one day can be predicted from the deaths the day before. This meant that the whole wave could be sketched out based only on data from the first few days or weeks as long as there was an accurate guess of the total deaths in a wave. The formula used to predict the next value can be plotted as a straight line. [119] Had interventions altered the trajectory this straight line should have deviated. There was no deviation from this sketched trajectory due to interventions, anywhere in the world.

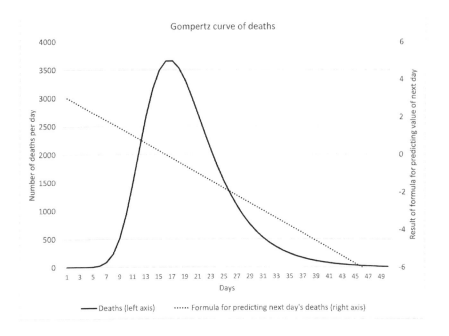

Figure 1: *Example of a Gompertz curve showing the rise and fall in death numbers in solid black using the left hand axis and the formula demonstrating the constant slowing in growth of deaths seen in the dashed line against the right hand axis*

If the limiting factor for spread had been human behaviour then there would have been notable differences in growth rates and the timing of the peaks in different places.[120] The predictable curve of these waves supports the hypothesis that everyone was exposed but only a proportion were susceptible.[121] Based on who developed antibodies, somewhere between 85 percent and 95 percent of the population were irrelevant bystanders who were not susceptible to any one variant wave. The variant worked its way through the remainder in a predictable fashion.

This is exactly what happens with influenza. (I am not saying covid is influenza, just that it behaves like it). For example, in 1957, public health officials feared that there may be a 1918 style pandemic due to a novel influenza virus, Asian Flu. It was a bad winter for deaths with 15 percent catching influenza.[122] With the arrival of spring and summer there was a hiatus until the following winter. The following winter a new variant caused another winter wave and so on until, after around a decade, almost the whole population had developed antibodies and that particular strain of influenza was replaced with another new one, Hong Kong flu, in 1968.

WHY DOES A WAVE END?

There have been numerous waves of covid now throughout the world. Each pre-Omicron wave has resulted in confirmed infections that generated antibodies in between 5 and 15 percent of the population, which is what was also seen with influenza waves in the past.[123] Over the course of two years, the levels of restrictions in place at the point when waves peaked and fell, in country after country have varied enormously. However, the peaks have occurred at predictable times. For example, deaths peaked in January throughout the Northern hemisphere, as was the case with other respiratory viruses pre-covid. Waves end regardless of human behaviour.

HOW MANY WERE SUSCEPTIBLE?

The clearest evidence that not everyone was susceptible to each variant comes from studies of transmission within households. When measured across a large sample only one in ten household contacts were infected.[124] This proportion was constant for each variant but at the beginning of a wave this figure was marginally higher and by the end marginally lower.[125] The UK public health authorities measured this carefully. In the UK data, when two members of the same household tested positive within a day of each other, these were assumed to be separate cases caught outside the household.[126] Other researchers included such cases as transmission, resulting in higher rates for household transmission.

It takes quite a leap of imagination to continue to make predictions on the basis that everyone was susceptible when there was plentiful evidence that the vast majority did not catch it, despite sharing a home or even a bed with an infected person. The proportion of a household that was susceptible was slightly larger than the proportion of people who developed antibodies to each variant. It is difficult to quantify how much of that difference was due to a proportion avoiding exposure and how much was due to antibody testing failing to identify every genuine infection. Either way, with pre-Omicron variants[127] nine out of ten people were not susceptible and the Omicron wave figures were similar.[128]

The low household transmission rate presented an additional quandary for the official narrative. The 10 percent household transmission rate would mean that an infected person would need to live with 10 others in order to guarantee the spread of an infection to one other person. To spread to one other person outside the household would require 20 contacts. If on average people had only three household contacts (which is still much higher than the actual average) they would need to have 14 other contacts during their infectious period in order for the infection to not just die away. These numbers far exceeded averages in reality. This was further evidence that close contact transmission was not the driver of spread. The surges could only have occurred with aerosol transmission.

SUPERSPREADER EVENTS

Although the big picture story is clear that not everyone is susceptible there were case studies of outbreaks where the attack rate was much higher, leading to claims that a much higher proportion must have been susceptible. For example, the outbreak on the French naval carrier *Charles de Gaulle*, resulting in half the sailors having symptomatic covid.[129] The highest attack rate I am aware of is a 73 percent symptomatic infection rate out of 111 people at a Christmas party in a Norwegian restaurant, who *"were young and fully vaccinated."*[130]

The ten percent of contacts who catch *SARS-CoV-2* is an average. A random sample of the population, particularly when the groups are small and when they may have been exposed to similar other infections in the past, could have a higher proportion of susceptible people. It is also critical to understand the importance of the dose of virus that a person is exposed to. Occasional superspreaders emit virus at extraordinarily high rates leading to these extreme accounts.[131] The higher the dose the higher the proportion who will be infected. The fact that there are occasional superspreader events with high attack rates should not detract from the key point that overall only around ten percent of household contacts will catch each variant. The figure for contacts outside the home is only five percent. The remaining population are not susceptible to average rates of exposure.

Hector Drummond, author of *The Facemask Cult*, points out that 90 percent of virus emission in one study of 37 people came from just 2 of them. *"The great majority of the subjects produced very small amounts of virus copies in their fine aerosols relative to these two."*[132] Such evidence confirms the concept of the superspreader.

The claims made about the importance of superspreader events seem, in retrospect, to have been greatly exaggerated. The fact that chains of transmission die to nothing relatively quickly could not explain how waves surged, but superspreader events could be used to explain the surges. Superspreader theory suggests that such events accounted for 80 percent of transmission. If true, in 2021 alone, one in every nine people would have caught covid in a superspreader event. Even though smaller superspreader events may have become old news, the small proportion of larger events would surely have remained newsworthy. Yet, where were the reports? The only such reports were of events with large numbers of attendees, like Glastonbury Festival, where the rate of infection among those attending was similar to the infection rate in the background population at the time.

Even those superspreader events that were reported were exaggerated. A database of global superspreader events compiled by the London School of Hygiene and Tropical Medicine overestimated the numbers in several ways.[133] Firstly, instead of counting those infected at a particular venue they included every subsequent case in the community that resulted from the consequential chains of spread e.g. to family members at home afterwards. For example, 1,200 people developed symptoms after attending two huge church gatherings in South Korea and this was exaggerated to 5,016 cases once every associated case had been included. The worst example was an inclusion of what may indeed have been a superspreader event at a football match in Bergamo which was reported as having involved every one of the people in Bergamo who had ever tested positive at the time of the report, all 7,000 of them.

Entire outbreaks that occurred in prisons and nursing homes were included as if every infection was caught within a day or two of each other. For example, an outbreak in a prison in Washington state that lasted *"just a matter of weeks"* was included as a superspreader event when the same profile in the community would be considered the natural trajectory of a wave.[134] Similarly, an outbreak over the course of a fortnight in a prison in San Luis

Obispo, California affecting 180 people was reported as a superspreader event affecting 614 people, which included all the inmates and staff who had tested positive for the whole of 2020.[135]

Finally, the reported numbers of infected individuals were also exaggerated by including everyone with a positive test result, even if they did not have symptoms of infection. Positive test results, particularly in those who never develop symptoms, can reflect contaminated air in a healthy person (see *Belief Six: If you test positive you have covid*).

One well publicised superspreader event amongst the Mount Vernon Presbyterian Church choir in Washington led to singing being banned globally even in nursery school. The story was of a rehearsal with sixty members, on 6th March 2020. They used hand sanitiser and remained socially distant but forty five members of the choir, with an average age of 67 years, contracted covid and two died.[136] A retrospective analysis of the outbreak revealed that many members of the choir were coincidentally already infected from community transmission in the days leading up to their rehearsal.[137] The presence of multiple infectious individuals among the group likely led to an unusually high dosage of exposure. People who may have been immune with a lower dose may well have been susceptible to this higher dose.

VIRUS VS IMMUNITY

What difference does dosage make? There is a balance between immunity and viral attack. In theory, an infection occurs when a virus overwhelms immune defences. In theory, there exists a big enough dose that anyone's immune system would be overwhelmed. For a more typical dose there will be a level of immunity which is just sufficient to provide protection. If immune protection is less than this threshold then a person would succumb. People with adequate immunity to protect them from average exposure to a particular variant could be overwhelmed with high levels of exposure.

There is one source of evidence (albeit incidental) on dosing and susceptibility. In a human challenge trial, scientists deliberately infected young people in order to test a nasal covid vaccine. To reflect the real world the right dose had to be selected. The dose was decided based on how many cells could be infected in the laboratory and there was a plan to increase the dose if it was insufficient.

In the end the starting dose proved to be huge. A natural dose results in several days with no symptoms while the virus replicates exponentially, doubling in quantity every six hours.[138] The average time to become symptomatic after natural exposure was five days, or twenty doubling periods. At that point, each virus particle will have become over a million new particles giving enough virus to damage sufficient cells to cause symptoms. During the human challenge experiments more than a third of those infected developed symptoms within a day and the mean time to symptoms was only three days, or twelve doubling periods.[139] By three days there had only been sufficient time for each virus particle to have produced two thousand new particles. That would imply the infected dose given was 250 times higher than a natural exposure. A very low quantity of virus, perhaps only a single particle[140] for the most susceptible, is the infective dose in a natural environment.[141] Despite the enormous dosage only half of the participants tested positive and developed symptoms. Half were immune to infection even when exposed to a huge dose of virus, as likely happened on the *Charles de Gaulle* ship.

The importance of the amount of virus that people were exposed to was also evident on a population wide level. Densely populated cities had much higher attack rates than rural areas. London and Stockholm both saw 17 percent of the population develop antibodies in spring 2020 compared to 7 percent of those countries as a whole. The sheer number of infected people in a small space breathing out millions of viral particles every hour would have meant a much higher likelihood of being exposed to a sufficient dose to cause a symptomatic infection and the subsequent development of antibodies. In New York City, one of the most densely populated cities in the world, nearly a quarter of the population were infected in spring 2020 and areas with more overcrowded living conditions were worst affected.[142]

In the UK, there was no difference in infection rates among non-healthcare key workers compared to other workers who stayed at home.[143] Even healthcare workers in non-patient facing roles had similar infection rates to people who were not key workers. The exceptions were healthcare and care home workers who were patient facing and were more likely to have been exposed to large amounts of virus. However, even for these groups six out of seven had not been infected by autumn 2020.[144] Moreover, a proportion of those healthcare workers who had not been infected in early 2020 went on to be infected in each subsequent wave. The evidence that only a proportion of the population were susceptible to each variant was studiously ignored. This enabled people to ignore the extent of exposure and the role of immunity and to hold on to the belief that close contact transmission was key and that behaviour change had stopped spread.

Early observations of both the severity of the virus and the proportion who were susceptible came from reports of an outbreak on a cruise ship, the *Diamond Princess*, which was quarantined off

Japan in February 2020. It is a shame no-one has kept track of the people who were on the *Diamond Princess* to see how many avoided infection with the Wuhan variant but have since succumbed. This would have shown if they were immune originally but susceptible to a different variant.

WHY WASN'T EVERYONE SUSCEPTIBLE?

The government model of a fully susceptible population was based on the idea that immunity to a partly novel virus could only be achieved through exposure to that virus or vaccination. The WHO said in December 2020, *"WHO supports achieving 'herd immunity' through vaccination, not by allowing a disease to spread through any segment of the population, as this would result in unnecessary cases and deaths."* [145]

The WHO were assuming that only immunity specific to *SARS-CoV-2* could confer any protection. This exposes an embarrassing failure to understand how the immune system works. Our immune systems do not recognise whole organisms, ticking them off on a checklist of what we have and haven't been infected by. Nor do our immune systems recognise particular genetic sequences that are unique to particular infections. Instead, they recognise tiny parts of foreign organisms by their shapes. Rather than recognising the whole shape, different elements recognise tiny, chopped up fractions of the whole. In this way, even an apparently novel pathogen can be recognised as foreign. As long as there is a fragment of a virus or bacteria that our bodies can detect as different to ourselves then a specific immune response will be generated.

The beauty of this system is that our immune systems can be educated by one virus or bacteria in a way that protects us from a completely different one. For example, a study demonstrated clearly that antibodies to mumps provided significant protection from *SARS-CoV-2*.[146] This applied to antibodies developed after infection but was even more protective if the antibodies were acquired after MMR vaccination.

The education of our immune systems over the course of our lives is likely to contribute to the age differences in infection rates. Antibody levels from infection were substantially higher in the young than the old.[147] The difference between children and grandparents could be attributed to high mixing in the young and shielding in the old, with the working aged population falling between the two extremes. However, there is likely to also be a contribution from older people having more years of immunity that provides cross reactive protection. Being old is a double edged sword. As well as having a more educated immune system, immunity starts to decline as we reach the end of life, leaving the most vulnerable susceptible and at risk of dying.

Although antibodies have often been emphasised, the immune cells most critical to our viral immunity, the ones which kill cells infected with a virus, are called T cells. Our immune systems remember which T cells were successful. Patients who had *SARS-1* still had a strong T cell response to the virus seventeen years later in 2020.[148] T cell responses to other infections were shown to cross react against *SARS-CoV-2* providing protection for a large proportion of the population, from any one variant.[149] It is impossible to reconcile that fact with the idea that this virus was totally novel and therefore everyone was susceptible. The vast majority had immune systems capable of recognising it in some way and mounting a protective immune response.

HOST FACTORS

The immune system clearly plays an essential role in protecting people from any particular variant but the story of virus vs immunity is not as simple as claimed. There are conflicting views about the events that lead to an infection. The conventional view is that we rarely encounter viruses and when we do our immune system comes to our defence. One alternative view is named terrain theory. This theory holds that we are constantly being bombarded by numerous viruses and it is a dip in our ability to fight them off that results in an infection. Anyone who has cold sores, caused by the herpes simplex virus, will be familiar with how the immune system can keep the virus in check for years before a flare up when a person is 'run down'. People who have had a bone marrow transplant which requires a temporary obliteration of the immune system need to take prophylactic treatment to ensure they do not succumb to infection from viruses, bacteria or fungi. A major risk of bone marrow transplantation is dying from these infections so demonstrating the importance of the immune system in tipping the balance towards an infection. The terrain theory of susceptibility would predict high rates of infection among the population hospitalised for other conditions, which is what was seen.

A curious piece of evidence from hospital data supports terrain theory. Covid cases acquired in hospital peaked before covid admissions from the community.[150] If viral attack resulting in hospital transmission was key then the cases acquired in hospital would peak after the numbers admitted from the community had peaked. The data appeared to indicate a factor making people susceptible to infection that faded at a similar time in the hospitalised population as it did in the community.

On the other hand, if immunity were the whole story we would expect to see the highest infection rates in people with impaired immune systems. Although certain groups of adults with decreased immunity did have worse outcomes,[151] there was no increase in infection risk.[152] Perhaps that was due to shielding behaviours but there is also no increase in risk for children who are immunosuppressed and far more likely to be interacting with others.[153] If terrain theory were the whole story then there would not be numerous examples of healthy fit people who became infected and symptomatic with covid.

In reality both susceptibility and presence of virus are required for an infection. However, presenting disease spread as a simple battle between virus and immunity misses several paradoxes about how both *SARS-CoV-2* and influenza spread.

SEASONAL PATTERNS

Hope-Simpson, the GP we met earlier who dedicated his life to studying influenza, noted that there were several unexplained mysteries of influenza waves which he attributed to an unknown *"seasonal trigger"* saying, *"The seasonal influence often operates contemporaneously at places lying at the same latitude whatever their longitude."* [154] These included the fact that people who would be susceptible to a particular variant in the winter, did not succumb to it the preceding summer, even when it was already circulating. Something caused the virus that was already circulating to grow exponentially once autumn arrived. The trigger causes a small fraction of the population to become susceptible to influenza where they were not before. In his writings Hope-Simpson never claims to understand what that factor was.

The seasonal trigger comes from conditions for spread being ripe. It has been likened to tomato or potato blight spreading and destroying crops within days. The fungal organisms responsible are present all the time but it takes the right conditions for it to create a dramatic surge in disease. Several components from environmental factors, like temperature and humidity, to human factors affecting immunity and even differences in the virus itself may all play a part. Viral mutation may contribute to this in the time it takes to reach a point where it can evade immunity in a segment of the population and trigger a new wave. Hope-Simpson comments on how researchers found it easier to infect animals in the winter even with a controlled level of humidity and temperature suggesting there is a factor in addition to the environmental conditions.[155] Not knowing the details of why it occurs should not stop us acknowledging it as an important factor and studying it.

The seasonal trigger effect on influenza has been evident for hundreds of years and there is more than one trigger each year. Each winter season would start with one dominant influenza variant surging in autumn. As herd immunity built to the first variant the second one became dominant. Occasionally, both halves of the season were dominated by a single variant but with two separate surges. Often the dominant variant from the second half of one winter was responsible for the first peak the next winter. There were unpredictable years where a third spring surge would also occur. The reason why these seasonal surges happen yet the virus has little impact in the summer months is poorly understood.

The conventional textbook explanation for surges was that *"In winter, people huddle together indoors, promoting the transmission of airborne and droplet infections."* [156] Such a theory seems credible in the UK where we fling open the windows and take every opportunity to enjoy the mild summer weather. However, in numerous parts of the northern hemisphere there is plenty of huddling indoors in the summer months to keep out of the heat with no consequent rise in respiratory infections. Moreover, places with a mild climate, like Australia and California, also saw winter influenza waves.

Although influenza is a separate family of viruses to coronaviruses, they both have seasonal surges. Therefore, it is worth considering whether covid was responding to a trigger in the spring. Winter 2019/20 saw an influenza virus as the dominant seasonal virus which perhaps, prior to February 2020, prevented *SARS-CoV-2* from dominating. Influenza disappeared globally, first from China in January, then Japan in mid-February and three weeks earlier in Italy than in Sweden and the UK, coincident with the *SARS-CoV-2* surge.[157] This suggests a reciprocal relationship. In 2022, several countries saw a reciprocal relationship where influenza would re-emerge at times of low covid but disappear again on the arrival of a new wave and vice versa. Further investigation of the similarities and differences between influenza and *SARS-CoV-2* certainly warrants investigation.

There is good evidence of *SARS-CoV-2* circulating globally in autumn 2019. The illness has such characteristic symptoms that many people are confident that they were infected in 2019 despite the lack of a positive test to confirm their belief. One early recorded death from covid in the UK occurred in January 2020, implying the infection was caught in December 2019.[158] The Military World Games were hosted in Wuhan in October 2019 and many of the nine thousand attendees reported covid like symptoms subsequently before returning to countries all over the world. Mike Gallagher, US congressman wrote to the Secretary of Defence and Chairman of the Joint Chief of Staff in the USA asking them to investigate: *"Athletes from Italy, Germany, Sweden, and Luxembourg may all have become sick with symptoms consistent with COVID-19 at the games... One athlete reported 'nearly empty streets. 'It was a ghost town.'"* [159]

However, data on covid deaths, covid cases and the rise in symptoms all show that the surge began in February at the earliest.

SARS-CoV-2 was around before February, it was not surging because of a lack of a seasonal trigger.

How do we know there was a genuine surge? An Imperial College University group was set up, separate to Neil Ferguson, to carry out mass population screening and surveys, called REACT. More than ten thousand people, from right across the country, who developed antibodies to *SARS-CoV-2* were asked when they had their symptoms.[160] The responses provide a symptom based measure of the trajectory of the epidemic. There were cases reported throughout the preceding winter. However, there was a definite and precipitous rise in symptomatic cases starting in the last few days of February 2020. The key question is why exponential growth did not happen from autumn 2019 or in December and January and then did occur in February 2020. A response to a seasonal trigger would explain this.

Similarly, random sampling of patients seeing their doctor with influenza-like symptoms showed only a handful of positives until suddenly, in mid-March 2020, half of the samples sent came back positive for *SARS-CoV-2*.[161] Again, this indicates that although *SARS-CoV-2* was present prior to March 2020, another factor meant that susceptibility rocketed in February and March.

What could cause a surge? Human gene expression and vitamin D levels also show a marked seasonality in the northern hemisphere. Tests of which genes our bodies express show a massive switch between mid-winter and mid-summer, particularly in genes related to immunity.[162] Whether this is a cause or an effect of our relationship with respiratory viruses is not known. Vitamin D levels also show a strong seasonal variation and there are several studies showing that people with low vitamin D levels are more susceptible to infection and to complications from covid.[163] However, Vitamin

D levels peak in September when covid levels begin to rise so there must be other factors at play too.

The fact that deaths peaked in January as they always have due to other respiratory viruses, suggests that it was the seasonal trigger that was determining the extent of the winter wave. If differences in seeding patterns and behaviour were the key determinant of a wave then the fact that every northern hemisphere country saw peak deaths each January seems like quite an incredible coincidence.

There is no doubt that covid is seasonal in the same way influenza was even though as late as December 2021, Vittoria Colizza, director of research at the French Institute of Health and Medical Research, was one of several co-authors of a paper claiming, somewhat improbably, *"containment measures are estimated to have a larger impact on the epidemic compared to seasonal effects only."* [164]

Seasonality doubters often seem to define 'seasonal' to mean 'only in winter'. That is not what it means. It simply means there is a predictable pattern based on the time of year. Coronaviruses have been known for years to have a seasonal pattern.[165] There was an established body of work showing that other coronaviruses have a winter surge and are only present at low levels in the summer. Because of the emphasis that the media made about each wave being driven by viral variants the seasonal nature was denied. However, covid hospitalisations and deaths showed a very clear low each summer and high each winter. The levels in autumn and spring were more variable but were intermediate between the winter high and summer low.

What about the summer waves of covid? Studying hospitalisation and death numbers is crucial as they provide a basis for comparison with other respiratory viruses such as influenza, and are not as influenced by differences in testing. In 2009 and 2010, there was aggressive testing for 'swine flu' influenza, as has been done for *SARS-CoV-2*, which revealed four peaks of infection occurring at the end of December, March, July, and October in some countries in the Northern hemisphere.[166] Similar to covid, influenza cases in the summer mainly affect the young and result in fewer hospitalizations and deaths than cases in the winter. Therefore, it is important to analyse hospitalisation and death rates when assessing the severity and impact of respiratory viruses, as these measures provide a more reliable comparison than other factors such as testing.

Some counter the claim of seasonality by pointing to Israel and South Africa and show that they have had waves outside of winter. Israel has shown two distinct waves each year. The first starts in July and peaks in autumn and the second starts in December and peaks in January with few cases in spring. The records for influenza in Israel historically show a similar pattern. Hope-Simpson, the GP and influenza expert, demonstrated that although influenza death waves were distinct winter phenomena in the Northern hemisphere they continued for many more months the closer a country was to the equator. *"The conclusion seems inescapable that, viewed on a global scale, epidemic influenza is moving annually south and then north through the world population, a smooth yearly scanning of the world very different from the local episodic picture of the disease."* [167] *SARS-CoV-2* appears to have adopted a similar pattern.

South Africa is the other example often cited because they have had summer waves of covid. In 1918, South Africa had two spring waves in September and October with the following July winter wave being much less impactful. Influenza in South Africa also showed a clear second wave in September in 2011 and 2013.[168] Again, the pattern for covid has been two waves a year with peaks in August and January. Admittedly, January is later than the spring influenza peaks of the past but it has been replicated each year with covid. Seasonality is therefore not as clear cut as winter versus summer but more a reflection of the timing for when a particular country has conditions ripe for a surge.

What these two counterexamples tell us, is not that Israel and South Africa have unusual covid patterns but that they had unusual influenza patterns which covid copied.

Despite plenty of evidence that covid waves recur over time and each variant reaches the peak after a similar time, there is continuing denial of this fundamental truth by those who have made the most catastrophic mistakes. Matt Hancock, continued to claim in July 2022, that the forecasts were right and there would have been 500,000 deaths without lockdown, saying, *"Making forecasts of what would happen if you did nothing and then doing something and the forecast not coming true does not disprove the forecast."* [169]

He and others were never asked why they believed the Wuhan variant would have behaved differently to every subsequent variant.

GLOBAL PATTERNS

Some anomalies were clearly geographical yet believers in close contact transmission tried to justify these anomalies as due to differences in human behaviour. For South East Asia and the Antipodes at the outset every country was spared and this was attributed to mask wearing and efficient test and trace work. Not one country was an exception. Roll forward to spring 2022 and every country in South East Asia and the Antipodes had a large covid wave with significant hospitalisations and deaths. Not a single country was spared. Deaths reached as high or higher than the European average seen in spring 2020. How did interventions that had apparently worked so well before, all fail at the same time in every country in spring 2022?

Similarly, all of Eastern Europe had only minimal covid in spring 2020 and this was again attributed to interventions even though these same interventions, did not have the desired effect in Western Europe. The fact no country was spared in the affected regions and every area was spared in unaffected South East Asia and Eastern Europe, suggests a geographical environmental difference in transmission or broad differences in population susceptibility. A lack of seasonal trigger in multiple countries over a broad geographical area explains it. The same likely applies to the band of states in the centre of the USA which saw only low levels of covid in spring 2020 but were more badly affected by later waves.

To be fair, none of this was known at the outset. However, a fair amount of it could be assumed. We knew that coronaviruses had a similar seasonal course to influenza.[170] We knew the winter spikes peaked without intervention. We knew about the ubiquitous nature of influenza spread. We knew that influenza did not behave like measles continually spreading to anyone not previously infected. Instead it took around a decade to pass through a population, (sometimes skipping a year), only affecting a small fraction of the population each winter. We knew that influenza filled the air indoors in winter such that exposure was inevitable.[171] We knew that once infected, people were well protected from future variants of that particular strain.

With influenza people had a more fatalistic approach. Yes, a proportion of the population died every year from influenza. A number of these people were young and otherwise healthy. However, there was an acceptance that if you were vulnerable to a particular influenza variant then there wasn't much you could do about it. We all breathe the air and, at least in winter, indoor environments are full of infectious doses of influenza virus. Given the extent of exposure all that avoidance would have achieved would have been to delay the inevitable. Somehow, people have been unable to adopt the same fatalistic approach to covid. It is impossible to avoid breathing the air. If there is a variant that you are susceptible to then it is very unlikely that you will be able to avoid being exposed to it during the course of a wave and before the wave is finished you will succumb.

With any one variant, there is at least an eight out of ten chance that your immune system will protect you completely. For those who are susceptible to that variant there is little that can be done to avoid catching it. Attempts can be made to delay exposure. Authorities were honest about this in spring 2020, making clear that infections could not be prevented but aiming to delay them in order to *"flatten the curve"* of hospitalisations. Patrick Vallance, Chief Scientific Officer was clear that there was nothing that could be done to prevent infection, *"It's not possible to stop everybody getting it."*[172] Preventing infection by a ubiquitous virus in the susceptible is not achievable. If you have not been infected you either have durable prior immunity or there is likely a variant with your name on it that is yet to come.

	THE EVIDENCE MANIPULATION TRIAD
EXTRAPOLATE	Despite all the similarities with influenza and despite knowing that coronaviruses have a seasonal pattern, it was assumed that SARS-CoV-2 would behave like measles. This assumption was extrapolated to form the belief that only altering human behaviour would stop the virus working its way through the entire population and infecting 85 percent of the population in the first wave.
EXCUSE	The evidence for the seasonal trigger has been excused for years as being caused by people huddling indoors in winter.
EXCLUDE	All the key questions have been ignored completely. Why do waves peak independently of restrictions? Why does exponential growth occur even in strict lockdowns e.g. Australia and Shanghai? Why does a surge in infections occur with a variant that had been present for months without exponential growth? Why are people who were immune to a previous variant susceptible to a subsequent one?

TOP THREE MYTHS
1. Everyone was susceptible to the Wuhan variant
2. There was no seasonality
3. Surges were due to human behaviour rather than the conditions being ripe

5

RISK

The fear of death follows from the fear of life. A man who lives fully is prepared to die at any time. **Edward Abbey, 1991**[173]

Humans, even those comfortable with numbers, are notoriously bad at assessing risk. Even when our emotions are calm and we are able to focus the numbers can be hard to make sense of. But the real challenge happens when our emotions are engaged. Humans overemphasise anecdotes which cause us to engage emotionally while struggling to conceptualise very large or small numbers that lie beyond those we use day to day. If someone tells us the risk (in the abstract) of something horrific happening it is hard to disengage our emotional response to the horror in order to assess the likelihood of the hypothetical risk and this leads us to overestimating dangers. Engage a fear response and the rational side of our brains don't stand a chance.

From an evolutionary perspective this makes sense. Our brains are wired to be oversensitive to danger. In terms of evolution, only three things matter: food, safety and sex. Those of us better able to ensure survival, via finding food, keeping safe, successfully reproducing, and keeping offspring fed and safe, will make a larger contribution to the human gene pool. Genes that influence the way we think, feel and act with regard to danger have been selected and our brains hardwired to keep us safe. Some have argued that people's many talents and characteristics that contribute to the rich layers of life's tapestry only survived because of their contribution to securing food, sex and safety.

Because these evolutionary drivers are so fundamental they act through our emotions not through rational processes. Unfortunately, emotional drivers can override our rational thought processes. Overcoming depression or phobias takes more than a rational decision to not feel that way.

Dangers that are a threat to life and invisible can have a devastating impact. Any danger that is a threat to our lives can create an overwhelming fight or flight response.

Threats to life that we encounter more frequently, like those experienced by someone who regularly goes hang gliding, become familiar enough that we can control our emotional response and remain relaxed while contemplating them. Our rational brains provide logical reassurance that the threat is tiny as they do every time we cross the road, meet a stranger or see a lion in a zoo. When the risk we are presented with is outside of our everyday experiences we are further disabled by not being able to use past experience as a way of rationalising the risk and putting it into perspective.

Physical, real world threats can be dealt with and the danger overcome. Momentarily mistaking a stick for a snake is a good trade-off, designed to keep us safe. However, reassurance about an invisible danger can never be complete. Every reminder of it, say from walking past masked strangers or one way marker stickers, can evoke a further response.

Only a few were able to engage their rational brain in winter 2020 when the country was plastered with posters of sick, frightened people, with zombie-grey skin and piercing eyes captioned *"look into her eyes and tell her the risk isn't real."* No one has ever said there wasn't a risk, only that people had failed to appreciate how low their risk was. A fear propaganda campaign was designed to make that even harder.

One shortcut to assessing risk is to judge how people in positions of authority portray the risk. In spring 2020, Chris Whitty rightly emphasised that, *"For the great majority of people this will be a mild or moderate disease, anything from a sniffle to having to go to bed for a few days rather like with mild flu."* [174] His concern was that the low level of risk to an individual could – on a population level – lead to pressures on the NHS. By 22nd March 2020, the behavioural science advisors, SPI-B (Scientific Pandemic Insights Group on Behaviours), were saying, *"The perceived level of personal threat needs to be increased among those who are complacent, using hard-hitting emotional messaging."* [175]

The SPI-B team of 37 predominantly behavioural psychologists were set to work. Their remit was to 'nudge' or manipulate the public's behaviour. Their chosen methodology was to terrorise the public with fear messaging. Unfortunately, they were very good at it. (Other governments were so impressed that a company which was a spin off from SPI-B's original incarnation, the Behavioural Insights Team (BIT) or the 'nudge unit' is now active in 35 countries[176] and was sold for £15.4m in 2021). [177]

For a member of the public to manage to derive a correct interpretation of their personal risk they would not only have to quell their fear and crack the maths they would also have to handle the confusing situation of the government messaging seriously conflicting with reality. Yes there was a risk, but it was not what was being portrayed. Even the exaggerated worst case scenarios in terms of individual mortality had tiny risks for the young. The population had demonstrated that they would make huge sacrifices for the sake of others and to stop the NHS being overwhelmed. Why not continue to use those reasons as motivation to minimise spread? Why scare people as well? Not many were able to step away from the fear and accept that the government was behaving in a very odd manner.

By June 2020, three in ten people avoided public transport and nearly a quarter avoided public places altogether.[178] For those of us who did go out, it was easy to believe that life was returning to normal, because we never bumped into those who had been most terrified and were hiding at home. For those that were crippled with anxiety, many did not seek help as the daily news feed led them to believe their responses were perfectly healthy and rational.

BELIEF THREE: COVID WOULD LIKELY KILL ME

We are in a war against an invisible killer and we have got to do everything we can to stop it. **Matt Hancock, March 2020**[179]

We were told it was unprecedented. Our worlds were put on hold to prioritise it. There were posters, stickers and masked faces everywhere one turned as a reminder. So surely it was a significant threat? When a threat is amplified and reinforced by authorities then people's estimates of risk can go awry. Dr Colin Foad led a UK study surveying attitudes to covid and found, *"people judged the threat of COVID-19 via the magnitude of the policy response."* [180] Americans under 40 years of age consistently overestimated their risk thinking they had a 10 percent chance of dying if they caught covid[181] while research from Cambridge University would suggest in reality it was less than 0.024 percent.[182] Australians, on average, estimated their risk of dying at 38 percent, which was many fold higher than reality.[183] Trying to get people to judge their risk based on the actual numbers rather than an emotional reaction to what is happening around them is very hard.

No one was immune to the fear distorting their perception of risk. As Mark Woolhouse, professor of epidemiology on SAGE in his book, *The Year The World Went Mad*, commented that in summer 2020, *"not only politicians, journalists and teachers but many scientists and even some doctors... were unaware of the huge variation in risk across the population."* [184] Those who believed that younger people were at significant risk had their beliefs confirmed when senior government minister Michael Gove, said *"the virus does not discriminate... we are all at risk."*[185] BBC news then regularly reported rare tragedies among low risk people as if they were representative.

Understanding the risk of dying was not helped by people treating life expectancy as an entitlement rather than an average. By definition half of us will not reach our life expectancy. That works out at a quarter of a million British people and one and a half million Americans who die every year *'before their time'*. Overall more than one in five deaths of adults are in those under 70 years of age and one in ten are under 60. (For men alone, a quarter of adult deaths are in under 70 year olds and 1 in eight are under 60).

An increasing proportion of deaths have been from the baby-boomer generation. These were post-war babies for whom new schools were specially built. Their teenage rebellion gave us rock and roll and sexual emancipation. The economy grew with them such that disposable income and lives improved as they aged. They have reinvigorated retirement with volunteering, new careers and studying. In 2020, they were reaching a stage of life where they had to confront their own mortality. The stage was set for covid to trigger an existential crisis among baby boomers that affected us all. Ken Dychtwald is a gerontologist and author and has studied the baby boomer generation for four decades, *"Boomers have never been a stoic bunch. They're not going to allow their last chapter in life to be an extended period of loss, fear, pain, and suffering."*[186] The baby boomer generation has influenced the world as they progressed through every life stage and now they are changing the world one last time as they approach death.

The fear of dying from covid was only a symptom of the dysfunctional relationship our society has with death. Our society is not comfortable with dying. Historically, this was not the case. In Victorian times, the body of the deceased would be washed and then dressed up and displayed in the coffin surrounded by flowers.

Now bodies are removed promptly on death, making it harder for us to confront the reality of it. Victorians designed cemeteries with recreation in mind and people would meet and picnic there, relaxing. Yet for many, the threat of covid meant an unwelcome contemplation of mortality. If people already had a strong sense of their prospects as to their risk of dying each year then that could have provided an anchor with which to compare any increased risk from covid. However, there was no anchor. People were not prepared to face their own mortality. The result was a misinterpretation of personal risk.

WHAT WAS THE RISK TO AN INDIVIDUAL?

The average risk from covid was well understood in late spring 2020. It did not change until 2022 when it halved for both the vaccinated and unvaccinated with the arrival of the Omicron variant. A useful handle on the threat from dying if you caught covid was that the risk was almost exactly the same as your overall risk of dying that year from other causes. However scared you were of dying this year, is how scared you should have been about dying if you caught covid. If you had a realistic view of your own mortality this was a useful yardstick. The same information could, however, bring home to older people a sense of a high baseline mortality they had previously managed to ignore and rather than anchoring their fear of covid to a realistic low risk, make them more scared of dying.

The government reacted to the threat in an extreme way. The public look to the reaction of those in authority as a guide to how serious a threat is. Based on their reaction, it was only reasonable to deduce that the threat must have been severe. Being told that your risk of covid was similar to your underlying risk of dying this year was therefore not reassuring. Rather it led to an existential crisis. Somehow, the idea that those with least life left should be living it to the fullest was totally lost.

The risk of dying if you caught covid is shown in Table 1. Estimates vary. These risks are based on data from pre-Omicron variants provided by Cambridge University's Biostatistics Unit and are not the lowest estimates.

	CHANCE OF DYING IF YOU CATCH COVID (ADD A ZERO TO ACCOUNT FOR THE FACT MOST DID NOT CATCH ANY ONE VARIANT)	SAME ORDER OF RISK AS	NUMBER OF SEQUENTIAL HEADS TOSSED IN A ROW IN A COIN TOSS
UNDER 5 YEAR OLDS	1 in 270,000	Dying this year from a fire	18
5 TO 14 YR OLDS	1 in 77,000	Dying from a general anaesthetic	16
15 TO 24 YR OLDS	1 in 29,000	A clover is three times more likely to have four leaves and an oyster to have a pearl.	15
25 TO 44 YR OLDS	1 in 4,000	Four times less likely than the chance of finding a double yolk when you crack open an egg.	12

45 TO 64 YR OLDS	1 in 560 to 1 in 280 (at peak deaths)	Picking two aces in a row from a pack. During peak death, it was as likely as drawing four cards in a row from a pack and them all being Kings, Queens or Jacks.	9
65 TO 74 YR OLDS	1 in 120 to 1 in 43 (at peak deaths)	In summer, you would have been more likely to win after placing money on the horse with the worst odds in the grand national than to die if you caught covid. However, in December 2021 it was more likely but still only as likely as placing your money on zero in roulette and winning.	7
75 YR OLDS AND OVER	1 in 29 to 1 in 5 (at peak deaths)	In summer, the risk was of flipping a coin 5 times and it coming up heads every time. At peak deaths four in five survive	5

Table 1: *Risk of dying with covid based on Infection Fatality Rates from the Medical Research Council's Biostatistics Unit at Cambridge University*[187]

The risk for older age groups has not been constant over time. At peak deaths in winter the proportion of cases dying also peaked. In the intervening summer months the proportion dying dropped to a low. Part of the reason for this has been that transmission in care homes and hospitals has also varied over time with winter transmission seemingly impossible to control. The plethora of measures deployed, from testing, protective equipment, banning visitors and vaccination, all appear to work well in the summer before collectively failing to work at all in winter (making all the measures pointless).

HOW DO WE KNOW THE RISKS ARE OVERESTIMATES?

Had 85 percent of the population been infected based on these risks the total death tally would have been somewhere between a quarter of a million and over one million deaths. These risks equate to an infection mortality rate similar to or higher than the 0.9 percent used by Ferguson. Therefore these risks represent the threat as a worst case scenario before a full understanding of the actual risk could be measured.

It is important to emphasise that the risk of dying of covid is zero in any one wave if you do not catch it. Only around 7 to 15 percent of people have caught each variant and this has been true for every variant in numerous countries. The chance of catching covid in a particular wave was the same as picking a card from a deck and it being an ace. For Omicron, it would have been an ace or a king. Therefore, for any one (pre-Omicron) wave, the figures in the table above showing the chances of dying would be ten times less likely than stated e.g. for 25 to 44 year olds it would be a 1 in 40,000 chance.

The government said in August 2021 that the overall risk of dying was not 0.9 percent but 0.096 percent so another zero could be added to the risk of dying figure in the table if this was believed.[188] I think this lower estimate was too low as it was based on flawed PCR testing which estimated twice as many cases in the population as antibody testing did (see *Belief Six: If you test positive you have covid*). The risk was reduced still further with the arrival of the less deadly Omicron variant. If Omicron had arrived without any of the fear built up around previous variants it would have been totally unremarkable. Deaths were lower than in an average winter.

A high proportion of those who did die were already at high risk of dying. In March 2020, Neil Ferguson said, the proportion who died from covid that would have died soon anyway, *"might be as much as half or two thirds of the deaths we see, because these are people at the end of their lives or [who] have underlying conditions."*[189] In fact 95 percent of deaths with covid were in people with pre-existing conditions.[190] With increasing age, a growing proportion of the population have a condition that the NHS would classify as a pre-existing condition. Nevertheless, the risk for the totally healthy is vastly lower to that which was presented by authorities and even in the table above both of which relate to the *whole* population.

The counter, of course, is that people who are obese or have comorbidities including hypertension, dementia and diabetes are at higher risk. People from ethnic minorities and lower socioeconomic groups were also at higher risk of dying. Someone with multiple comorbidities would adopt the same risk as a person ten years older than them (but remember how exaggerated the figures in the table are).

How much difference did comorbidities make? Based on spring 2020 estimates, the chances of a healthy 35 year old woman dying having caught it would be less than the fatality risk from a general anaesthetic.[191] For a 55 year old man with comorbidities the risk, having caught it, was less than the risk of a woman dying in childbirth. For a healthy 75 year old woman the risk, if she caught it, would be less than the risk of her being injured in a car accident this year. Even for an 85 year old with comorbidities the risk of dying from covid would be less than the chances of dying while in a care home for one year.

A disproportionate number of covid deaths occurred as the result of infections acquired in hospitals and care homes. Including these deaths as part of a risk calculation is disingenuous when talking about healthy elderly people in the community. For example, a sample of 30,000 patients who caught covid before spring 2021 only after a long stay in hospital had a mortality rate of more than one in four.[192] Those closer to death by virtue of age, comorbidities or current illness were all more at risk from covid and extrapolating their risk to the whole population was a way of inflating fear levels.

RISK TO CHILDREN

The risk in children was comparable to seasonal flu. It was never common knowledge but a few children and young people, even healthy ones, would succumb every year to influenza. When I was a medical student, I met an otherwise healthy man in his twenties who was struggling to stay alive in intensive care and this had a dramatic impact on me. I was upset by his situation but also surprised at the realisation that the public merrily ignored this risk from influenza. This was the equivalent for me of the *"Look him in the eyes and tell him the risk was not real"* poster. A single example is not a good way to judge risk. Influenza contributed to the deaths of a large number of people each year and a proportion were young and healthy and their deaths are an even greater tragedy than a death in the old. However, worrying about this would not help anyone. In the grand scheme, the public's relaxed attitude to 'flu' was the right approach all along.

In all of 2020 and 2021 there were 41 covid attributed deaths in under 15-year olds which was slightly less than the average number of deaths attributed to influenza over a two year period in the past.[193] Even then that is the risk of dying 'with' covid. The Paediatric Intensive Care Audit Network, PICANet commented, *"It is not possible to say that any of the deaths on PICU were as a direct result of Covid-19, merely that these children had a Covid-19 positive test prior to or during their PIC [paediatric intensive care] admission or at post-mortem."* [194] That does not mean that all the children had pre-existing diseases. For example, a healthy child who caught meningitis and needed intensive care might later have tested positive for covid before dying. Clearly the risk of dying would be higher for a child who caught it when already in intensive care for another condition. The risk for a healthy child at home is therefore far lower than the 1 in 77,000 figure shown in the table.

PUTTING THINGS IN PERSPECTIVE

By the end of 2021 there had been 800 deaths of under 40 year olds and fewer than 150 of them had no pre-existing conditions. Have you ever had the slightest concern about being bitten by a snake in this country? It is quite reasonable not to have been concerned; the chances are incredibly rare. The NHS reports that there are 100 snake bites from adders every year. Had every snake bite made it into the newspapers each day then your concern levels may well be raised but the underlying risk is just as negligible. In terms of risk of dying, around 650 people die falling downstairs every year, a similar number are murdered and another 200 people drown. Just like the government poster, you could not in earnest *"look these people in the eyes and tell them the risk was not real,"* but that risk certainly was negligible. It was not worth you changing how you lived your life in order to avoid the very real but very low risk.

For older populations the risk was higher but in 2020 and 2021 six out of seven deaths still had nothing to do with covid. There is no point being so afraid of dying that you forget to live. All balance seemed to be lost in the restrictions placed on care homes for the elderly. People in care homes have on average one to two years of life left to live. When there is only a small amount of precious life left, the priority must surely be to fill that time with love rather than spend it trying to hide from an inevitable and fairly imminent death.

RISK TO WHOLE POPULATION

The overall risk of dying was comparable to previous bad influenza seasons. January 2021 had more deaths than any year since 2008. However, the proportion of the population who died in January 1990, 1997, 1999 and 2000 was higher than in January 2021. If all winter deaths since 1983 are plotted together it is almost impossible to pick out the covid winter. The exception is spring 2020 where the peak was a similar magnitude to a bad January but at a much later time of year. The later timing of the 2020 deaths was unusual.

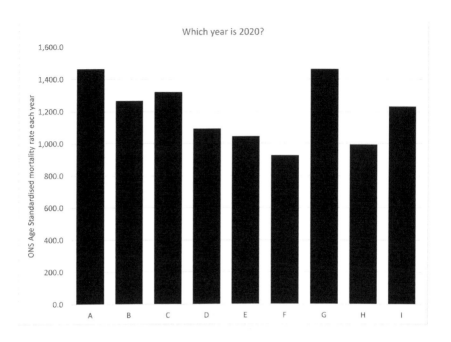

Figure 2: *These bars show the* Office of National Statistics *mortality rates195 after adjusting for the age and size of the population, for a selection of years since 1990. Can you pick out 2020?[196]*

Because aggressive testing for influenza was never carried out, many influenza deaths which occurred in winter were never diagnosed as such but instead were recorded as cardiac or other causes of death. They are thought to be due to influenza partly because there is a plausible biological mechanism and partly because good and bad years for cardiac deaths would follow the same pattern as good and bad years for influenza deaths. In contrast, for covid, unlike for influenza, healthcare could not be accessed without extensive testing. That meant there were very few deaths where covid contributed that would not have had a covid diagnosis. The measure of influenza deaths is usually given as deaths caused directly by influenza whereas deaths 'with influenza' give numbers more than three times as high even without aggressive testing.[197]

Whereas influenza deaths are counted for a single winter season before the counter returns to baseline, for covid the counters have never reset. At the end of 2022, the Office of National Statistics were continuing to report on the total covid death tally over the preceding three years.[198] They also chose to compare covid deaths to deaths from other infectious diseases excluding influenza and other respiratory virus deaths resulting in their ridiculous claim that *"COVID-19 was the underlying cause of more deaths in 2020 than any other infectious and parasitic diseases had caused in any year since 1918."* [199]

A fair comparison to make is to look at total deaths in a covid winter versus an influenza winter. Covid policy resulted in unexpected deaths that were not all due to covid which makes this comparison harder but it remains fairer.

Because influenza disappeared when covid arrived, the risk from covid should not be considered an entirely new additional risk. The risk needs to be considered in comparison to the risk that would have been there from influenza in the absence of covid.

A key contrast with influenza was that the number of people requiring lengthy stays in intensive care was significantly higher for covid. In the end it wasn't just the total number of patients requiring care that overwhelmed intensive care as the length of time that they each required treatment. The result was that total intensive care demand was higher than seen in previous years and an expansion to capacity was needed to accommodate the extra demand. However, with respect to deaths a bad influenza year is a fair comparison.

WHY HAVE PEOPLE OVERESTIMATED THE RISK?

Because absorbing the meaning of a low risk is difficult, most people had an understanding that was not sophisticated. The risk was worse than seasonal influenza but not as bad as Ebola. The mortality from Ebola if you catch it is 90 percent. For a healthy 35 year old, Ebola is 100,000 times more deadly than covid. Although people are very ready to point to evidence that covid is slightly worse than seasonal influenza, that comparison is at least in the same ballpark.

How have so many people overestimated these risks? In addition to people being bad at assessing risk and governments deliberately increasing the fear of dying, there was a third factor that contributed. Chinese videos were widely shared which portrayed a much deadlier disease.

Social media was flooded with images of apparently young, healthy people collapsing suddenly, dead in the street. Apparently such sudden deaths were so common that they were frequently caught on camera. Such collapses in the street were not seen in other countries. More astute people than me spotted the game that was afoot. On slowing one video, it was possible to see the collapsing actor stretching his arms out at the last minute to break the fall. One person spotted that the patient in several images was the same person as the doctor in other similar images. The most ridiculous video, a real life Scooby-Doo episode, was one of a team of police with 'SWAT' written in English on their backs, stopping a man and then catching him with a butterfly net when he removed his mask. There was also extensive footage of people sick in intensive care. Such footage could be filmed any day in any hospital using patients sick from other conditions, but normally, most of us remain oblivious to the drama of other people's struggle with illness.

The source of these videos is not clear. Michael Senger, a San Francisco lawyer and author of *Snake Oil: How Xi Jinping Shut*

Down the World[200], makes a compelling case that the Chinese Communist Party set about to encourage the West to lockdown and that these videos were a part of that strategy. Whereas writer and philosopher, Ben Irvine, puts forward a case in his book *The Truth about the Wuhan Lockdown* that these videos originated from activists.[201] The Chinese Communist Party strategy at the time appeared to be to down play the virus. However, a significant part of the population of Wuhan had lost faith in authority following bad handling of safety concerns regarding chemical exposure. There were therefore activists already in Wuhan who were primed to criticise any further failure of authority to ensure public safety. It is possible that some of these videos did originate with these activist groups. In all likelihood both sources were responsible. It is hard to believe activists would risk arrest by impersonating police officers for a video, for instance.

Another striking image that induced fear was one of mass graves in Hart Island in New York. The island had been used for many decades as a burial site for the poor with relatives only allowed to visit from 2015. Footage from above of a trench where bodies are stacked three deep, was used to amplify fear. New York City Council member, Mark Levine, tweeted, *"Soon we'll start 'temporary interment.' This likely will be done by using a NYC park for burials (yes you read that right)."*[202] He later retracted the comment. Bodies were indeed buried on Hart Island. By the end of 2020, 2,334 bodies had been buried on Hart Island[203] which was more than in previous years, for example in 2018, a bad year, there were 1,213 burials.[204] To put it another way, in 2020, 2.8 percent of deaths[205] resulted in a burial on Hart Island compared to 2.2 percent in 2018.[206] The implication that mass burial was the only way the city would cope with the number of deaths was not true.

In addition to the fear inducing imagery there were news reports that amplified fear. When doctors were interviewed about the pressures intensive care units were under they would often make claims that the data could not support. There undoubtedly was a large increase in admissions to intensive care that did put immense pressure on doctors. Intensive care doctors have always been at the frontline for rationing healthcare managing a finite resource. Whenever accepting a patient on to an intensive care bed they have to be mindful of who the next patient may be that might need it more. The next patient may have been the doctor's colleague or even themselves.

Doctors had a belief that the number of people needing treatment was doubling after a predictable number of days and would continue to. It is easy to imagine how doctors might predict the situation after one more doubling and report as if that had already happened. By the time behaviour changed as a result of their fearmongering there would indeed have been another doubling – had the doubling carried on indefinitely (which of course it didn't).

Even if someone can be persuaded of the evidence that their risk of dying is low, there is a problem. There is a much stronger belief that conflicts with this evidence: the belief that covid is something to be feared. There was already sufficient fear before lockdown to cause people to work from home and to halve the numbers attending hospital emergency departments. Nevertheless, fear of covid was then deliberately amplified by the government and the media. Once an emotional response is established a rational belief will struggle to overcome it. Perhaps, the best approach is to treat such fear as irrational and ignore it, in the same way we do with our fear of rollercoasters.

TOP THREE MYTHS

---

1. The extreme response to covid means it must be really dangerous

---

2. Total deaths were in a different league to previous influenza seasons

---

3. Every covid death with a mention of covid was because of covid

---

6

FAITH IN OUR BELIEFS

'Tis with our judgements as our watches, none. Go just alike, yet each believes his own. **Alexander Pope, 1711**[207]

Even when we know we are falling for an optical illusion, our brains still tell us that what we see is right. Given that we all act as if we are omniscient it is worth taking time to consider the foundations of our beliefs, why we believe new 'truths' and what the flaws in that system are. If we want to prevent mistakes we must understand how we choose what we believe, why we believe those things are true and what commits us to a belief.

CHOOSING WHAT TO BELIEVE

It is easy to claim that in choosing what to believe we first gather all the evidence, assess it dispassionately and logically and then draw conclusions. In reality people constantly make shortcuts in deciding what to believe. The philosopher Descartes' said our errors of judgement come when we abuse our free will to believe things without sufficient evidence. However, to constantly assess the world for 'sufficient evidence' would require a level of doubt and effort to trawl through all the evidence that would be utterly exhausting. That is not how we decide what to believe.

Our brains, unlike computers, do not work purely on logic. Instead we rely heavily on preconceived notions. Our beliefs are based on the most likely scenarios. When we open an envelope we expect to find a letter because that is the most probable outcome. But it is not the only logical, possible outcome. Logically, anything small enough could be in there, seeds, feathers, stamps. Rather than work through each of the logical possibilities we take an educated guess and assume it will be a letter. That system works really well in the real world.

Our experiences teach us what to expect. Familiarity has a profound effect on what we believe which can be mimicked when there is repetition of information. When we hear a statement repeatedly, we are more likely to think it is true, even if it is incorrect.[208] There was no shortage of repetition of covid slogans by politicians and public health authorities. Repetition makes something familiar but it does not make it right. Isa Blagdon, a novelist and poet, wrote a passage in 1869, often shortened to *"If you repeat a lie often enough it becomes the truth."* [209] When the lie is presented from trusted authorities that certainly seems to be the case. (The counter to this might be that a truth repeated often enough only by people not in positions of authority becomes a conspiracy theory).

There is an inbuilt bias whereby our first impressions of something new have a disproportionate impact on our beliefs. For many of us, our first exposure to covid came from fear propaganda produced in China of young people dropping dead in the street. Young people do not drop suddenly dead of covid in the street; it was fabricated. Being aware of how that may have influenced our subsequent thinking might help with piecing together a more honest picture of the risks from covid. Fortunately, there is also a bias towards information we have recently taken onboard, so it is not too late to read and learn more.

WHY WE BELIEVE WHAT WE BELIEVE

Multiple factors influence what we believe including our upbringing, our society, the source of the information and incentives. In addition to these biases, there is also a bit of a quirk in human wiring which leads us to believe stories we have invented ourselves to explain our own behaviour.

Our experiences play an integral role in our expectations of the world, how we were brought up and the culture we live in shape our beliefs, particularly our religious and moral beliefs.

Overlaid on this foundation are countless things we believe that we have not experienced. Every fact from history and all sorts of truths as relayed by family, friends, your doctor, the news. In these cases, we have no direct evidence of what happened and instead use our faith in the source to verify the belief. Even extraordinary things can be believed when all our trusted sources concur that it was true.

News sources are relied upon to an unhealthy extent. Many of us have read an article in the news in an area we have detailed knowledge of and recognised that many of the facts and the subsequent interpretation was plain wrong. However, we then continue to read the rest of the news, reporting on areas about which we know less, and we believe all of the contents, as if our area of expertise is the only one where journalists will be hopelessly wrong.

Added on to that age old problem, since 2019 there has been coordination of the news stories that are reported through the 'Trusted News Initiative'. This is a cartel of media establishments who control the official narrative under the guise of stopping fake news. Organisations signed up include the BBC, Facebook, Google, YouTube, Twitter, Microsoft, Reuters, *Financial Times* and *The Wall Street Journal* among others. Any agreement to speak in unison destroys the role of the media as an error correction mechanism. In the USA, the pharmaceutical industry provides a significant income stream to news channels with nine out of ten drug companies spending more on marketing than on research and development.[210]

Announcements from Public Health Officials concurred with what the news was reporting. In a normal situation that would be because the news were reporting what the Public Health Officials were saying. However, when the CDC director Rochelle Walensky heard a CNN news report reproduce Pfizer's own press release of their trial results almost verbatim, she said, *"I can tell you where I was when the CNN feed came that it was 95 percent effective, the vaccine. So many of us wanted it to be helpful, so many of us wanted to say, 'Ok this is our ticket out.'"*[211] The fact she was relying on CNN to inform her about covid rather than the other way around means that there are powerful official sources who need to be doubted as much as the journalists.

Against the backdrop of cultural beliefs and what we have learnt from trusted sources our beliefs can change based on incentives. Although we might hope that overt incentives would not change our beliefs, they can do so on a subliminal level. These incentives may not just be financial. Career progression or job security can all affect what we believe.

For the small group of people who dedicated their lives to the study of epidemics, when one finally comes along they are incentivised to believe it to be more deadly and extensive to justify the hours they dedicated to the subject. The work carried out by these teams of epidemiologists was the definition of a solution waiting for a problem.

Incentives can also be negative. The scientific advisors would have taken the full blame had they underestimated the extent of the threat. However, harmful policies undertaken as an overreaction to the threat would be blamed on politicians. There was therefore a perverse incentive for the advisors to communicate worst case scenarios in the hope they would be covered from any future blame.

EXPLAINING OUR ACTIONS

The actual reasons policy decisions were taken at the time may never be fully known – even to the people who made the decisions. A distorted view of history occurs because we can explain our decisions, especially ones we have acted on, by inventing narratives.

Psychologists Richard Nisbett and Timothy Wilson devised an experiment, published by the *Psychological Review* in 1977, where people shopping in a department store were asked to choose the best quality, nylon stockings, from four identical pairs. Not only did shoppers confidently make a selection they also, when asked, provided reasoning for their selection. The shoppers post-rationalised, attributing their choice to the quality of the knit, weave, sheerness, elasticity, or workmanship of the stockings, coming up with a total of eighty reasons. When there are that many reasons it is because none of them are right. We like to explain things even when the explanation is beyond us. As the stockings were identical there could be no genuine reason and these reasons were invented to justify their decisions after the event. One factor that did seem significant was that shoppers most often chose the pair they viewed last, on the right of the display, but only one of the shoppers (a psychology student) articulated this. Worse, when asked directly about the possibility of a position effect, participants denied it *"usually with a worried glance at the interviewer suggesting that they felt either that they had misunderstood the question or were dealing with a madman."*

When we cannot explain how we have acted, our brains go into overdrive inventing narratives that could be true. On occasion we do not know why we behaved a certain way. In order to recognise that, we need to consider and reject each of the stories that we invent to explain our behaviour. That's hard work that most of us do not bother with.

So, for example, incentives such as pizza slices, doughnuts and even cannabis joints were used to increase vaccine uptake. The fact that such incentives work shows there were people who only made the decisions to be vaccinated based on a trivial bribe. However, it is likely that those who took the vaccine in return for the bribe would have justified their decision to be vaccinated based on other reasons such as their desire to protect others. The latter belief would therefore have been adopted and internalised, at least for some of those who accepted the incentive, thanks to a slice of pizza or a doughnut.

WHY WE STOP BELIEVING

As well as understanding our worryingly fallible basis for choosing what to believe it is also worth considering what makes us stop believing something. Kathryn Schulz sets out in her book, *Being Wrong: Adventures in the Margins of Error* what makes us discard or keep beliefs.[212] Children toy with how we rely on trusted sources when teasing their friends and accusing them of being gullible. It is very easy to drop a belief when you realise you have been tricked into it briefly e.g. as a prank. It is also easy to correct a simple mistake, such as an incorrect historical date. However, once we act on a belief then we own that belief in a different way. It is possible to remain questioning and open minded but it becomes harder.

A small action will make reversing a belief difficult, even simply writing down an idea. Communists during the Korean war have used this fact to indoctrinate people including prisoners of war by making them write essays giving the pros and cons of a particular argument and then quoting back their own words out of context. Cults are also known to have done this.

Those in power have to reconcile taking action that may have been the biggest decision of their lives and have impacted massively on the lives of those around them. Even if they now doubt the benefits, it would be a painful admission of being wrong and of having caused unnecessary harm. Perhaps, rationally, they were not hearing dissenting voices at the time and had no other source of information. Perhaps they were scared themselves. Rationally, they should be able to understand that others may have behaved similarly if they too had failed to listen to dissenting voices. However, emotionally, it is less easy to escape. This was their defining moment and will be their legacy. They will want to defend it.

Even as late as May 2021, both Dominic Cummings, a key government advisor, and in June 2022 Matt Hancock who had been in charge of health in spring 2020, continued to express in public that they believed there would have been 500,000 deaths in the absence of intervention. The implication that this was the imminent threat of the first wave was blatantly wrong – see *Belief Two: Everyone was susceptible* and *Belief eight: Lockdowns saved lives*. Had they done too much on the basis of that figure to be ready to let it go, even while it was evidently so wrong?

Once committed to a belief it is unclear what tips people into changing their mind. People fall out of love, leave cults and switch political allegiances but trying to predict when that might happen is impossible.

7

THE POWER OF FEAR, SHOCK AND SHAME

I am not so much afraid of death, as ashamed thereof; 'tis the very disgrace and ignominy of our natures. **Thomas Browne, 1643**[213]

Humans are complex but there are aspects of the way we think that are simpler than might be ideal. An overly simplified but useful model of our brains describes layers of control.[214] At the most basic there is the reptilian derived centre that demands our basic needs are met: hunger; thirst; and libido. Above that we have our emotional centres which gives us a rich experience of the world through love; happiness; sadness; surprise; anger; fear and disgust. Thankfully as well as these more basic areas we have a modicum of control from our cognitive centres which are able to learn, think and rationalise. Unfortunately, our belief that we are rational actors is often an illusion and our emotional centres have a great deal of power. Using the rational, thinking side of our brains is slow and hard work, whereas the emotionally driven response is our default situation. It's our emotional brain that takes a trip to the biscuit tin when we are distracted and everyone is aware how hard it can be to engage our rational brains to make us stop that kind of habit.

Beliefs about covid matter because they impacted behaviour. In summer 2021, fear led to four in ten people avoiding touching anything in public with nearly a quarter avoiding public places altogether.[215] An utterly distorted view of reality was held by the majority of the population, with a dangerously awry understanding of their own personal risk. People who worked as scientists, doctors, politicians and lawyers and even the behavioural science team producing the propaganda were themselves not immune to fear. There was therefore a risk of a positive feedback loop such that scared people were driven to frighten the public still more.

FEAR DISTORTS BELIEFS

Fear led to a massively exaggerated belief about the risk of the young dying. Surveys carried out in November 2020 demonstrated that on average people believed the mean age of death from covid had been 65 years old.[216] In reality, the mean age of a covid death was 80 years old and the median age was 82 years old, one year older than the average age of death in the years prior to covid.[217] That does not mean that young people did not die. For the youngest their chances of dying over the course of a year were lower than for influenza which *SARS-CoV-2* had replaced.[218]

As well as who was dying, the perception of the numbers dying was utterly distorted. A survey asked people to estimate the percentage of the population who had died of covid in spring 2020.[219] The average estimate was that one in fourteen people had died which was a hundred times higher than the real figure. Assessments of personal risk were also exaggerated with no accounting for the much lower risk in the young. Counterintuitively, anxiety levels were higher in those in their 30s, who were at minimal risk, compared to older groups, whose risk was higher.[220]

ADDICTED TO FEAR

Fear has its own positive feedback loop. Hunting and finding frightening information to reinforce a fear results in a surge of the brain reward chemical, dopamine. Even though there is no pleasure to be had from fear, like many things that cause dopamine surges there is a risk of addiction. Dopamine is designed to motivate us to meet our needs, whether we are quenching a thirst, feeding a hunger or falling in love. When dopamine is stimulated in the absence of a need that can be sated, there is a continuing desire to try for more and therefore there is a risk of addiction arising.

Like fear, shock and surprise also stimulate dopamine and our addiction centres. When an experience does not fit with our expectations we experience an emotional reaction of either surprise or shock. Our response to either surprise or shock is to focus all of our attention on the situation. If we open an envelope and find, not a letter, but several crisp banknotes, we would be pleasantly surprised. If we found white powder the negative connotations would make us feel shocked but the way our brain responds in either case is similar. There is a dopamine surge but there is also a surge of noradrenaline.[221] The latter gives as a few moments of sharp focus, which is why when an event is particularly shocking, say in a car accident, every detail can be remembered, as if in slow motion. Our facial muscles totally relax producing a, sometimes amusing, blank expression or even a jaw drop, as all our concentration is given over to taking in information on the shocking situation.

The shock addict would be hooked on the news waiting for the latest death tally or some new alarmist horror story. The fear resulting from this new information might lead them to try to take action to keep themselves safe from the perceived danger, whether that was through avoiding other humans, wearing a mask or getting a vaccine. However, they would return to the news only to find that all their efforts were not making them feel safer and there continued to be a frightening invisible threat. They were getting the dopamine and noradrenaline hit but could never fulfil the need of feeling safer. Short bursts of noradrenaline give us amazing concentration enabling us to absorb detailed information. However, high levels of noradrenaline for extended periods just lead to stress and anxiety.

The reports became more hysterical and exaggerated, feeding the shock addiction. Any recent shocking story would be brought to mind when surrounded by posters, stickers and, of course, masked faces. It was telling how often the media chose to present frightening anecdotes rather than a discussion of the merits or otherwise of government strategy. On 29th December 2020, I heard an interview of a London nurse on BBC 5 Live's breakfast programme, in which she portrayed a horror-film-like scenario,

"It's so contagious, lots of pregnant women are coming in, newborn babies are catching it. It's horrendous... It is literally anyone and everyone in its path... You can be the most healthiest person in the world and it will literally make you its victim. You can't breathe properly, it affects your kidney function; it may affect your heart. We've had healthy colleagues that have, like, been diagnosed with cancer or healthy colleagues that have now got heart problems or... are now kidney patients. This is what covid has done to them. This is the first wave and this one is more infectious and contagious and we don't know what this strain can do, so it's very worrying." [222]

This kind of broadcast was so common that the interviewer, Helen Skelton, did not even question the idea that covid was causing sudden cancer. A news broadcast designed to cause the most shock possible could not have done a better job.

It was not just the news providing shocking stories which continued the fear. The threat was invisible and every person, especially children, were portrayed as potential 'vectors' of disease, like fleas or mosquitos. The covid posters and stickers were constant reminders of the invisible threat as were the masked faces. A sense of control could be gained by donning a mask, crossing a pavement to avoid a stranger and sanitising hands. The general and media environment and these behaviours perpetuated the fear. Even as an onlooker who was no longer scared it was easy for me to sense the shift in fear levels as mask wearing and covid posters in the street waxed and waned. Unfortunately, scared people were less good at recognising the manipulation.

FEAR KILLS

It was broadly possible to measure how scared the population was based on the attendance rates in hospital emergency departments. Rates fell to half of normal levels in spring 2020, did not fully recover over the summer and then fell again over the next two winters to 80 percent of normal levels. Attendance for life threatening surgical emergencies and heart attacks dropped as well as well as for more minor cases.[223] Only if you thought healthcare was fairly ineffective could you think this would not impact on mortality levels. Indeed thousands of deaths from non-covid causes occurred in the community and there were fewer non covid deaths in hospitals than normal.

Fear among staff also impacted on care. Nursing home staff have to work hard to frequently encourage their patients to drink in order for them to remain adequately hydrated. However, fear coupled with absent colleagues and staff eking out the rationed protective equipment, meant that keeping patients hydrated was no longer possible. A lack of hydration has been cited as a contributory factor to deaths in care homes at this time along with neglect due to staff shortages and lack of care from visitors.[224] Care home mortality rocketed but these deaths were from non-covid causes as well as covid causes. For the whole of April and May 2020 the total number of care home deaths above expected levels (the average of 2015-2019) from causes other than covid were about equal to the covid deaths.

I had covid in July 2021. It made me more sick than I had been in years with fever, cough and unpleasant eye pain. A good friend who was sick at the same time as me was hospitalised, spent a long time in intensive care and many more months to recover and I was fearful for him at that time. While I was ill I would occasionally have catastrophising thoughts about me being one of the unlucky ones. Rationally I knew how ridiculously low the probability was but fear does not hear reason. I had been subject to the full fear campaign too and it affected me at that time despite my best reasoning. Yet even so it remained intuitive to me that a state of fear would be detrimental to my prospects of physical recovery from covid. Indeed, covid patients who had higher levels of anxiety or the stress hormone cortisol were in fact more likely to die.[225]

SHAME

The fear of dying was accompanied by the fear of being accused of causing others harm. I heard many rational thinkers prepared to do irrational things entirely to reduce the potential shame of being accused of having caused spread. It was not a reduction in spread they were hoping to achieve, just protection from the accusation that they were irresponsible. The fear of being one of a tiny number of pre-symptomatic superspreaders, much publicised in the press, chastened people into compliance. For some, even where the fear of dying was realistically low, the fear of being called a 'covidiot' was greater. For them the focus was imagining dying with the embarrassment of not having carried out any number of allegedly preventative steps. Did the fear of shame, of wanting to ensure they avoided the accusation that they were responsible for covid deaths, also impact policy makers' decisions?

Fear and shame have both had a large detrimental impact on people's beliefs and behaviour. Fear not only formed beliefs about covid but, combined with shock, it sustained them. Fear among both the public and healthcare staff contributed to the rise in mortality seen in lockdown.

BELIEF FOUR: DEATH CERTIFICATES ARE NEVER WRONG

Doctors are expected to state the cause of death to the best of their knowledge and belief, it is not required that the cause must be proven. **General Medical Council, May 2020**[226]

The public has much faith in medicine. Doctors are trusted to predict our futures through the medium of diagnosis, to help us defer death and to smooth our arrivals and departures from this world. Generally doctors do know why their patients died and death certificates reflect that knowledge but death certification is not black and white.

The combination of faith in death certification and faith in medical testing created a perfect storm. Death certification is more of an art than a science. A doctor who has seen the patient recently must collate all the evidence of disease in the lead up to death and from that information deduce a sequence of events that led to death and factors that contributed to it. The same death can be interpreted differently by different doctors.

For example, say Abigail had lung cancer. Her doctor did not expect her to live for very much longer and sure enough she developed pneumonia and died. Lung cancer predisposes people to pneumonia and in the past the organism causing the infection would have been considered of only secondary importance. However, Abigail wanted hospice care which meant compulsory testing. While in the hospice she tested positive for covid. Was Abigail going to die regardless of covid? Yes. Would she have died that day regardless of covid? That is harder to say.

One doctor might write her death certificate as saying that pneumonia due to lung cancer was the direct cause but it may have been contributed to by covid. However, another doctor may say pneumonia due to covid was the direct cause which was contributed to by lung cancer. The argument about which is right depends entirely on whether a doctor would believe that in the absence of covid Abigail would have continued living with her lung cancer. Alternatively, it hangs on the question, would Abigail, without the presence of covid have still died of her lung cancer and when. These questions are more philosophical than scientific.

The WHO released guidelines that when covid was listed as the direct cause of death, other causes could not be listed as having contributed.[227] That is correct in terms of strict causation meaning certifiers would have to select which of covid or the underlying disease would be classified as the direct cause and which were described as a contributing cause. The WHO went on to recommend, *"A manual plausibility check... in particular for certificates where COVID-19 was reported but not selected as the underlying cause of death,"* which would maximise deaths attributed to covid. The guidance went on *"always apply these instructions, whether they can be considered medically correct or not."*

Our system of death certification has stood the test of time because it served the purpose it was designed for. That purpose was to offer an explanation of death for the family and to collect a broad brush view of causes of death for public health purposes. It is not the rigorous system that people seem to think it is, capable of determining the exact death tally from a virus. Joy Fritz, a death certification clerk said, *"Aside from tracking age, gender, and place of death of the deceased, using death certificates for anything beyond closing bank accounts is a disservice to society."* [228] Certification has some uses but its accuracy should not be relied on when making decisions that overturn people's lives.

The problems with defining a covid death were brought into sharp relief when the US CDC started denoting covid deaths in the vaccinated as being *"asymptomatic"* or *"from a cause unrelated to COVID-19."*[229] Such deaths before vaccination were all considered to be covid deaths. By redefining deaths in this way, the number of covid deaths in the vaccinated could be minimised. Such shifting of the goal posts benefits pharmaceutical companies but not the public.

Suggestions that a proportion of deaths may have been misdiagnosed are worth examining and do not mean that every death was mistakenly labelled. Considering deaths from covid merely as numbers may seem callous and uncaring. That is not my intention. Many people have died from covid and there is no intention to be disrespectful to those that lost their lives or the families they left behind. However, it is important that questions relating to accuracy of the diagnosis of death as a whole can be asked and the evidence examined.

HOW COULD DEATH DIAGNOSIS BE WRONG?

The UK Coronavirus Act states, *"Coronavirus means severe acute respiratory syndrome coronavirus 2 (SARS-CoV-2); coronavirus disease means COVID-19 (the official designation of the disease which can be caused by coronavirus)."*[230] Despite drawing this clear distinction between the virus and the disease, every death after a positive virus test was counted as a covid death in the government statistics. It took until mid-August 2020 before only deaths within 28 days of a positive test became the measure and over five thousand deaths were removed from the data in England alone.[231]

The number of deaths within 28 days of a positive viral test tallied well with the number of death certificates with a covid diagnosis. Public health authorities pointed to this as evidence of the accuracy of their measure. However, it could equally be interpreted as evidence of the inaccuracy of death certification. With Omicron two thirds of mentions of covid on death certificates did not include it as a direct cause of death, whereas prior to Omicron 90 percent did, despite us knowing that for the majority of patients who tested positive it was a mild or moderate infection. It was estimated that 16 million people in England had developed antibodies after infection by the end of 2021[232] (pre-Omicron) and only 3 percent (just over half a million) of those had required hospital admission.[233] Of those in hospital, some had a positive test and no symptoms at all. Some had covid some time before but still tested positive on admission to hospital despite their symptomatic recovery. It is quite reasonable to say there were deaths that were incidentally after positive test results. Where were the coincidental deaths in such people prior to Omicron?

A death diagnosis is entirely dependent on the information available. Where there was a positive covid test result that would have formed part of the evidence. In the knowledge that a patient's death had already been recorded as a *"death within 28 days of a positive test"* by the government, doctors were in a bind. Could they record a death with no mention of covid? What evidence could they provide that covid, with its myriad symptoms, hadn't contributed to the death? One paper looked at the accuracy of covid diagnosis and concluded that *"neither absence nor presence of signs or symptoms are accurate enough to rule in or rule out disease."* [234] It was far easier to include covid on the death certificate than exclude it.

The WHO have declared that differences in the definition of a covid death resulted in *"a substantial lack of comparability between countries."* [235] Comparing trends in data between countries might have meaning but comparing totals does not, especially when there were also differences in the pressures doctors were under to diagnose covid death. In the US, hospitals were paid two to three times more in Medicare payments for covid patients than patients with other respiratory illnesses.[236] Surviving families were also paid up to $9,000 of death benefit where covid was listed as the cause of death.[237]

One UK doctor friend told me of an incident where a man died overnight in the emergency department from a massive brain haemorrhage as a result of a ruptured blood vessel, an aneurysm. When my friend came to relieve the doctor after his night shift, he asked him to do the death certificate. In his exhausted state he responded, *"but we haven't got his covid test back yet."* Deaths from bursting aneurysms should not be attributed to covid, even if a patient tested positive and even though covid can affect clotting. No matter how good your clotting ability is, it will not save you from a massive burst aneurysm in your brain.

In a hospital where doctors were reminded of covid at every turn, thanks to masking, the likelihood of them thinking of covid rather than other causes of death was much higher. Ideas that come to us more readily affect how we think. Psychologists refer to this as the availability heuristic. The same logic is why we think there are more words that begin with k than have k as a third letter. We are more likely to take notice of token words like 'kite' or 'kettle' than to acknowledge the other five words in this sentence with a k.

OVERESTIMATING COVID DEATHS

If someone is diagnosed as having died of covid when a covid positive test was incidental, then their death will not be recorded as the true underlying cause. The result would be a deficit in deaths from other causes. That is what was seen in the first covid winter from December 2020 to February 2021.

Each week more covid deaths were diagnosed and each week there were more deaths from other causes that did not happen. As covid deaths began to fall, there were fewer missing deaths from other causes. In total there were more than 40,000 deaths from non-covid causes that were expected to occur during that first covid winter which were not seen and I strongly believe they ended up being labelled as covid deaths.

Aside from covid being overdiagnosed and deaths that had nothing to do with covid being categorised as covid deaths, there are three ways in which this could have been interpreted:

> 1. Deaths caused by flu had in the past been attributed to heart disease and other causes. Now that we were mass testing we were proving the underlying viral cause of these deaths and re-categorising them as covid deaths.

> 2. Covid killed the already dying.

> 3. People who would have died in winter had already died in spring 2020.

No-one else was putting forward the first hypothesis that countered our overdiagnosis argument so Jonathan Engler, Joel Smalley and I published the idea.[238] Many of the extra deaths seen each winter were not diagnosed as due to influenza but might have been associated with it. Perhaps we have underdiagnosed the contribution to cardiac and vascular deaths from other respiratory viruses in the past by not testing as much? The 2020 to 2021 winter season was just beginning when we published this idea. Subsequently, covid deaths rocketed accounting for almost all the deaths above summer levels. A further spike occurred in January 2021. Throughout that winter, almost no deaths were attributed to other respiratory causes. The graph shown uses the summer low of deaths as a baseline from which to plot the weekly deaths through each winter. Only a fraction of the extra winter deaths were attributed to respiratory disease in the past but almost all were once we started testing extensively for covid. It is clear that spikes in respiratory deaths (solid black area) coincide with spikes in non-respiratory deaths (black dotted line) suggesting a common factor causing both.

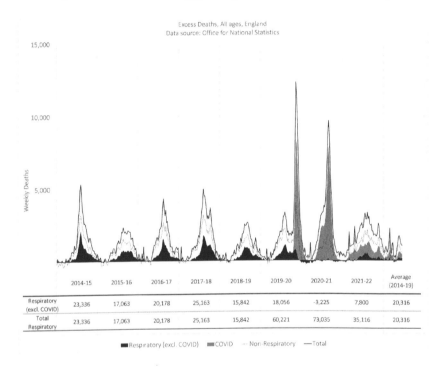

Figure 3: *Deaths above summer baseline levels since 2014*

Although that theory is compelling, the evidence does not support it. Other countries diagnosed deaths more accurately. In these countries deaths attributed to covid approximated the number of deaths in excess of normal levels. Also, in the second covid winter (2021/22) in the UK there was no repeat of missing deaths from other causes suggesting death diagnosis improved.

Furthermore, a surrogate of covid deaths could be used. Deaths above expected levels that were attributed to any acute respiratory infection could be used as another way of measuring covid deaths. Both the first covid spring and the second covid winter saw a similar proportion of covid deaths also being recorded as acute respiratory infection deaths.[239] However, in the first covid winter, a far smaller proportion of covid deaths were also recorded as having had an acute respiratory infection, further suggesting overdiagnosis.

The second point – covid killed the already dying - is a philosophical one. Is it really possible to blame a death that was going to occur anyway on a test result?

The frail were certainly at greater risk from covid death. A study carried out in a mortuary in Spain in 2017, showed that when testing elderly people who had died from a huge range of causes half of them tested positive for respiratory viruses.[240] Of those, one in seven were coronaviruses. There are likely many viruses that it is not possible to test for which may have meant more than half had virus onboard. Only one in fourteen of them were diagnosed as having had a respiratory infection before their death. These were 'infections' that were going unnoticed while people died of other causes. Therefore, there was a huge bias here in susceptibility to testing positive on dying. It becomes hard to unpick whether these viruses contributed to the demise or were just bystanders at deaths. There are a number of situations where virus could be detected but it could be argued that attributing the death to covid is overcalling. What if life was only shortened by a matter of weeks or even days? What if the virus was just a signal of a dying person having a weak immune system prior to death? What if the virus was an innocent bystander?

The problem of diagnosing the dying with a positive test result is exemplified by hospice deaths. Nearly three thousand covid deaths have occurred in hospices. By definition these people were dying of another cause and yet when they tested positive their deaths were attributed to the virus.

The third point – that those who would have died from non-covid causes were already dead – should have meant a lack of non-covid deaths that would begin to return to normal levels at some point. If that were the reason for the lack of non-covid deaths then it would have been quite a coincidence that the peak in covid deaths occurred at the same time as the number of non-covid deaths started to recover. A similar effect was not seen in other countries nor in the following winter.

Inaccuracy of death certification is also evidenced by the fact that a higher proportion of covid deaths were in men. The ratio of deaths in men and women could be used as a separate way of measuring covid death. At certain periods, including through the first covid winter equal numbers of men and women had covid labelled deaths suggesting that there was significant overdiagnosis or a factor making women as susceptible to dying as men just for this period.

UNDERESTIMATING COVID DEATHS

There were attempts by lockdown sceptics to minimise the impact of covid. For example, a report from Italy which showed only 3 percent of Italians with a covid death certificate had no pre-existing conditions was used to claim that 97 percent of the covid deaths had been over-diagnosed.[241] Similarly, there were two Office of National Statistics reports which gave low figures. One stated that fewer than 6,183 deaths had had only covid mentioned on the death certificate, up to the end of December 2021.[242] A separate Office of National Statistics report that stated only 17,371 covid deaths had not had another underlying condition mentioned as part of the direct cause of death, up to September 2021.[243] The difference between these two low figures can be explained by there being thousands of deaths where covid was indeed the only direct cause but there were other conditions mentioned as contributing factors but not direct causal factors.

However, this lockdown sceptic interpretation is also wrong. Genuine covid deaths in the otherwise healthy, if certified thoroughly, would have had other conditions mentioned on the certificate, for example, pneumonia, septic shock or conditions caused by clotting problems. Claiming such certificates were not genuine covid deaths results from not understanding death certification. Also, it is quite wrong to say that anyone with a pre-existing condition could not have died because of covid.

INVESTIGATING DEATHS

The fact different doctors will certify the same death in different ways shows that there is uncertainty about the cause of death. In the medical world, being uncertain about the predominant cause is very different to not knowing the cause of death. Where the cause is not known the doctor must refer for investigation by the coroner who might request a post mortem. The coroner's system is in place to investigate preventable causes of death. Therefore, where the cause was natural, even when there are remaining questions, a coroner's referral is not appropriate. However, a post mortem with the consent of the next of kin is still an option to investigate the death further.

In England and Wales fewer than one in seven deaths have a coroner's post mortem and less than 1 percent of hospital deaths are investigated[244] which compares with a rate of only four to eight percent in the USA.[245] The proportion having an autopsy fell further in 2020.

Given the importance of death diagnosis to the response to covid and the impact the response had on society there was a strong case for tightening up death certification. Instead the requirements were loosened. The Royal College of Pathologists issued guidance saying, *"In general, if a death is believed to be due to confirmed COVID-19 infection, there is unlikely to be any need for a post-mortem examination."* [246] A few pathologists did carry out post mortems and the Royal College reported on the findings in twenty eight cases in May 2020. There were other pathology departments who stopped carrying out post mortems altogether.

The Coronavirus Act removed the usual safe guards for death certification.[247] A doctor who was not the attending doctor, even where the patient never saw a doctor, could sign the certificate if they felt able to *state to the best of their knowledge and belief the cause of death."* [248] NHS guidance stated, *"If before death the patient had symptoms typical of COVID-19 infection, but the test result has not been received, it would be satisfactory to give 'COVID-19' as the cause of death."* [249] The NHS also provided for dedicated certifiers who were, *"medical practitioners whose role does not usually include direct patient care, such as some medical examiners, to provide indirect support by working as dedicated certifiers."* [250] These certifiers could speak to staff in a care home, who were struggling to care for their residents because of staff absences, then view the deceased and approve a death certificate and cremation on that basis.

HOW CAN WE UNDERSTAND THE REAL IMPACT OF COVID ON DEATH LEVELS?

The National Covid Memorial Wall on the south bank of the river Thames is a 500 metre display of 150,000 hearts to represent each person with a death attributed to covid. A number of bereaved families have personalised a heart with a touching story of grief. When I see it, I cannot help but think of those people who contacted me desperate to have their relative's death certificate changed after an overly eager doctor had ascribed covid to an otherwise explainable death. I also wonder how long the wall would need to be to cover the influenza deaths over the course of two or three years or, for that matter, deaths that resulted from covid policy.

There is no doubt that policy itself will have led to non-covid deaths. Healthcare was not accessed in the usual way thanks to fear of attending hospital and restrictions on face to face contact for people with a cough or fever. Staffing levels collapsed due to isolation requirements including for those staffing ambulances and emergency departments. There were many more factors that I cover in *Belief Nine: Lockdowns are not harmful.* Lack of staffing and fear of attending would both rise and fall as a covid wave passed. Resulting deaths would coincide with the timing of covid waves. Separating these preventable deaths from deaths caused by covid over and above what would be expected from any other respiratory virus becomes very hard to unpick.

In theory, the impact of covid could be calculated by measuring the number of deaths above normal levels. The difficulty is knowing what normal would have looked like in the absence of covid. The average over previous years is often used but there is considerable variation year to year with a bad year of deaths often followed by a year with fewer deaths and vice versa. Furthermore, after a period of significant deaths, as we saw in spring 2020, it would be fair to assume there would be fewer subsequent deaths but it is hard to know by how much.

One way to better understand the impact would be to audit the deaths that have occurred. The Covid19 Assembly proposed to do just that and persuaded me to help.[251] Several doctors and other healthcare professionals came forward to help. Audits were begun looking at side issues such as the extent of in-hospital spread or the impact of changing protocols. However, each of these audits were blocked by more senior doctors or by management and data has not been forthcoming, as yet.

Audits were carried out elsewhere and resulted in huge shifts in the total covid deaths. Two California counties reviewed covid deaths resulting in a 25 percent reduction in covid deaths. Health officials in Alameda County described these as *"clearly not"* caused by covid.[252] However, the remaining 75 percent still included deaths where *"covid could not be ruled out"*. Similarly, there was a 22 percent reduction in covid deaths in Santa Clara after a review.[253] Applying a federal definition of covid death to nursing home deaths in Massachusetts[254] resulted in a third of previous covid deaths being reclassified.[255] A review of death certificates in two regions in Sweden found only one in six were decisively due to covid a further one in seven were totally unrelated to covid and the remainder were frail patients where covid was considered a *"contributing factor."* [256] In the USA, over 77,000 deaths were removed from the covid tally in March 2022 because, *"its algorithm was accidentally counting deaths that were not covid-19-related."* [257] This also resulted in a reduction in child covid deaths of 24 percent[258] right after those inflated figures had been used to persuade parents to vaccinate their children.

The reason the death data was so important was that it should have been an absolute number not subject to bias. Exaggeration of covid death numbers could easily be called out by comparing to the lower numbers of deaths in excess of expected levels. However, there were undoubtedly additional deaths caused by policy interventions and quantifying these is very difficult.

	THE EVIDENCE MANIPULATION TRIAD
EXTRAPOLATE	Modelled data is the perfect ingredient for extrapolating to produce wild claims. In terms of covid death one bad offender was a *Lancet* paper published in March 2022 which aimed to model the world's covid deaths.[259] They managed to claim three times as many covid deaths as had been officially recorded. Rather than believe any country was capable of correctly counting their dead, they used a model to *estimate* the total deaths in each country. Yes, they invented the number who had died. A large part of their model was reliant on an assumption that covid killed more people in countries nearer to the equator, which is as counterintuitive as it sounds. The result was to inflate covid deaths in southern states of America while reducing them in the northern states. The model also managed to claim that Japan was so incapable of counting corpses properly that over 110,000 deaths above expected levels had been missed.
EXCUSE	Overdiagnosis of covid deaths in the first covid winter led to missing deaths from other causes. The discrepancy was dismissed with the excuse that the missing deaths just had not been reported due to a delay in the recording of the data. Given enough time, numerous non-covid deaths would be recorded and make the discrepancy disappear. They never did.

EXCLUDE	Other evidence for overdiagnosis of covid deaths has been ignored. As well as the missing non-covid deaths in the first winter there was evidence from audits where deaths were reviewed and many deaths reclassified as being due to other causes. There has been no official response to this evidence.

TOP THREE MYTHS
1. Most people who died with covid had their lives cut short by many years
2. Death certificates are a reliable way to measure the impact of either influenza or SARS-CoV-2 on mortality.
3. A virus identified around or after death must have contributed to death

8

CHANGING YOUR MIND

The man who never alters his opinion is like standing water, and breeds reptiles of the mind. **William Blake, 1790**[260]

As we saw in *Being right and feeling omniscient*, although we like to believe we are constantly weighing up the facts, in reality changing our minds involves much more than just the presentation of new evidence. Pinning down what makes us change our minds is really hard. In retrospect, it can be hard to identify the moment where one switched opinions on something. If there was such a moment there were likely multiple steps that led to it. The same is true on a societal level. Political and social historians spend careers writing and debating about what caused cultural shifts because, even in retrospect, they cannot agree.

Anyone who cannot change their mind will end up being proved wrong. Information in an epidemic has been likened to the 'fog of war' and so being flexible as evidence emerges is essential. It is much easier to achieve this for an individual than an institution.

Nevertheless, institutions globally managed a dramatic shift in March 2020 from following their pandemic plans to instigating damaging lockdown policies. (The emails from Fauci show him to be taking a measured approach and questioning draconian measures until doing a sudden u-turn and recommending all of the US shutdown. It is unclear why.) There was also an international shift from saying masks did not work to implementing mandates in summer 2020. Quite how these dramatic u-turns took place in every country in unison deserves an explanation and I am not sure I can provide one.

UNCHANGING INSTITUTIONS

In normal times institutions are very resistant to change. University funding comes from a small number of funders. They and the universities seek reassurance that scientists will be able to demonstrate outcomes to justify the support they were given. It is therefore risky for scientists to challenge the bounds of knowledge where new techniques are required which may not work, meaning results may not be forthcoming. Alongside the funding issues, scientists work in ever larger teams which are bound to moderate the most disruptive thinkers. There are several signs that progress in science has slowed.[261] The age at first innovation is increasing while the number of patents per capita has been decreasing. Fewer Nobel prizes are for fundamental breakthroughs. There has also been a steady decline in publications that are disruptive i.e. prove past hypotheses were wrong.[262] All this shows that exploration at the bounds of our knowledge is limited. Instead, the focus has been on increasing our depth of understanding, adding in details to what we already know. Scientists plough in familiar furrows hoping to reach deeper knowledge.

Any research that might upset the careers or legacy of those in powerful positions is unlikely to find a funder willing to embrace it. Funding is triaged by peer review but, the leaders of the funding bodies ultimately decide where the money is allocated. Those making the decisions benefit from their own patent royalties and will not want to support any research that might lead to potential competitors to those patents. The disincentives are not just financial. Those allocating funding likely have bodies of work that have contributed to the consensus beliefs in their field. Researchers who threaten that work will be seen as a threat to the reputation of these gatekeepers. This setup benefits mediocre thinkers who understand how to play the system. The bureaucracy around maintaining a university position producing publications and the reward for compliance over brilliance also puts off the best intellectuals who seek work elsewhere.

Over time more funding from industry has flowed into universities and with it can come restrictions on publications, as well as patents, as the sponsors own the intellectual property. Some industry sponsors have delayed or banned public release of research results.[263] A particular problem with pharmaceutical funding was highlighted 20 years ago by previous editors of the *New England Journal of Medicine*, Arnold Relman and Marcia Angell, who said *"The medical profession is being bought by the pharmaceutical industry, not only in terms of the practice of medicine, but also in terms of teaching and research. The academic institutions of this country are allowing themselves to be the paid agents of the pharmaceutical industry. I think it's disgraceful."* [264]

Max Planck, Nobel prize winning quantum physicist is credited with the sentiment that *"science progresses one funeral at a time."* [265] There was truth in that. Studies have shown that on the death of prominent researchers, publications by their collaborators rapidly declined and the number of new researchers entering the field rose.[266] Increased life expectancy may have put the brakes on scientific progress. For example, there are doctors in the USA who are nearing retirement and have had their whole careers during the

38 years that Anthony Fauci has been director at the National Institute for Health. In that time, researchers whose work has challenged Fauci's have struggled to progress.

WHAT STOPS PEOPLE CHANGING THEIR MINDS?

It is much easier to change your mind if you're not committed to an idea. The more we act on an idea the more we own it and the harder it becomes to let it go and change course. Once a belief in an idea starts to form part of a person's identity then it becomes even harder to change their mind.

It is important to be aware of this. I made a case for continued overdiagnosis in autumn 2020 which most people believe was wrong. I continue to think there was truth in it and could show you evidence for why I think that. However, I am aware that, having pronounced on it in public, I have created a bias such that I do not judge that evidence fairly anymore.

The Millerites believed the second coming of Christ would happen on 22nd October 1844. They told the world and when the date passed uneventfully they adjusted their belief about the date. This allowed them to continue their underlying belief that the second coming was imminent that had become part of their identity. In 1954, a study was carried out of a UFO cult in Chicago who believed a flying saucer would rescue them from a flood on a specific date. When it didn't occur some left the group but the others rationalised why it hadn't happened and became more committed and evangelical about their beliefs. The researchers also noted that recruiting others to the belief system provided additional reassurance that it was believable. The study was criticised because up to a third of the supposed cult members were undercover observers and these people and the media may have pressured the cult to extend their commitments. However, similar patterns have been seen with other cults. One of the researchers of the UFO cult study noted that *"The jeering of non-believers simply makes it far more difficult for the adherents to withdraw from the movement and admit that they were wrong."* [267]

CHANGING OTHER PEOPLE'S MINDS

Being presented with opposing views can often lead to a position of aggressively defending existing beliefs. A clash between opponents creates entrenchment where both sides set out to prove the other wrong by choosing as evidence the most extreme examples. These clashes result only in highly emotional exchanges where no analysis of the evidence presented is achieved.

In the real world there is plenty of conflicting evidence to present. Scientific papers can be contradictory because science is messy and researchers introduce their own biases from their own beliefs. Even where the evidence only supports one side, the opposing person will instead find some associated part of their narrative and present evidence to support that. Where attention is successfully drawn to the actual issue and where the evidence is not easily challenged, there is usually an attempt to discredit the messenger as a shortcut to being able to dismiss the evidence as false.

In order to win an argument someone must be persuaded and that can only be achieved if a rational conversation can be had, where beliefs are challenged, without creating this emotional defensive position. It also requires a willingness on both sides to discover the truth and learn. Unfortunately, the covid discussion was so polarising that any indication that a person may have held an opposing view on some aspect of the narrative meant that their voice would be dismissed altogether on other aspects. Many of the zero covid proponents blocked dissenting voices who had only ever politely questioned what was being said. (Blocking on social media is a way to control what you see and who can comment on what you say).

I had a policy at the outset that I would not block people but I reversed that when I was under incessant personal attack. I changed my policy to blocking anyone who attacked me personally or who shared illegally hacked materials. There are a number of people who have continually challenged me throughout my time on Twitter who I have not blocked. Very often they were far from kind in the way these challenges were made and they did often feel like attacks. But where there was an attempt to challenge the content using evidence I decided I needed to hear their voices. Science requires debate, even when the debaters can be rude. Even when it hurts.

David Charalambous, a communications expert and the founder of reachingpeople.net, taught me and many others how using questions can avoid defensive emotional positioning. I already used questioning frequently in situations where there was uncertainty. David's questioning was much more thought provoking. One of his most powerful questions was, *"If the government asked you to do something immoral, would you do it?"* This disabling question meant that a person's thinking could no longer be outsourced to others. They must take their own ethical position which can then be compared to the government's.

Changing people's minds was most successful when their friend listened patiently to the other person and then simply asked a challenging question and let them mull for a few days. I am not claiming in any way that this is easy to do. I have been credited with helping people change their mind but, if that is true, I have had far more of an effect on strangers than on people close to me.

I have witnessed friends and family come round but other factors seem to have played a larger part than anything I said.

There remains much we do not understand about how and why people change their minds but we do understand some aspects of what makes it harder. The sheer length of time that people have believed covid lies along with the public commitment to them will mean it will be very hard or perhaps impossible for everyone to admit they were duped. An important step to achieving that will be institutional change and political change but these are far harder to achieve than persuading an individual and that is hard enough.

9

THE OPPOSITE OF SCIENCE

> *In science it often happens that scientists say, 'You know that's a really good argument; my position is mistaken,' and then they would actually change their minds and you never hear that old view from them again. They really do it. It doesn't happen as often as it should, because scientists are human and change is sometimes painful. But it happens every day. I cannot recall the last time something like that happened in politics or religion.* **Carl Sagan, 1995**[268]

If science is the exploration of uncertainty and the correction of wrong thinking, then the opposite is having and maintaining a set of unshakable beliefs with exaggerated certainty. Perhaps, the opposite is religion. My second book (forthcoming) *Spiked: A Shot in the Covid Dark* explores this further looking at the strange parallels between Covidianism and religion. It's not just religion where certainty can be exaggerated. Perhaps, the opposite of science is politics?

Politicians must decide what they think will work best, act on that decision and be judged on the results. Scientists, in contrast, propose a way the world might work. They then set out to show why that model is wrong in order to improve their understanding. If scientists are not proving theories wrong they are not doing their job.

Politicians are, in part, salesmen who need a pitch, which requires at least the illusion of certainty. Talking openly about uncertainty in politics can be made to look like dithering, indecisiveness or a lack of confidence. Having made their pitch they will want to defend it even as the cracks appear. For covid, where there was doubt this was spun to maximise fear levels. We don't know how many people are susceptible – perhaps it will be everyone. We don't know how it spreads – perhaps every person is acting as a potential vector. We don't know how the next variant will behave – perhaps it will be far more transmissible and deadly. Doubt was always presented as the possibility that things were about to get much, much worse.

Those in power presenting suppositions as certainty were not allowed to be questioned. That is not science. Science requires curious minds to exchange ideas, debate and challenge. Covid was not to be debated, whether in parliament, which made itself redundant, or even among friends and family. Science is not banned from polite conversation but politics and religion are.

When a topic is banned from polite conversation that implies that people do not want their beliefs questioned. Why was that? Perhaps, it was a result of fear. Fear of the virus meant that comfort could be found in believing that the authorities knew what they were doing and had everything as controlled as could be. There was also fear of the work they would need to do if they were proved wrong and had to research to restructure their thinking. Perhaps, people knew the authorities had misled them but admitting that meant having to face the frightening question of why they would do that.

Either way the politicians were presenting philosophy driven by hysteria and the public believed what was said with a religious fervour. Questioning was not allowed in public and even among friends. A professor told me he was instructed to be more careful in what he shared in public. His boss said he knew everything stated could be backed up with evidence and even said he shared the same views, but the truth was not to be spoken in case it risked the reputation of the university.

I was faced with medical friends who would not engage in conversation about covid. The first time was when I emailed a group of friends to seek their views on a paper presenting the evidence for overdiagnosis. One replied immediately (she could not have opened the attachment let alone read it) saying, *"We'll have to agree to differ on that one."* Another time I was trying to engage a medical friend on issues around the vaccine and just raising the topic resulted in this friend, who I'd never known raise her voice, angrily responding *"Don't, Clare. Just don't."* Thereafter, the topic became a taboo.

While debate was not allowed among friends, it was continuing online. The belief, on both sides of the debate, that lives were at stake has meant that passions were raised and polarisation followed. The polarisation of views on covid has meant that the debate has been heavily politicised in a tribal way. When people present an argument in a politicised environment they do so with exaggerated certainty. Science cannot thrive in such a politicised environment. Politics requires the illusion of certainty and the scientific method relies on uncertainty which can then be tested and measured to step closer to the truth. Science moves forward in increments with evidence that leads to people conceding they were wrong about something. The heavily politicised environment around germ theory 150 years ago resulted in germ theorists exaggerating their case. Rather than sticking to the facts around germs they went a step further denying that any spread through the air was a possibility. The ramifications of the politicisation of that debate is still causing harm today.

Despite the politicised nature of the debate in the public realm, parliamentary debate has been minimal with all sides agreeing or with the opposition and the media pitting Government policy against even harsher hypothetical approaches. The argument that the interventions have caused more harm than benefit has not been properly heard or debated. A number of doctors and scientists have risked their reputation, knowing they will be viciously attacked, to try to make the case publicly. These people have nothing to gain from taking such a position and everything to lose. Many more doctors and scientists share their views but have been too afraid to speak out. Those that did, faced attempts to discredit, shame and have careers terminated. Such an environment is not compatible with a journey of discovery towards the truth and away from error.

Politicians do not always tell the truth. The truth is often too complicated, uncomfortable or tedious. Instead politicians use narratives that are easy to communicate – a key skill for any politician. Presenting simple narratives enables evasion of complexity, ignoring truths that may harm their cause and reframing to try and keep public support. There is power which comes from selling simple narratives because politicians who admit to complexity and uncertainty might find it harder to win the crowd.

At the heart of epidemiology beliefs about the response to an epidemic was the idea that the public were irredeemably ignorant and reckless. It was not possible for the public to comprehend the danger of exponential spread and they would respond to the threat too late. Any reassuring words would cause people to alter their behaviour, spreading disease. In reality, the public restricted their behaviour well before any requirement to, falsifying that core epidemiological belief and there was no evidence of harm from altered human behaviour.

Public health officials have to bridge the gap between being scientists and politicians. It turns out, in terms of what level of certainty to portray, the simplicity of the narrative and the level of hysteria they all had both feet firmly in the politicians' camp. The history of public health doctors presenting simple narratives with too much certainty stretches right back to the evangelical Dr Charles Chapin, who we met in *Belief One: Covid only spreads through close contact.* He was the doctor who founded the myth that all infectious disease spread only through close contact. The opposite of science is a combination of projecting certainty where there is none, like politics, and holding on to beliefs with religious zeal.

BELIEF FIVE: A NEW VARIANT SPELLS DOOM

Why do some die and some live? The answer was clearly, that on the whole the best fitted live. From the effects of disease the most healthy escaped; from enemies, the strongest, swiftest, or the most cunning; from famine, the best hunters or those with the best digestion; and so on. Then it suddenly flashed upon me that this self-acting process would necessarily improve the race, because in every generation the inferior would inevitably be killed off and the superior would remain – that is, the fittest would survive. **Alfred Russel Wallace, 1908**[269]

Viruses can mutate. A mutation could, in theory, make the virus more transmissible or even more deadly. Perhaps mutations could even mean that immunity from infection stops working. As if these invisible threats were not enough to scare people, fear could be amplified with hypothetical future variants that would be even worse. And so it was.

As soon as the spring 2020 wave started to fade, scare stories began about the potential for a new variant that would spread more easily and be more deadly. This story was rehashed with every variant associated with a wave and many more, like the *"Killer from Manila"* as the *Daily Mail* dubbed a Philippine variant, that came to nothing.[270]

MORE TRANSMISSIBLE

The conventional narrative says that if a new variant is to replace its predecessor and become the dominant variant it must by definition be more transmissible. However, reality was not as simple as that. The transmissibility, measured as the percentage of household contacts that end up infected, changed during a wave. It started relatively high at the outset shifting to low at the end. It was not a constant. Despite this being clearly demonstrated in summer 2021, public health authorities continued to compare the declining transmissibility of previous waves with the rising transmissibility of the next variant.[271] Will Jones, who wrote from the outset about covid, tirelessly and always in a carefully evidenced way, demonstrated that when variants were compared over the entire course of a wave, the differences were miniscule.[272]

It was widely believed that Omicron was indeed more transmissible. However, the transmission rate proved to be no higher when measured as the proportion of contacts infected. The reason the surges were more dramatic was because the incubation period shrank. The time to become infectious had reduced. The time until the next generation of infections also reduced and they would therefore become infectious sooner. The shortened time between infections did lead to a sharper rise in the trajectory of the wave. But it was not true that people were more likely to become infected and the wave fell just as sharply.

A different way to explain the sharper surge was simply that there was a stronger seasonal trigger meaning more people were susceptible in a short time. Looking at the trajectory of deaths over past winters in the UK there are some years with gentle autumn and winter surges and some where the rise is much steeper and shorter as with Omicron. It was the winter seasonal surge that led to deaths in the UK peaking in January every year. In reality, it was the seasonal surge that caused variants to become dominant. The government reported in December 2022, that a variant that closely matched the Alpha variant was circulating 6 months before the

Alpha surge and a close match for Omicron was circulating a whole 18 months before the first Omicron surge.[273] Although viral changes may play a role in which variant becomes dominant, their role is exaggerated when ignoring the impact of the seasonal surge.

MORE DEADLY

Viruses evolve. The result of all evolution can be summarised in terms of securing food and sex and avoiding danger. For a virus, these translate into successfully entering cells to hijack resources, reproducing and spreading and avoiding the body's immune response. The world is therefore full of viruses that have optimised their ability to infect and spread including all the viruses that cause a common cold. Dead people do not cough and sneeze. Sick people do not wander around sharing air with others. Viruses therefore succeed over time by making people less sick not more sick. In fact, respiratory viruses switch off certain genes in a temperature dependent way such that they become inactive when they enter the warmer air of the lower lung but remain active in the colder upper airway.[274] These upper airway infections benefit the virus by not making their host too sick, helping them evade the immune system so reinfection is possible and causing coughing and sneezing to help with spread.

How deadly a variant is depends on more than just the virus. In periods of overdiagnosis of death a variant will appear more deadly. In a population that has recently had many of its vulnerable die, it will appear less deadly. Having had a mild winter with Omicron in 2022-23, there will be more people who are close to death who did not succumb and will therefore be more vulnerable to a future covid variant. There may be a 2023 variant that appears more deadly but part of that will be due to there being more vulnerable people in our population rather than the virus itself being different.

The genetic mutations that are seen in a 'new variant' are changes in the genetic code for a particular protein. Such changes often result in a vital viral protein being unable to function such that the mutation will not be passed on. Only mutations that allow continued functioning survive. In terms of function, the variants cannot differ significantly from each other and therefore the behaviour of subsequent variants can largely be predicted from previous variants. Omicron was the exception to this rule.

OMICRON

Omicron was significantly milder and had a different symptom profile to previous variants. Early on it was evident that infection was more likely to be restricted to the upper airway and not involve the lung. For most people it was more like a cold, stayed in the upper airway and was less likely to create lasting immunity.

If there had been no testing and none of the drama over previous variants, it is highly likely that Omicron would have gone unnoticed in the UK. Omicron was a blessing. As Bill Gates put it, *"Sadly, the virus itself - particularly the variant Omicron – is a type of vaccine. That is, it creates both B cell and T cell immunity. And it's done a better job of getting out to the world population than we have with vaccines."* [275] (No, I do not know why that made him sad).

The very first cases were reported in Botswana after four people, from an undisclosed country, were diagnosed while on a *"diplomatic mission."* [276] South African doctors reported repeatedly that they were seeing fewer hospitalisations and deaths than for previous variants. [277] In the meantime, the WHO tried to keep fear levels high, with the Director General saying, *"There is this narrative going on which is 'it's milder or less severe', but we're undermining the other side, at the same time it could be dangerous, because the high transmissibility could increase hospitalisations and deaths."* [278]

Part of the spike genetic sequence looks similar to the dangerous bacterial protein that causes toxic shock syndrome. Toxic shock syndrome happens when a particular bacteria produces a toxin which binds directly to white blood cells. These cells then release chemicals that leads to an overreacting immune cascade, a cytokine storm, that can ultimately lead to circulatory shock, organ failure and death. The virus had this same sequence enabling it to bind these cells too and cause the cytokine storms that led to severe illness or death in pre-Omicron variants. In the Omicron spike sequence this area was made non-functional with mutations.

Omicron had lower hospitalisation and death rates in South Africa, where only 35 percent were vaccinated. However, Omicron was as deadly in South East Asia as the original Wuhan variant had been in Europe despite extensive vaccination. The deadliness was clearly related not just to changes in the virus but also either to the susceptibility of the host population or the degree of panic. Two years in, Europeans had been exposed to several variants such that the population had a primed immune system and many vulnerable people had already died. Europe had also tired of responding to a covid wave with panic. A calmer approach, with fewer people scared of accessing healthcare, may have helped keep mortality lower.

IF YOU HAVE HAD COVID DO YOU NEED TO FEAR A NEW VARIANT?

While vaccine immunity can be evaded by a variant, is the same true for naturally acquired immunity? Prior to Omicron the answer was that between 98 and 100 percent of people who had had an infection previously would be protected from infection in a further outbreak.[279] For all positives detected in England, prior to Omicron 99 percent of people who had tested positive by PCR had not had another positive once they had had time to recover.[280] By March 2022 that figure had fallen to 95 percent because Omicron was more likely to cause reinfections.[281]

In the pre-Omicron era a study from Qatar demonstrated that for someone who had previously been infected, *the risk of having a severe reinfection is only approximately 1% of the risk of a previously uninfected person having a severe primary infection.*[282] Any subsequent infection would therefore only present a negligible risk even in the vulnerable. The same authors repeated this work after Omicron and found that protection against hospitalisation was maintained at 90 percent and there were no critical or fatal covid cases in the reinfected.[283]

Panic that antibody levels would wane to nothing was unwarranted. All antibody responses fade when not needed but they stay at low stable levels in case they are needed again. Antibodies and cellular immune responses are maintained as part of immune memory with follow up for 15 months showing stable responses.[284] Immune memory for the original SARS virus was maintained for at least 17 years.[285]

If covid continues in the ecological niche previously occupied by influenza there may yet be more variants to come. A good year for deaths is often followed by a bad year so the winter following Omicron may have more deaths which will be blamed on a *more dangerous variant*" but will likely have more to do with the vulnerability of the population who escaped death in the previous winter. Ultimately the transmissibility and deadliness will be in the same ballpark as previous variants and any further attempts to scare people will be unjustified.

TOP THREE MYTHS
1. A new variant is capable of being markedly more transmissible
2. New variants might be far more deadly
3. Immunity from one variant will not protect from the next

10

IN THE BEGINNING WAS THE WORD

The purpose of Newspeak was not only to provide a medium of expression for the world-view… but to make all other modes of thought impossible. **From 1984, George Orwell, 1949** [286]

The narrative has been shaped by the way words have been misused. Important words had ambiguous definitions leaving them open to abuse whereas the official definition of other terms were brazenly changed.

The philosophical debate about how language restricts or influences our ability to express ourselves is complex and will no doubt continue to be the subject of academic studies for years to come. Even if language does not restrict expression, there is good evidence that what a language forces us to say has an influence on how we perceive the world. For example, when people who used gendered nouns in their language were asked to assign random adjectives to inanimate objects,[287] people associated masculine objects with adjectives such as dangerous, strong and towering whereas feminine objects were described as elegant, fragile and pretty. This was also the case even for words like apple, wars, rain, clocks, forks, mountains and tables which were associated with a different gender in Spanish than in German. Even when only pictures were shown and the participants of a study[288] were asked to assign voices to animate the objects they chose male or female voices to match the genders assigned by their language. What have we been forced to say recently and how has that influenced thinking about covid?

REASONABLE WORST CASE SCENARIO

In the aftermath of the swine flu 'pandemic' reports were written about the mistakes that had been made. A major mistake has been overly pessimistic models predicting severe loss of life. Neil Ferguson presented the government with a scenario where 65,000 people would die of swine flu. This was presented as a *reasonable worst case scenario.'* In a report on the lessons to be learnt from the swine flu response the European CDC commented:

"Modellers offered a range of possible scenarios. Despite a variety of scenarios, decision-makers and politicians had a tendency to focus on the worst-case scenarios, assuming that by alerting the public to the worst case, unpleasant measures would look more acceptable. The phrase 'reasonable worst case scenario' caused confusion because 'reasonable' was interpreted as 'likely' when in fact it meant 'still very unlikely'.[289]

The word *'reasonable'* is critical here. The National Risk Register,[290] which outlines risks to the UK, includes as 'reasonable worst case scenarios' the *"worst plausible manifestation of that particular risk."* In 2017, the same register defined plausible as having *"at least a 1 in 20,000 chance of occurring in the UK in the next five years."* [291] That would mean a 1 in 100,000 risk for each year. To put it into context, 20,000 years ago an ice age ended and humans have only been farming for just over 10,000 years. In common parlance, something that would only happen once in every 100,000 years would be considered implausible and certainly not a *'reasonable'* expectation for the year ahead and worth planning for. A 1 in 100,000 risk is surely an *'extreme,' 'apocalyptic'* or *'fantasy'* scenario and should never be described as *'reasonable'* or *'plausible.'*

Neil Ferguson could have fallen back on the above excuse and explained that his 500,000 figure was an extremely unlikely scenario but worth considering as an outside case. He did not. He repeated the figure in public, three days after lockdown began, saying, *"Without those controls, our assessment remains that the UK would see the scale of deaths reported in our study (namely, up to approximately 500 thousand)."* [292]

In June 2020, Ferguson claimed that half the lives lost could have been saved if we had locked down a week earlier.[293] Even late on in 2021, the claim was that the figure of 500,000 had been reasonable and would have happened in the absence of lockdowns and vaccination.[294] However, it is clear from the minutes of the SAGE meetings that *only* the *'reasonable worst case scenario'* of 85 percent of the population being infected and 0.9 percent of them dying was discussed.[295] No other figures were discussed as possibilities prior to lockdown. In the most favourable light possible, this was an utter failure of communication and it must have been clear to the modellers that their work was being interpreted in a way that would lead to overreaction and potentially destructive policies.

INFECTION

You might think that we would have had a clear definition of an infection. It is a term used frequently and seems simple enough. However, medicine has never had the vocabulary to distinguish between a disease caused by a microorganism in the absence of risk to others and a disease that makes you infectious to others. Both are called an infection. An *'infection'* was used to describe someone who had tested positive for the presence of a virus. The way testing was calibrated meant that numerous people tested positive with viral levels too low for them to have been infectious to others.

Even if they were not infectious, did they fulfil the first definition of having a *"disease caused by a microorganism"?* Is it enough that virus has been detected in the respiratory tract? Many people seem to have a cartoon-like image of immunity as if it produces a force field protection from infection. In reality the air we breathe is full of viruses and other nasties. Our nasal hairs filter out most and the rest must penetrate the mucus lining the airway and escape the immune attack from antibodies secreted onto the surface of the airway. If that succeeds the virus can bind to a receptor on the cell surface and enter a cell. Once in a cell, it can hijack the cell's apparatus and replicate. Unlike most viruses, *SARS-CoV-2* and influenza do not result in the bursting open of the cell to release the virus particles but instead they are thought to leave the cell by *'budding'* one at a time from the cell surface. The idea is this next generation of virus particles can then infect neighbouring cells repeating the process of replication. All the while, the infected cells and virus particles will be under attack from the immune system which will kill infected cells and remove free virus for destruction. Only once multiple cells have been turned into virus factories can sufficient virus be produced to make the person an infection risk to others. The cell destruction that accompanies this phase of the illness means that the infectious period coincides very closely with the symptomatic period. Irritation of the airway results from immune attack and cell destruction leading, inevitably, to a cough, which expels the virus particles.

The risk of infecting others is not reached until just over a day before symptoms appear. In an epidemic, the ability to diagnose people who are pre-symptomatic and in the incubation period before symptoms develop could be important if altering behaviour before the moment of infectiousness would reduce spread. (That theory is based heavily on the assumption that close contact is a requirement for transmission, see Belief One). The zealousness with which covid was diagnosed meant that many people who would never become infectious were also diagnosed as positive cases. People whose immune system defeated the virus before the viral levels were a risk to others and those for whom virus was present but never successfully entered a cell were all given positive test results and described as having an infection. There were also people for whom a positive test was an error. All of the above groups would have been asymptomatic (see *Belief Seven: 1 in 3 people with covid spread it while asymptomatic*). These non-infectious groups made up a much larger proportion of the asymptomatic population than those who were in the short lived pre-symptomatic infectious period. Anyone with no symptoms and a positive test result was very unlikely to have been an infection risk to others.

Describing a person as having an *'infection'* would inevitably have led to the incorrect assumption that they were *'infectious.'* This misunderstanding formed the basis of the fear around asymptomatic spread and the exaggeration of the case numbers.

CASE

The word 'case' itself was misused. In an epidemic it is believed that the number of cases must be measured to track the spread of disease, measure its growth and, ideally, limit its impact. The traditional definition of a case in an epidemic was based on symptoms with or without a confirmatory test result. China used that definition throughout. The rest of the world abandoned this fundamental principle of how a disease is diagnosed. This was at least in part the result of attempts by public health authorities to have centralised diagnosis. By *'diagnosing'* patients based only on a laboratory result, they could easily count numbers and please the politicians and media who wanted a daily *'case'* count to publish. The latter was not compatible with a system where a doctor who had seen the patient face to face was allowed to express an opinion as to whether this was a genuine *'case'*.

The presence of symptoms and diagnostic assessment of a doctor were disregarded and every positive test result was described as a *'case'*. One perverse consequence of this was that rather than the disease state being defined as a classic clinical presentation based on symptoms, the test became the definition of the disease. Once that inversion had taken place the list of associated symptoms began to grow and included having no symptoms at all. Gastrointestinal symptoms were added to the main symptom list and for many doctors there was no persuading them that these symptoms were not a genuine presentation of the disease. Symptoms like cough, fever, loss of smell, toe rash were predictive of who would test positive in a study from the pre-Omicron era.[296] However, symptoms such as headache, sore throat, diarrhoea, nausea, vomiting and abdominal pain were just as likely to be seen in those who tested negative as those who tested positive. Gastrointestinal symptoms did not predict who would test negative or positive.[297]

As well as every person with a positive test result being described as a case, in USA[298] and Canada[299] even some with negative test results were included in the data. Having defined cases too broadly, the pressure to produce a figure for the number of *'cases'* each day meant that no sense check could be employed. There was no feedback from the treating doctor and no opportunity to question the diagnosis or appeal it where the patient never did develop symptoms. In part, this was due to the belief around what asymptomatic positivity meant. For some people, the idea of an *'infection'* had become so distorted that the presence of a handful of virus particles in the air being breathed that happened to be picked up on a swab was sufficient to meet their definition.

A more nuanced approach to calling a 'case' with a series of different definitions would have been a huge help. Someone who was asymptomatic but positive could have been referred to as merely *'screening test positive'* and would not be considered a *'case'* unless they went on to develop symptoms. A *'suspected case'* of a patient with symptoms and a positive test result would become a *'definite case'* if they also had a positive antibody test.

There was a very strange lack of information about the numbers of people who were symptomatic and tested positive. Not only did public health authorities avoid the issue but so too did numerous research scientists. Huge numbers of papers published during covid based their analysis on positive test results and very often failed to mention symptoms at all. When something so fundamental to diagnosis is omitted from a paper it does make one wonder what the paper would have shown if they had included that information.

While using the word *'infections'* led to assumptions about infectiousness, it was the 'case' numbers that led to assumptions about extent of the problem. If we had had a term to describe someone with a virus on board who was not an infection risk, then overdiagnosis, exaggeration of case numbers and ultimately fear could have been reduced. A term to describe someone who had tested positive but did not have a clinical picture of covid disease would also have been useful. Imagine how different the public's fear would have been if only patients with raised antibodies after infection were described as having had covid. I make this suggestion since that was how influenza has been measured in the past. People with influenza may have been tested and diagnosed in hospital or by their doctor and otherwise the extent of disease was only measured in retrospect. Those people who were infected and never particularly sick would have shrugged their infection off as a cold and moved on.

PANDEMIC

As with the word 'infection' the word 'pandemic' was poorly defined. The WHO did not have an official definition prior to 2009. The WHO Pandemic Preparedness homepage from 2003 said, *"An influenza pandemic occurs when a new influenza virus appears against which the human population has no immunity, resulting in several simultaneous epidemics worldwide with enormous numbers of deaths and illness."* [300] In 2009, the WHO announced the onset of the swine flu pandemic. A month before the words *"enormous numbers of deaths and illness"* were removed from their website.[301] Loss of life was no longer a feature, only the presence of a novel virus. This new definition could be used to describe seasonal influenza every winter which mutates and returns as a different, novel variant each year. In fact, their new definition would mean that we are currently still living through the 1918 influenza epidemic as there remains global circulation of descendants of that virus. From 2009, there was a marked mismatch between the potentially inconsequential viral spread that WHO were prepared to describe as a pandemic and the pandemic plans which were designed for implementation only when there was a catastrophic risk of societal disruption from illness with huge loss of life.

In the absence of a definition, its meaning could be deduced from how the term was used. Pandemic plans, internationally, were written about hypothetical situations in which the loss of life would be 4 to 30 times higher than in a seasonal influenza season.

The government published a plan for emergencies generally, The National Risk Register of Civil Emergencies 2017[302] where the campaign to *"Catch it, bin it, kill it"* with tissues was the only recommendation to the public for a pandemic flu with consequences of *"up to 50 percent of the UK population experiencing symptoms, potentially leading to between 20,000 and 750,000 fatalities and high levels of absence from work."*[303] Even the advice to use tissues, while polite, may not have had any significant impact on an aerosolised pandemic flu. The 2014 Pandemic Plan defines a pandemic as causing illness *"substantially higher than even the most severe winter epidemics"* with mortality *"increased in younger age groups."*[304]

There was an implicit assumption that the lives lost would include the young. Even with these expectations of huge loss of life, there was an understanding in pandemic plans that actions to limit the spread of an airborne respiratory virus would be largely futile.

In the same way that there is a hierarchy for what would be called an infection, there is an equivalent for describing a pandemic (see table 2). The measure of significant virus was reduced from infectiousness, to any virus detected and the pandemic measure from being a threat to life, to any outbreak which is not contained regardless of how serious it is.

HIERARCHY OF INFECTION	HIERARCHY OF PANDEMIC
1. Virus detected	1. Virus exists
2. Virus has replicated in cells	2. Human to human transmission
3. Patient is infectious to others	3. Spread between countries
4. Symptomatic disease	4. Sickness
5. Threat to life	5. Threat to life

Table 2: *Hierarchy of meaning for terms infection and pandemic*

There were other terms used in a manipulative way. The word 'asymptomatic' was used to conflate people with genuine infections in the pre-symptomatic period with those who had an adequate immune response and would not be infectious (see *Belief seven*). Further words had their meaning changed including 'herd immunity', 'vaccine' and 'unvaccinated' and these will be discussed in *Spiked: A shot in the covid dark.*

Where does that leave us? Terms were changed that meant a pandemic could be declared and sustained where previously the threshold would not have been met. Imagine if the definition hadn't been changed and the attempts to instil fear would have been for a *"novel respiratory pathogen with a mortality rate similar to bad influenza but a higher intensive care requirement.*" It really would not have packed the same punch. The potential case numbers were exaggerated with use of a *"reasonable worst case scenario"* and the actual case numbers were inflated with misuse of the terms 'infection' and 'case'. Whenever language was simplified or redefined it was always to present covid as more frightening and the solutions as more straightforward.

BELIEF SIX: IF YOU TEST POSITIVE YOU HAVE COVID

> Clinicians are also driven to make a diagnosis, sometimes forgetting what is actually important to the patient. Not every test or treatment will actually help – in fact, some may cause harm – and many interventions are not even evidenced based. ***Uncertainty in Diagnosis, Royal College of Pathologists, 2018***[305]

There is rightly much faith placed in medical testing by both doctors and patients. It is justified, not because tests are perfect, but because doctors use them in a reliable way. To prevent underdiagnosis, tests can be repeated when there is a clinical suspicion. To prevent overdiagnosis imperfect tests can be cross-checked in two ways. Firstly, tests are primarily used where there is already a clinical suspicion of a disease, reducing the risk of error. Secondly, tests done in an apparently healthy population, e.g. cancer screening tests are always followed by confirmatory testing. Covid regrettably has been defined such that symptoms and the doctor's opinion have been ignored and testing of the apparently healthy has been used in the absence of confirmatory testing.

WHAT IS A COVID CASE?

The justification for saying people with no symptoms were *'cases'* was that there was an incubation period where a person would develop symptoms within days. Calling these pre-symptomatic positives *'cases'* is justified. However, a high proportion of people described as *'cases'* never developed symptoms. There is no evidence of them producing antibodies either. Using circular logic, these never-symptomatic people were defined as having had covid on the basis of the test result, without ever demonstrating that they were an infection risk to others. (See below for how infectiousness can be demonstrated).

If we took the same approach of calling everyone with positive screening tests a *cancer case* there would be huge numbers of people *with cancer* and the cancer mortality rate would plummet. In reality the vast number of patients who are positive on cancer screening are given the all clear with subsequent testing. For covid the more positive test results that were described as *cases* the less deadly it was per case. Ironically, calling out the overdiagnosis that leads to covid looking *less* severe than it was, results in being called a *"covid-denier."*

In theory, PCR was perfect. Utter faith in medical testing resulted in a belief that, unlike any test in history, PCR testing never produced erroneous positive results. That led to a series of errors including, exaggeration of the covid problem, overdiagnosis of covid cases, hospitalisations and deaths and disparaging of other tests.

TESTING ERRORS

In what way are tests imperfect? A useful way to think about testing is to think of a spectrum. For example, take the height of dogs. Height can be used as a measurement to determine whether or not a dog is a Great Dane. There is a height above which all dogs are Great Danes. There is also a height below which none are. In between there is a mix of dogs. If height is used as a test, a choice must be made. Do you want a test that finds every possible Great Dane or only the definite ones? Whatever height cut-off you choose for the test will be a compromise between these two decisions. Either you choose to emphasise how good the test is at finding definite cases and miss possible cases or choose to maximise the possible cases and end up including more dogs that are not Great Danes. This choice of cut-off must be made for every test.

Tests that can detect every possible case at the expense of including the innocent are used for screening e.g. cancer screening tests. For example, even low levels of abnormality seen in the cells from a cervical smear or low levels of calcification seen on breast cancer screening triggers further investigation. Both these findings are a spectrum that signal a possible cancer. Tests that will give a definitive answer but are not suitable for testing a population, like a tissue biopsy, are used to confirm the diagnosis.

Doctors know that tests are imperfect. However, the process of diagnosis involves a doctor forming the kernel of a belief about what the possible diagnoses might be. Testing can then be carried out to confirm their suspicions. Once they have weighed up the available evidence the doctor increases their belief in one particular diagnosis in order to convincingly sell the idea to the patient. Faith in testing translates to belief in test results. There are times when both can be misplaced.

TESTING HARMS

Underdiagnosing patients could lead to increased community transmission or result in an infectious patient being placed on a hospital ward with vulnerable people. Overdiagnosis could lead to uninfected patients being placed on a covid ward and then actually catching covid or being given inappropriate medical treatment as well as their real reason for being sick going undiagnosed. The impact on an individual of overdiagnosis could mean delayed surgery and a prolonged miserable hospital stay. On a population-wide level overdiagnosis resulted in isolation of people and their contacts who were never an infection risk to others with all manner of financial and psychological implications. The exaggeration of cases and inflated death figures from overdiagnosis led to inflation of the fear narrative.

Because there was a risk of real harm either way, there was conflict in what people prioritised. Those who wanted to find every possible case at any cost for fear that unstoppable exponential growth and huge amounts of death would otherwise result were the Hyperbolic-Hawks. Those who wanted to minimise the harm caused by overdiagnosis were the Data-Driven-Doves.

The Hyperbolic-Hawks feared a genuine infection being missed and going on to cause harm. In reality, if a genuine infection was missed symptoms would develop and other testing could be offered. (The significance of spread during this period is addressed in *Belief Seven: 1 in 3 people with covid spread it while asymptomatic*). If, on the other hand, a healthy person was incorrectly described as a case there was no recourse.

POST-INFECTIOUS POSITIVES

In August 2020, three men in their twenties went to Italy to teach at a summer school.[306] Two of them fell ill with covid and isolated in Venice.[307] They then headed to Florence hoping to get home from there. In Florence they all tested positive and they were put into isolation in separate rooms of a hotel. They were not allowed into the corridors, food was brought to their doors in plastic tubs and they only saw each other when let out for more testing. The rules stated they needed two negative tests in order to be sent home. Two months later and they were still in the same situation having all tested positive repeatedly. Finally, the Italians changed the rules so that a single negative test and 21 days of isolation were considered sufficient. They were released and made it back to Blighty just in time to not have to undergo a further 14 days of isolation that had been introduced in the UK.

How did these tests keep coming back positive? Were these samples being contaminated by a member of staff between the hotel and the laboratory or perhaps in the laboratory? Given it was a problem for all three of them this may well have been the case. Could there be another reason?

Dr James Barnacle, a doctor at Imperial, noticed a shift in who was testing positive.[308] In spring his covid patients had been very sick but by May many of his covid patients did not have acute covid at all. In fact by 13th May 2020, 80 percent of them were admitted for reasons other than covid. Most of the remainder were being treated for complications of covid rather than acute covid. Nevertheless, all these patients were being added to the tally of covid hospital admissions.

There seemed to be an issue with testing giving positive results after infections had cleared.

WEAK VS STRONG TEST RESULTS

The amount of virus produced changes over the course of an illness. Imagine a graph showing the amount of virus produced over time and you are travelling along the trajectory. It starts as a path just riding above sea level, soars exponentially as it climbs a virtual mountain during the early symptomatic phase when 100,000 particles a minute are being emitted and then plummets rapidly downwards before the illness is over. PCR tests measure viral RNA which is the equivalent of DNA and, when a cell is hijacked, is the code that it makes the cell produce the virus. The PCR results do a similar sharp ascent but the descent is very gentle and a patient will not arrive back at sea level for up to 90 days.

Patients who are sick and in the mountain area, generally have clear cut test results combined with symptoms and these results are easy to interpret: they are very likely to have covid. The confusion has arisen around results from the foothills both before and after the mountain. The trajectory of viral emission is almost always plotted on a logarithmic graph which squashes the peak right down. As a result, a false impression is given of the size of the difference between weak positives at the foothills of the mountain and true positives from the mountain itself. The public health authorities have found people standing on the equivalent of molehills and declared that they must be on the mountain. Any viral RNA, even in someone not infectious, must indicate that they were a ticking time bomb of infection, right?

HOW WAS TESTING GETTING THIS SO WRONG?

As well as assuming PCR testing was infallible there was another flawed assumption. PCR testing is presented as a measurement of the quantity of virus present but it is actually not that precise on two counts. Firstly, it does not measure virus particles themselves but the viral equivalent of DNA, its recipe for life, RNA, and only parts of that sequence. RNA is very unstable and it was assumed it would be undetectable shortly after symptoms end. The assumption was that PCR positives would relate to the illness and the period that someone was infectious. However, with the way testing was calibrated, a PCR test would be positive before and potentially long after the infectious window.

Secondly, it is also not a direct measurement. PCR works by first copying each letter of the sequence into its DNA equivalent. This DNA is then repeatedly copied, doubling what is present until it is detectable. This is done through several steps which double the DNA present and these steps are repeated. At the end of each doubling a measurement is taken to see if there is sufficient relevant DNA present to reach a target quantity called the threshold. The result is therefore highly dependent on the amount of virus collected in the sample, the efficiency of the doubling process and the choice of threshold at which a positive is declared.[309]

It is widely accepted that sampling difficulties and other issues can lead to one in ten (or more) samples from genuine cases being inadequate.[310] The result is a negative test largely because enough virus was not collected (a false negative result). A good thorough rub, where the swab has plenty of contact with the throat, is more likely to get to the cells that are infected which will clearly affect the quantity of virus detected.

The machines that carry out the testing do not give a binary result or even a spectrum of numbers from which, like the dog heights, a cut-off must be chosen. The machines produce a graph. The graphs must be interpreted to decide if they show valid positive results. For covid PCR results, instead of skilled scientists, artificial intelligence was used to interpret the graphs. UgenTec secured a contract for this role. A competitor, Diagnostics AI, proved that UgenTec's graph interpretation was inadequate, and sued the UK government on the basis that cases would be missed, winning an out of court settlement for £2m.[311] Aron Cohen, Diagnostics AI's chief executive commented, *"The government is paying out a lot of money. And they're paying this out, you know... to avoid having to have these issues aired in court, and to have discussions over the accuracy of the testing."* [312]

Despite these clear challenges, many in the scientific community treat the result produced after interpretation of the graph as an accurate measurement of the amount of virus present in that patient. PCR can be a very useful test but this overinterpretation of it leads to too much faith being put in the results. Like any test things can go wrong. Strong positives are more likely to reflect a true positive but there is room for error in all the results, as we shall see.

The public health authorities had set ideas: that PCR was infallible at detecting virus and that viral RNA had a short half-life. Once the virus was dead there should be no detectable RNA. Based on this model of the world, the presence of any viral RNA must mean virus was present. Carl Heneghan, Director of the Centre for Evidence Based Medicine in Oxford, was a key voice in demonstrating that the real world was not complying with the model that public health authorities were using. The data was demonstrating that something unexpected was happening. The post-infectious positives were generally weak positives where doubling of the DNA had continued beyond the point at which infectious virus could be recovered and the results were not clinically meaningful.

In December 2020, a team from Harvard demonstrated that the virus's RNA could end up as DNA within the cell nucleus of the respiratory epithelial cells and that *"the integrated sequences might account for PCR-positive tests."* [313] The cells would keep producing viral genes until they died, months later. The young men stuck in isolation in Florence may have been waiting for their cells to die.

The problem could have been solved in two ways: involving doctors in diagnosing patients or ensuring testing was only calling infectious cases. Both these options were dismissed.

CALIBRATION

The spectrum of numbers produced as a test result needed to be interpreted. The Hyperbolic-Hawks were so afraid of missing a single case that even the weakest positive was treated as genuine.

The Data-Driven-Doves wanted to use a cut-off that would identify only those who were an infection risk to others. No-one knew what level that was at the outset.

As with any machinery that takes measurements calibration is required to ensure the measurements are accurate. The numbers produced by the test needed to be calibrated to ensure that interpretation reflects something clinically meaningful. The laboratories that designed the testing protocols did not do this work to make sense of their recommendations.[314] By summer 2020, the testing had finally been calibrated in a handful of laboratories. The measure used to standardise the interpretation of the test was *viral culture*, which demonstrates the presence of intact virus particles capable of entering a cell. Viral culture tests use cells in culture and measure how many cells were deformed in one way or another by virus infection. For *SARS-CoV-2* and influenza, these changes are more subtle making the test more subjective.

The WHO had warned in 2012 of the danger of winding down viral culture capacity to replace it with easier, cheaper PCR testing saying that PCR *"should not distract from the crucial role of virus isolation."* They also noted that China was seeing a *"rise in the number [of laboratories] performing virus isolation only."* [315] The danger the WHO was pointing out is relevant because it is usually considered essential that new PCR tests be calibrated against the gold standard of viral culture testing.

Such calibration experiments were few and far between. Those that were published in summer to autumn 2020 showed a range[316] for when[317] PCR positives were too weak to infect cells. The cut-off these calibration experiments determined could be translated back into the amount of times the original material had been doubled in the PCR test. Samples were only capable of infecting cells when they contained enough RNA so that it reached the cut off threshold after being doubled by between 17 million and 17 billion times the original amount. Authorities_were calling tests positive when they had doubled the material up to 500 billion to a trillion times which means the starting amount was way smaller (29 to 58,000 times smaller) than a sample with intact virus.[318] Even in these extreme circumstances, results were described as 'positives' from 'cases' without any caveats. Going back to the dog detector analogy, the public health authorities were finding miniature poodles and saying they must be Great Danes. The wide range (17 million to 17 billion) demonstrates the importance of calibration in each laboratory. The result in one laboratory can have a very different meaning to the same result in another laboratory. This may have been due partly to the subjective nature of assessing a positive on viral culture as well as other differences between laboratories in their PCR setup.

The laboratories themselves were tested as part of external quality assurance. These checks ensured that laboratories described samples as positive which were diluted so much that they were near a *"detection limit"* of only 3 to 4 virus particles in a 5 microliter extract of the sample.[319] A paper on laboratory quality assessment in Glasgow concluded, *"Laboratories that were unable to detect low-concentration samples… should strive to improve the sensitivity of their molecular assays to prevent false-negative results."* [320] Such focus on ensuring every laboratory called such weak results 'positive' meant that people would be described as 'cases' when they had a thousand times less virus present than any infectious person. It would also be easy to accidentally contaminate a sample with so little virus even from aerosols in the laboratory air.

The Office of National Statistics provided confirmation that the viral culture test was in the right ballpark for determining who was infectious.[321] They carried out random screening of the population and found they had two types of positive test results and described a cut-off. People whose tests were weaker than the cut-off were distributed randomly among the households tested and those people whose tests were stronger were clustered in households. This clustering was evidence of transmission of infection. The cut-off they found tallied with the cut-off from viral culture work. It represented somewhere between 800 and 5,000 virus particles per sample extract tested, depending on the laboratory doing the testing.[322] The amounts needed were far higher than the 3 to 4 particles laboratories had to call positive. These results were available as raw data but the Office of National Statistics never wrote an accompanying report and never drew attention to their findings.

When the Office of National Statistics were using PCR testing of a random sample of the population to estimate how many were infected at any one time, they included weak results as meaningful. They did not use the cut-off that their own work had demonstrated was the cut-off for infectiousness.

A weak positive result does not indicate an infection risk or sickness and are therefore meaningless and by definition they are false positive results. However, I have often called them *"clinically meaningless positives"* or *"incidental positives"* to avoid the backlash that calling out *"false positives"* created after summer 2020, (apparently people using that term were far right apologists with blood on their hands). By any name, such a testing regime was failing individuals and society.

By July 2020, the people who had been so derided for speaking out on this issue had their first vindication. Dr Anthony Fauci admitted the problem on a podcast saying how frustrating these positives were for patients and doctors alike and that many of these results were *"just dead nucleotides, period."* [323]

The NHS conceded that there was a problem in September 2020 and said that staff and patients should not test during this 90 day post-infectious period *"because fragments of inactive virus can be persistently detected by PCR in respiratory tract samples for some time following infection."* [324]

In December 2020, the WHO warned about overdiagnosis saying, *"Users of PCR reagents should read the [instructions] carefully to determine if manual adjustment of the PCR positivity threshold is necessary to account for any background noise which may lead to a [weak positive] being interpreted as a positive result."* [325]

WHAT CAUSES TESTING ERRORS?

The opinions of physicists were heard loudly not just in the realm of modelling but also regarding testing. The only diagnostic expert on SAGE was a biochemist who was Chief Scientific Advisor for Northern Ireland. In contrast, in April 2020, there were seven epidemiologists, six mathematicians along with a computer scientist, two experts in artificial intelligence, an engineer and an astronomer. By July a further five epidemiologists and five people with expertise in either maths, physics or engineering were added. I am all for multidisciplinary input, but by July 2020, experts in food security, EU and anti-discrimination law, adolescent sexual health and a vet had all been added while there was still no expert in mass diagnostic testing. Professor of Public Health, Allyson Pollock tried to draw attention to the fact that the UK National Screening Committee, responsible for advising the government on mass testing, was never consulted. She said, *"Extraordinary that UK National screening committee not involved. They have the experts. All mass testing of symptomless people is screening."* [326] The Royal

College of Pathologists also commented on their lack of involvement saying, *"The capacity and capabilities of the pathology community and resources of the NHS and partner pathology services were under-utilised at an early stage in this...we consider that any expansion of services would have been more effectively undertaken in partnership with the health service from the outset."* [327]

Prioritising finding every possible case over finding definite cases and prioritising speed and volume of testing over quality of testing meant that problems with overdiagnosis were more likely to result. Speed and volume were made the key priorities. Over 2.3 million tests were carried out on a single day on 5th January 2022.

For a physicist, testing equipment in the laboratory, the false positive rate is deduced from the lowest ever positivity rate produced by that equipment. Medicine is more complex. The whole process needs to be considered in medical testing of this kind and there are multiple variables along the way that can contribute to an error rate. The equipment is the least of our concerns. Each laboratory will have its own false positive rate and it will vary day to day depending on a number of variables. For that reason, the rate is deduced not from a minimum but from a plateau as the rate of false positives fluctuates around an average.

There are four ways, apart from equipment errors, in which overdiagnosis in the absence of virus – a miscarriage of diagnosis – can occur: a profiling error, mistaken identity, contamination of the chain of evidence and using too low a burden of proof.

PROFILING ERRORS

Profiling errors arise because some groups within communities can have a higher baseline false positive rate. This is a problem seen with police facial recognition cameras pointed at the general public. They produce mainly false positive results which are disproportionately of young black men.

It is important to target testing at those with a high clinical suspicion of the disease being tested for. The most extreme example of bad profiling is where a subgroup are targeted for testing precisely because they have a high false positive rate. A false positive would lead to more testing of contacts who might also be from a subpopulation with a high false positive rate.

Remember the study from the Spanish mortuary?[328] Half of the people who had died of any cause had a detectable respiratory virus on board. Testing the dead for covid would be bad profiling, but it has occurred. The fact that the dead harbour detectable virus does not mean that it contributed to their death but it does suggest that people ill with other conditions may be more likely to have a small amount of unchecked virus in their respiratory tract even if they do not succumb to a full blown infection. Testing debilitated people who do not have characteristic covid symptoms is also bad profiling.

People who are sick are more likely to test positive for a virus. Think of how cold sores re-emerge when people are unwell. Patients in hospital would be more likely to test positive for any virus than patients in the general population. The Spanish mortuary study showed one in fourteen of the dead had a positive result for a coronavirus, almost all of whom were not diagnosed in life. That rather suggests that the balance between viral attack and our immune response can tip towards the virus when we are ill with other conditions. However, it also suggests that the way we carry out testing is so good at picking up tiny amounts of virus that a diagnosis can be made that is not relevant to the patient's condition or death. If it is not clinically relevant, then it is a false positive result. It is therefore nonsense to take a false positive rate for a predominantly healthy population, like the Office of National

Statistics random sample and assume it applies to a sick population like those coming forward for testing because of symptoms.

MISTAKEN IDENTITY

The second cause of error is mistaken identity. Although PCR testing is very good at detecting the unique genomic sequence it is designed for, it is not infallible. Near matches with other DNA or RNA could result in errors in a proportion of the tests. Testing for only one gene within the RNA sequence rather than three increases this risk. Human DNA has even been mistaken for a different coronavirus in a PCR test.[329] An apparent outbreak of *SARS-1* in a care home in British Columbia in 2003 turned out to be an illusion created by false positive test results.[330] Both PCR and antibody testing had overdiagnosed a common cold causing coronavirus.

Coronaviruses are a family of viruses and, although the spike protein of the covid virus is fairly unique, the rest of the virus has a great deal of overlap with other coronaviruses many of which have not been classified or sequenced. These similarities can cause mistakes in PCR testing. As coronaviruses are seasonal, this type of mistaken identity could cause a seasonal variation in the false positive rate. However, testing of the whole viral genome was a way of checking that the virus present was *SARS-CoV-2* in those samples of good enough quality and this was carried out extensively at huge expense. Overall where this problem did exist it must have been fairly minor.

CONTAMINATION IN THE CHAIN OF EVIDENCE

The third and a critical cause of error is contamination of the chain of evidence. Contamination can occur when the sample is taken, during delivery to the laboratory, during checking in of samples or when opening and working on them. It is worth remembering how tiny the virus is and how many particles can be emitted by the infected. This contamination may come from aerosols in the air, from the breath of individuals carrying out the work or from contact with other patients' samples once in the laboratory. Claims that PPE would be effective at preventing contamination from swab takers etc. is like claiming that wearing chain mail would prevent you from getting sandy on a beach.

A delivery driver who is post-infectious and shedding RNA could contaminate the containers the samples are transported in. Whoever opens those containers could then transfer the RNA to the contents. If the same gloves are worn when opening numerous patient sample pots, then the possibility for contamination between samples will be high. Disturbing images were recorded by an undercover Dispatches reporter showing how some samples were handled carelessly and how leaking tubes from some samples could contaminate subsequent ones.[331] Similarly, BBC Panorama exposed how a hanging trail of snot could be lifted from one tube as the machine removed the pipette it had inserted, and then dragged over all the neighbouring tubes.[332] When shown swabs being dragged across the sample tubes by the machine, Prof Chris Denning, Director of the University of Nottingham Biodiscovery Institute said, *"naturally, little droplets are going to spray off in all directions and they are going to go into all the neighbouring tubes... There is almost zero question that this would lead to contamination."* [333] In its defence the Milton Keynes Biocentre responsible said that, *"its test positivity rate closely tracks the UK average."* That comment rather suggests that there were similar problems elsewhere.

Contamination is an issue largely because of the nature of the test rather than sloppy handling. The multiplying of the DNA present by well over a billion times means that even with highly competent sample handling, the risk of contamination will remain because only the tiniest fragment of contaminant RNA can create a false positive test result. Ignoring results from samples with too many doublings reduces the chance of these errors, but even then, not to zero.

BURDEN OF PROOF

Finally, the burden of proof needs to be high to keep the false positive rate low. Testing for a long sequence and for multiple genes keeps the burden of proof high. However, this only holds true if a positive is only called when all genes are positive. Calling any positive gene a positive result makes the test even more likely to produce a false positive.

As well as those specific errors there is a bigger problem of overinterpretation when genuine virus is present. The presence of virus could be due to four possible situations and they should not all be described as 'positive test results' and a 'case' with the implication of infectiousness:

> 1. Virus causing a symptomatic infection

> 2. Presymptomatic virus infection which could present an infection risk to others

> 3. Virus beneath levels that cause symptoms of infection thanks to immune response

> 4. Bystander virus of no consequence

HOW COMMON WERE FALSE POSITIVES?

One aspect that the public and politicians found difficult to understand was how even a low percentage false positive rate can cause such havoc. The percentage refers to the percentage of *tests* that will return an erroneous result, not the percentage of positive results that are erroneous. Lord Bethell, a health minister in the UK wrote in a letter that:

"Independent confirmatory testing of positive samples indicates a test specificity that exceeds 99.3 percent, meaning the false positive rate is less than 1 percent." [334]

This meant they believed 0.7 percent of all tests would return a false positive result. Not many tests in medicine have such a low error rate. PCR testing is one of our best tests from this perspective. However, when carrying out 1.5 million tests per day, as we did in the second covid winter, that would mean 10,500 people being falsely told they were positive every day.

The Office of National Statistics carried out random population testing to estimate the proportion of the population infected over time. The estimates produced with their methodology were extraordinarily high. Overall, the Office of National Statistics estimated that twice as many people had had covid than were diagnosed by either government testing or antibody testing. Their criteria clearly included diagnosing incidental low levels of virus. The consequence of overdiagnosing people coming forward for testing could be up to 10,000 misdiagnoses a day. Overdiagnosing a sample of the population and then extrapolating from that to the whole population using modelling could result in many more misdiagnosed hypothetical cases that did not exist. That is what the Office of National Statistics did, massively exaggerating the number of cases in the community.

Claims that low false positive rates in other countries meant that it must be low everywhere were also wrong. These countries had low prevalence at the time and were carrying out confirmatory testing. Requiring more than one positive result before declaring a positive massively reduces the false positive rate. For example, the New Zealand government said a positive PCR result was not sufficient and, *"following further investigations such as serology, repeat testing, history, and symptoms, they [could be] deemed to not be a case (eg, a likely false positive)."* [335]

There were two easy ways to measure the day to day false positive rate. The first would be for known negative samples to have been fed into the system without laboratory staff knowing. The second was simply to follow up the positives and see how many developed symptoms and antibodies. Neither were done or if they were the results were kept very quiet.

PCR SHOULD NOT BE THE GOLD STANDARD

Utter faith in PCR testing led to disparaging of better tests like antigen testing. Antigen tests are the plastic tests that you can carry out at home which test for the whole virus. The introduction of antigen testing meant testing could be sped up massively. But before they could be approved they were tested for accuracy against PCR testing. How can you tell if a new test works? Assessing a test requires that it be compared to another test but the clinical features also help. Ideally antigen tests and PCR would have been compared to a gold standard, such as seeing who developed symptoms and a positive antibody result and who did not. Ideally, these comparisons would have been carried out at different times of year to check for cross reactivity with seasonal viruses. None of this was ever done.

However, the two tests were compared. Antigen tests were positive only for people who were infectious and up the mountain in the graph. PCR testing was calling people on the mole hills positive.

For *SARS-CoV-2* there was a large grey area where PCR testing was positive but antigen testing was negative. This could be interpreted in one of two ways. The PCR was accurate and antigen testing missed real cases or the antigen tests were accurate and PCR overcalled people who should not have been called 'cases'.

The Hyperbolic-Hawks had so much faith in PCR that they attacked antigen testing claiming it was missing huge numbers of genuine cases. Professor Alan McNally ran the Birmingham laboratory responsible for PCR testing and carried out a mass testing experiment on students right before Christmas. They used PCR as their gold standard and claimed that antigen testing had found only 3 percent of the PCR positive students.[336] On the other hand, government testing had shown that antigen tests found 77 percent of all PCR positives.[337] Using PCR as a gold standard resulted in a huge, nonsensical range in the estimate of cases that were missed. However, using antigen tests as a gold standard, which accurately detected the infectious, showed that PCR testing in Birmingham had a false positive rate of 0.8 percent, the same value as for estimates from a population of symptomatic people (who would have had symptoms due to a variety of causes) but higher than from healthy people. How many cases were antigen tests missing and how many were PCR tests overcalling? Ideally they would have followed the small number of positive students to see who became sick and who developed antibodies. This would have allowed the accuracy of both test types to be properly assessed. They did not mention this. Professor Alan McNally refused to acknowledge they had any false positives at all, *"I don't know about outbreaks. And I don't care about comparisons. I know for a fact there are not false positives affecting numbers."*[338]

The problems were summarised well by Rochelle Walensky, who was a Data-Driven-Dove in September 2020 when she and her co-author wrote:

"But what if we re-framed the narrative? What if we instead asked what possible prevention purpose a PCR test can serve in a pandemic if it returns positive results in people who have already cleared the virus and pose no risk of further transmission? What good is a PCR test that identifies non-infectious individuals as candidates for isolation and quarantine? Who would even think of using PCR for outbreak containment if it sends up so many false alarms, leading contact tracers on so many wild goose chases, and undermining public confidence in the role of testing to keep us safe?

For purposes of surveillance screening, those antigen-based negatives worrying the FDA aren't false negatives at all; those are true negatives for disease transmission. Far from being problematic, in the context of outbreak containment, the antigen test's limited window of sensitivity is a major asset. The antigen test is ideally suited to yield positive results precisely when the infected individual is maximally infectious." [339]

Her description here summarises well the value of antigen testing and its benefits over PCR. By January 2021, she was appointed director of the CDC and although she pushed for antigen testing in that role, she switched, allegiances becoming a Hyperbolic-Hawk, and using the PCR measure *"to really capture our case counts and really get a good view of where we are in terms of the epidemiology."*

ONE HUNDRED YEARS OF SOLITUDE

How much benefit did covid testing have? Between lockdowns people were forced to isolate themselves if they had been in contact with someone who had tested positive. The results of the first four months of the UK test and trace programme were published in

Nature.[340] At that point 1.7 million people had been identified as contacts and had been asked to isolate for around a week – a total of 32,600 years of isolation between them.

Only 6 percent of those who isolated eventually tested positive, in line with other measures of non-household transmission rates. In theory these 100,000 had been exposed and if infected their infections could not have been prevented. However, their isolation prevented them from passing the infection on to their contacts. It also prevented their contacts passing it onto others and so on. The study found an average of four contacts per person. Using these numbers we can calculate how many infections would have been prevented if everyone isolated as instructed and if isolation was a panacea. (The 100,000 people would have infected 6 percent of their 400,000 contacts which is 24,000. Those 24,000 who would have infected 6 percent of their 96,000 contacts which is 5,800 etc). The answer is 32,000.

Using the most pessimistic, Imperial estimate of fatality that would mean 320 lives were saved working out at 100 years of isolation for every hypothetical life saved. Even the one life saved for every *one hundred years of solitude* is dependent on the belief that isolation stops transmission and the assumption that transmission occurs through close contact. In reality, those hypothetical 320 people may well have gone on to encounter the virus later in the wave and have succumbed anyway.

The authors of the paper realised that these dwindling numbers could not explain the exponential growth in numbers seen in the real world. They therefore guessed how many cases were prevented as a result of isolation – using modelling – and decided the figure was around 280,000. If you believed in asymptomatic transmission then perhaps you could exaggerate the numbers like this but you would have to believe that people transmitted asymptomatically to complete strangers at a far higher rate than to people they knew. (We'll cover that in the next belief). If we accept that it is highly unlikely that close contact asymptomatic spread between strangers is more efficient than close contact symptomatic spread with people we live with then another explanation is needed. Aerosol spread between strangers at a distance fills that gap.

PCR testing, like any medical test is not infallible. When carried out on the scale it was used and when the priority was to never miss any possible case, there was always going to be a trade-off where healthy people were diagnosed as cases. The consequences of this, in terms of people having to isolate were immense. There was a total denial of false positive results and a fear from those responsible that if every person with the tiniest amount of virus in their airway was not diagnosed then there would be preventable deaths. The repercussions of that approach were an exaggeration of cases and deaths, perpetuating the fear and the damaging isolation of people who were not a risk to others.

THE EVIDENCE MANIPULATION TRIAD	
EXTRAPOLATE	The effects of mass testing and isolation were extrapolated by modelling to claim many more deaths were prevented than a reasonable analysis could conclude.
EXCUSE	Numerous convoluted excuses were given for the myriad of different data sources indicating overdiagnosis.
EXCLUDE	No attempts were made to measure the operational false positive rate in the laboratories.

TOP THREE MYTHS
1. Bystander virus is clinically meaningful
2. A positive PCR test means there is viable virus around
3. Mass testing can help us control spread

11

THINKING INDEPENDENTLY

Facts are stubborn things; and whatever may be our wishes, our inclinations, or the dictates of our passions, they cannot alter the state of facts and evidence.
John Adams, 1850[341]

From spring 2020, I could see that covid was not as serious as the influenza pandemic that pandemic plans were written for. There was a threat to the elderly and vulnerable and there were valid concerns over our insufficient intensive care capacity. However, beyond those thoughts I did not spend time critically thinking about the situation. I was working as well as being a busy mum of four and was happy to go along with the tide.

I have tried hard since to remember that state of mind. Why wasn't I more questioning? What would have woken me up? Most of all, how can I get through to people who continue to believe everything they hear on the news unquestioningly?

The closing of schools did not sit easy with me. However, as it was only for three weeks I just shrugged my shoulders and got on with it. Five months later I came up for air.

Two things changed for me. I had stopped working, 'because of covid', and my children went back to school. Only then did I have the time to investigate something that had been gnawing at the back of my mind. Through July and August we had seen a constant trickle of cases and deaths that never seemed to get closer to zero. These looked identical to what you would see from low level testing errors. Medical tests are not perfect as we have seen. Every diagnostic test has two types of error. They can either miss real cases or they can give someone healthy a false positive result. This latter phenomenon happens for a certain low percentage of tests done. With the amount of testing we were doing, false positive test results could have accounted for almost all the cases that were being diagnosed. The question of whether this trickle of cases and even deaths could be accounted for by test errors was gnawing at me constantly right through the summer. How much genuine covid was there in summer 2020?

FALSE POSITIVE TESTING ERRORS

The type of testing errors that I was concerned with are not just a function of the test equipment. I was worried about all the errors that can cause a miscarriage of diagnosis as we saw in Belief Six.

With the children at school, I had the time to properly investigate. I found the data that allowed me to calculate the percentage of positive test results each day. Sure enough since July the percentage of tests that were positive had been less than 1 percent. There was day to day variation which is to be expected with differences in the laboratories, different populations being tested and just sheer chance. However, while the percentage fluctuated up and down, it tended to a mean. The mean was 0.8 percent for the community testing and 0.4 percent for the hospital testing. The difference could be attributed to different laboratories doing the testing and the latter including testing of healthy staff as well as patients. Having 99.2 percent of your results avoid the false positive problem is a phenomenal result. Testing rarely gets better than this.

MEANINGLESS POSITIVES

It seemed odd that this wasn't being discussed as an issue, so I thought I was probably wrong. Nevertheless, I wondered how I could test the hypothesis, just in case. Covid has certain characteristics in spring 2020 that act as flags in the data. Covid disproportionately killed men, the old and ethnic minorities, especially people of black ethnicity. The key to testing the hypothesis was thinking about the difference between the population who would test positive randomly – i.e. the characteristics of those being tested – compared to the population who would be most likely to die from covid. When patients were given a meaningless false positive test result before their death, if their symptoms had overlapped enough with covid their doctor may have included covid on their death certificate. A diagnosis of death based on a meaningless positive test result would reflect the whole population that was tested and died and would not be sexist, ageist or racist in the same way.

To illustrate, let's imagine a blue sticker test. People would be tested and if they were 'positive' they would be given a blue sticker. It is a test for something that does not exist but the test comes back positive once in every 200 or so tests. The positives would be randomly distributed but only to people who were tested. A random selection of those who had the test would be allocated a blue sticker. If testing was mostly carried out in hospital then they would not be a representative selection of the whole community. The people who ended up with blue stickers would be a representative sample of people who had had the test i.e. the hospitalised population, older and sicker than average. The blue stickers are totally random so the demographics of the group with blue stickers would be identical to the demographics of the group who got tested.

It is worth remembering that being admitted to hospital is itself associated with a mortality rate. It is a lower mortality rate than for covid, but it means that some people would die with a blue sticker. The blue sticker had nothing to do with their death but because they were given out to a random selection of those entering the hospital, some would coincidentally have a sticker before their death. If the sickest patients were repeatedly tested then their chance of getting a blue sticker would increase as they kept rolling the dice. If there was repeated testing of the sickest then the mortality rate for those with blue stickers could end up being higher than the average mortality rate for the hospital as a whole. It would look like the blue stickers were deadly. But they're just blue stickers – they're harmless.

By summer 2020, every patient admitted to hospital was tested for covid. There were sufficient tests that if a test were negative, then a patient suspected of having covid could be tested again. Anyone who was sick enough to go to hospital and was particularly sick while in hospital would be tested more often and would be more likely to end up with a random positive test result. These randomly allocated positive test results would be in addition to any genuine positives.

EVIDENCE OF DIAGNOSING THE WRONG PEOPLE

If positive covid test results were mostly false positive then the covid test would be behaving like a blue sticker test. That means we can note the characteristics of the population who would end up dying with a blue sticker. These characteristics can be compared to those of the population who died of covid in spring 2020 and then summer 2020. Strikingly, of the first 200 patients admitted to intensive care with covid, nearly one in six were of black ethnicity. Of those admitted subsequently the figure was lower falling to only one in twenty of those admitted after September 2020.[342] This may have reflected outbreaks in different regions of the country but there were still areas affected with a significant black population so it was surprising for it to have fallen to a rate that was as low as a random sample of the population. It's what we would expect from blue sticker deaths. That would suggest covid was no longer more severe in black people. Certain groups are either more susceptible to a virus or not. The virus had not changed significantly in that time.

The age distribution was not helpful because it turns out that covid kills each age group in exactly the same proportions as life does. However, during spring 2020, men accounted for 60 percent of deaths. Throughout summer that figure reverted to 50 percent, as would be expected from blue sticker deaths. If an equal number of men and women were tested then an equal number would end up with a blue sticker. Why was the virus no longer killing more men? It could be explained by men being better at shielding than women such that the proportion of men dying would be lower, but that would be a rather odd thing to believe. The alternative explanation was that meaningless positive results had been given to an equal number of men and women. When some of these died with a covid positive test result their death could be misdiagnosed. Misdiagnosed covid deaths would be equally likely in men and women. So the people affected seemed to be more like a sample of the tested population than a typical covid population.

Let us turn from the people diagnosed, to the behaviour of the disease itself. During the course of summer 2020, it was clear that the fatality rate of covid was falling in every country. One theory was younger people were being infected in the summer and were less likely to die, but even in the older age groups the fatality rate had fallen to a third of its original value.[343] The early highs could have been due to a lack of testing. However, by May testing was so readily available, even for unlikely cases, that nineteen out of twenty tests done came back negative. Despite that, the fatality rate kept on falling.

In April 2020, a third of those diagnosed in hospital died. It could be argued that there were insufficient tests. If tests had been more freely available, more people with incidental covid, who were in hospital but less sick might have tested positive. If so, the proportion of covid hospitalised patients who died would have been lower. However, by May and June 2020, there were plenty of tests in hospitals. By August, the fraction dying had halved compared to April. (Although I did not know it at the time, the proportion dying would return to near April levels in October and then December 2020 before falling again). The fall from May, once tests were plentiful, to August 2020 needs an explanation. Either the disease was changing in severity over time, genuinely becoming less deadly, or the covid diagnoses were being diluted with patients who were misdiagnosed. If it were the latter, it would be impossible for the hospital covid mortality rate to fall below the blue sticker mortality rate with the existing testing strategy.

Covid deaths followed a predictable pattern, with death occurring around 20 days after symptom onset, varying with age. From the outset every death after a positive covid test was being recorded as a covid death no matter how long after the test. Prof Carl Heneghan campaigned for a change in definitions and won.[344] He had pointed out that, in July 2020, the death figures were increasingly inaccurate, *"Public Health England figures are about double the ONS*

[Office of National Statistics] figures because PHE are reporting anybody who has had a positive Covid test in the past." [345] He showed that two thirds of covid deaths had occurred more than 28 days after diagnosis at the beginning of July and this was true of five sixths of the deaths by mid-August. So the tight relationship between diagnosis and death for covid was also lost in the summer. If a frail patient was dying and was admitted with a covid diagnosis in March or April their death would have occurred within four weeks. A death that followed a meaningless false positive test result would be related to when people were sick or were admitted to hospital and got tested but death could be more drawn out. The longer time to death for these patients admitted in the summer 'with covid' suggests either that the disease was much less severe in summer or that patients were being overdiagnosed, or both.

After Carl Heneghan's publication in July 2020, Public Health England started to publish data for deaths within 28 days and deaths within 60 days. A total of 5,377 deaths (1 in 8 of the deaths recorded) were removed from the covid tally in this way. No deaths before May needed revising. The data for those deaths that occurred in the summer had to be significantly revised.

Rather than excuse these anomalies with a series of separate explanations and suggestions that the virus had changed in several ways, Occam's razor should be applied and the conclusion reached that these patients did not all have covid. How can we test this hypothesis further? The hypothesis could have been confirmed by checking the antibody status of those who were diagnosed with covid in the summer and survived. This was never done.

The government published data showing the capacity for antibody tests which reached 120,000 per day by July 2020 and continued at or above those levels. Antibody testing would have been an excellent choice for diagnosing people admitted with symptoms. The test can be done at the bedside and shows whether a person had been infected and how recently. The result would have been quicker than PCR and allowed for effective triage into covid and non-covid parts of the hospital. I do not know why they were not used more. The false positive rate for these tests is higher than for PCR and there may have been a fear that people would believe they were immune when they were not. Perhaps there was also a desire to reinforce the dogma that there was no substitute for vaccine induced immunity. Even without making antibody testing available for the general public there were numerous instances when it could have helped with accuracy of diagnosis and yet they were not used.

SPEAKING OUT

Carl Heneghan and I were in the same year at medical school in Oxford. He had since become Director of the Centre for Evidence Based Medicine there. He had spoken out about the false positive problem so I contacted him for the first time in twenty years and asked what I should do. I took his advice to just publish and set up a blog and twitter account and *"get it out there."*

On his advice I posted an article entitled *"Waiting for zero,"* on 7th September 2020, explaining how we could never reach zero covid deaths because of the false positive problem. I shared it on Twitter ending with a warning to plan ahead for an early family Christmas, *"Coronaviruses are seasonal. If you have vulnerable relatives I recommend putting your sprouts on for Christmas in November – but make sure you eat them before they go soggy."* [346]

That was the point at which I entered a parallel universe of the unheard. It transpired that there were scientists, doctors, lawyers and other experts who had taken issue with the government narrative around lockdown from the outset and had been fighting to be heard for months already. In the same way that I approached the problem from a testing perspective, numerous scientists from every discipline had identified issues in their disciplines, which on their own were only part of the puzzle. Only by working together and understanding what these issues were could the big picture be seen. Many of the people who understood the big picture were polymaths who perhaps did not fit in well to the hierarchical and restrictive world of academic science. However, they were excellent scientists who were comfortable juggling multiple hypotheses and only closing off possibilities where the evidence was conclusive. Each of them was caring, articulate and concerned about what they had seen.

It was an utter privilege to meet so many brilliant independent thinkers. Nick Hudson, a South African actuary had set up PANDA, an international collaboration of scientists investigating covid and invited me to join. I also met Jonathan Engler, a doctor, barrister and entrepreneur and we challenged each other's thinking trying to make everything we were observing make sense. By the end of the year, Narice Bernard, had brought together a number of professionals and scientists who had been speaking alone to form a body the Health Advisory and Recovery Team – HART. Jonathan and I would go on to become co-chairs. The aim was to speak as a group to reduce the attacks that we were *"outliers"* or *"lone wolves."* As well as a place to share and have ideas challenged it became a network of the calm and rational and a real source of strength. In the meantime, I was attacked incessantly on Twitter by people who thought I was wrong. It was a really odd reaction. Their offence at criticism of the testing and my calling for checks, such as antibody testing, was utterly out of proportion. That was my moment of realisation. Both the silencing of those with concerns and the attacks were red flags that free debate that lies at the heart of scientific inquiry and democratic accountability was under attack. Such debate matters even when people are wrong. To paraphrase the philosopher John Stuart Mill, it is important to hear other arguments for four reasons: you might be wrong; you might be partially wrong and learn something; you might be right and can crystallize your arguments by debating or you might be right but by denying free speech your view is undermined by its perception as dogma.

At that time, I believed we had over diagnosed covid since peak deaths but I still worried that we had a largely susceptible population. It was a common hope that the spring 2020 wave had brought us close to herd immunity. From May 2020, results of antibody testing led to huge disappointment. Most of Western Europe had only reached levels of antibody positivity of around 7 percent, implying that 93 percent of the country were still susceptible to covid. These results left the government very scared to release the lockdown and without a strategy for the future except the hope that vaccines would save the day. The promise of potential vaccines from summer 2020 meant that restrictions and lockdowns were then justified on the basis that freedom could only come with vaccination.

I was inundated with people trying to persuade me to look at the mathematics of the spring wave which showed that the trajectory was not impacted by behavioural changes and I became convinced. This led to the hypothesis that the majority of the population already had a level of immunity to this variant (whether from other related coronaviruses or indeed totally different infections like mumps)[347] which had kept covid at bay and that therefore the worries of a rebound were wrong.

Other people tried to persuade me of other theories e.g. that viruses do not exist, which I have not been persuaded by. It is a virus as much as influenza is a virus, which is to say they have important characteristics that are different to other viruses. There certainly is an illness with very characteristic symptoms associated with the presence of a unique genetic sequence. There have been plenty of instances of chains of transmission indicating a common agent in the environment and those infected develop antibodies to the protein that the genetic sequence codes for. It is true that unlike the majority of viruses it does not cause cells to explode when infecting them in culture, instead budding as new particles leave the cell surface. However, the same can be said of influenza. I don't doubt that there is more to learn about these diseases.

For each of these hypotheses I went back to first principles and tried to weigh up the evidence for and against. It was really hard work and it took me months before I felt I had a grip on the big picture. What was reassuring was that there was disagreement among the dissenters on all sorts of topics. It was the disagreement and debate that was so lacking in the official version of events.

My concerns about overdiagnosis in summer continued into autumn. Those that thought PCR was infallible had to adopt painful distortions of thinking in autumn 2020. Somehow the vast majority managed it.

Autumn 2020 saw covid deaths peak at half the numbers seen in spring 2020. Overdiagnosis meant that other measures of covid showed autumn 2020 to be far less of a problem than March 2020. Denial of overdiagnosis meant that a whole series of excuses had to be dreamt up.

The crowd who believed PCR was perfect, let's call them Toady, responded in all sorts of imaginative ways when presented with the evidence. I and others believed all the evidence pointed to one thing – overdiagnosis and were trying to point out the scam. Let's call us Ratty. A summary conversation of genuine excuses I was told would sound something like this:

Ratty: *PCR based covid case numbers were exaggerated by overdiagnosis. There were fewer people in autumn 2020 seeking non-urgent care for covid (111 calls and online searches etc) than in spring 2020.*

Toady: *Fool! March phone call data was artificially high because of people trying to get hold of tests. The drop in online NHS 111 searches was because the novelty had worn off by autumn. People knew all they needed to without needing to search the NHS website.*

Ratty: *You can also see PCR case overdiagnosis in other data. Fewer adults went to their GP with difficulty breathing than normal for the time of year. In addition, fewer went to the emergency department with an acute respiratory infection. Both were well below the levels seen in March.*

Toady: *You're not thinking! There was no influenza in autumn 2020. That is what led to a reduction in acute respiratory tract infections!*

Ratty: *But there was also no influenza in March 2020. So why were more people sick then?*

Toady:...

Ratty: *Total emergency ambulance calls in autumn were only one tenth of the level seen in March 2020.*

Toady: *In March 2020 everyone was so terrified of going to hospital that they did nothing until they were in urgent need of care. Whereas,* *by autumn 2020, patients were being admitted before the emergency arose and that means they did not need an ambulance. Also, the higher emergency attendances in March 2020 must have been due to panicking which flooded health services with covid like complaints that were not covid.*

Ratty: *That makes no sense. People were told to stay away from healthcare services from late February if they had a cough or fever. The result was a marked drop off in healthcare access.*

Toady: *Some people didn't access healthcare in spring but acute respiratory infections were an exception in spring. At that time people were panicking and overusing ambulances. By autumn people stopped panicking.*

Ratty: *Okaaaay. PCR overdiagnosis also explains exaggerated hospital admissions. How come the level of empty beds remained constant in autumn while more and more hospital beds were occupied with covid patients?*

Toady: *Non-covid patients were clairvoyant and avoided hospital, despite being sick, at exactly the required numbers for any given hospital to allow covid patients to take their place.*

Ratty: *Sure they did.*

Toady: *Besides, there was marked hospital transmission so that patients admitted for other causes caught covid in hospital.*

Ratty: *Covid patients need longer stays. If it were all genuine hospital transmission, in autumn, it would still have led to fewer empty beds. Even if it was all hospital transmission, wouldn't that imply very few patients admitted from the community?*

Toady: …

Ratty: *Have you seen how deaths from causes other than covid have fallen with the rise in covid deaths? It's as if people dying of covid would have otherwise died of something else. Or maybe they died anyway but had covid on their death certificates as the cause. After the peak, as covid deaths fell the non-covid deaths recovered at the same rate.*

Toady: *You idiot. You are sharing death data based on the date the death occurred and not the date it was registered. It takes time to register a death meaning the data has a lag. Given time the non-covid deaths will be registered as having occurred and this deficit will disappear.* [It never did]. *Besides, lockdown saves people from dying of non-covid causes too.*

Ratty: *The missing covid deaths are every cause, not just traffic accidents! What about the fact that in spring 2020, the number of deaths above expected levels in a particular region could be accurately predicted based on the number of covid deaths. Now the two are almost totally unrelated to each other.*[348]

Toady: *I've already told you -there are data lags! If only you would wait two more weeks you would see the excess deaths climb in line with covid deaths.* [They did not].

Ratty: *Look at the antibody levels across the population. Why haven't they risen if there is so much covid around? Only the North West had a rise before December.*[349]

Toady: *Antibodies wane. It just so happens they are waning at exactly the right rate such that for every person whose antibodies were no longer detectable there was a new infection to take their place. Also, the sample used was disproportionately drawn from regions that were less affected.* [No region was unaffected according to PCR testing].

The alternative to all of the above beliefs was simply that PCR had overdiagnosed covid, in autumn 2020. Perhaps the motivation for believing the complex set of 'Toady' arguments was that it meant you could continue to believe that the data driven individuals who you had called right wing apologists, who were trying to kill granny, were wrong.

THE MATHS

My concerns of overdiagnosis in the autumn were supported by a mathematical proof I had been taught in a lecture, as a doctor. There is something about the environment of a lecture theatre which lends great weight to the words of the speaker. I remember vividly the content but I regret that I cannot remember the name of the lecturer. The lecturer taught us how overtesting can make a disease appear both less severe and more deadly. The disease was becoming less severe through autumn, demonstrated by the number of hospitalisations or deaths per 'case' falling. At the same time, the disease was becoming more deadly, demonstrated by the number of deaths per hospital or ICU admission rising. Diseases cannot become both less severe, making individual patients less sick, and more deadly at the same time. Identifying this contradiction enables overdiagnosis from false positives to be identified. Overdiagnosis leads to this contradiction because an excess of cases dilutes the apparent severity of the disease. In the meantime, the illusion of a more deadly disease occurs when overtesting of the dying in hospital leads to overdiagnosis of death.

I had serious concerns that the covid data was no longer reliable because of overdiagnosis and that a good deal of the autumn wave was actually a *"pseudo-epidemic,"* where the illusion of an epidemic is created through overdiagnosis. It has happened in the past and can be very difficult to both prove and bring to an end as those making the diagnoses truly believe in the testing.[350] Alongside this overdiagnosis, there were pockets of genuine covid outbreaks at the same time with associated excess deaths. These were largely areas that had not been hit hard before. However, other regions including most of London had no excess mortality, above what respiratory viruses would normally cause at that time of year but an increasing, disproportionate number of apparent covid deaths.

INTENSIVE CARE PRESSURES

The main counter to the overdiagnosis data were the intensive care pressures. Intensive care physicians were passionate that they could diagnose their covid patients without testing. I am sure that was true for many of the patients. Like any doctors though, testing will sway diagnosis and once a diagnosis has been made the doctor becomes invested in it as a belief. The number of intensive care patients in spring 2020 did not reach feared levels. Nevertheless, more than a third of intensive care beds in use in April 2020 were beds that had been created to cope with the surge. Two thirds of total beds were filled with covid patients in April.

In April 2020 and in January 2021, there were more covid positive patients in intensive care than would normally be in intensive care from all other causes put together. This was partly because covid patients spent a long time in intensive care. The mean stay was more than double the average stay for non-covid intensive care patients.[351] In May 2022, there were still 35 patients in intensive care who had been admitted over a year before. Therefore, a build-up of covid patients over winter would have significantly contributed to the pressures seen in January.

Covid did increase demand for intensive care. However, something else was happening. There were fewer non-covid intensive care patients than normal. Operations that needed intensive care for recovery were cancelled. It is hard to know how many of the missing non-covid patients were patients who were not admitted to intensive care in order to preserve beds for covid patients. There are intensive care consultants I have spoken to who believe that care was not rationed and those who needed care got it. However, one reported that some sick patients who would have been treated in the past did not receive treatment[352] and two intensive care consultants published a letter explaining how centrally dictated criteria for admission meant that *"vulnerable groups become a self-fulfilling prophecy"* because criteria that were not predictive of who would benefit from intensive care were being used to exclude people.[353]

The other explanation for the missing non-covid intensive care patients would be that many of the patients on intensive care with covid would have been on intensive care for other reasons in the absence of covid. For some this could have been because they, or someone like them, would have succumbed to a different dominant respiratory virus at that time of year. For others it could mean that they were admitted because of other medical problems and then caught covid as an added insult. Lastly, there may have been patients who were already in intensive care whose covid test was positive because of the amount of virus present in the intensive care air.

Evidence for this latter hypothesis came from a Spanish study that looked for confirmatory antibodies in hospitalised patients.[354] Of those PCR positive patients who were in hospital for fewer than seven days, nearly 90 percent did not develop antibodies. Half of those who were PCR positive on intensive care never developed antibodies either.

The truth was there was overdiagnosis. With Omicron it was widely accepted that half of hospital patients with a positive test were coincidentally positive while being treated for something else.[355] Omicron was undoubtedly milder, but prior variants were still mild for most people. Yet, even in retrospect there was no admission that there had also been incidental mild positives in hospital with previous variants.

WINTER 2020/2021

Despite having called for early cooking of sprouts when I first published in September 2020, by November I had forgotten this warning. When the alpha surge appeared, in January 2021, I was sceptical. The first sign of something odd happened in December 2020 when the percentage of tests returning positive rose synchronously in every region of the UK.[356] Prior to December, each region had a unique course with urban areas dominating in spring 2020 and rural areas in autumn 2020. It was quite extraordinary to see the meandering coloured lines on the graph all turn a corner and rise in lockstep across the country, in December. No region was excluded. Even isolated islands including the Isle of Wight, Anglesey and the Shetlands, despite having had quite separate trajectories prior to that point, participated in this synchronous rise.

Shortly afterwards a few key indicators changed. Ambulance calls doubled to a third of their spring level. GP consultations for difficulty breathing again went above normal levels. Covid-like attendances in the emergency department returned to spring 2020 levels. Excess deaths and covid deaths again correlated and antibody levels leapt up.

There were still oddities that could be explained by continuing overdiagnosis particularly of the dying. In spring 2020 there were ten covid ambulance calls for every five hospital covid deaths but by January 2021 ten covid ambulance calls were matched with thirty covid hospital deaths.

As covid deaths rose the expected number of non-covid deaths from a wide range of causes again fell away. There were more covid deaths than missing non-covid deaths but it was as if a substantial number of covid deaths had replaced deaths from other causes. When covid deaths peaked the missing deaths from other causes also reached a maximum. Non-covid deaths returned to expected levels only once covid deaths declined. Hospitalised patients with symptoms that overlapped with covid could have been overdiagnosed by PCR. The combined effect of a positive PCR and a system which automatically registered such deaths on a government database as a covid death could easily have biased what doctors chose to put on the death certificate.

The Hyperbolic Hawks were fearful of missing a covid case with a false negative result and saw me as a danger. In their view, every missed positive result, even in someone with no symptoms, was an outbreak in the making and could lead to deaths. Isolating the healthy was, in their view, a small price to pay to prevent unnecessary deaths. I found myself under attack repeatedly for asking questions and trying to bring nuance to the testing strategy.

As it became evident in autumn and winter that more people were sick, the attacks on me increased. Two people who cared about me and were advising with the best of intentions suggested to me in autumn that I should delete my tweets. One argued for a fresh start as we joined forces in HART and the other couldn't see any benefit in keeping them but only retained the potential for certain ones to be used against me out of context. It did not seem a good idea to have tweets calling out problems with testing being shared out of context when people were clearly sick again and I deleted my tweets. As a strategy for not being attacked this utterly backfired. The result was a barrage of accusations that I was somehow admitting I was wrong or expressing guilt or else was trying to hide something. None were true. Of course, it also led directly to certain ones being used, out of context, against me. I continued speaking out aware that many in HART did not have the support or savings to enable them to risk their jobs. I entered the dark winter of 2020/21 partly self-censoring what I said in public and discussing concerns in private with HART members.

A VICTORY FOR THE HYPERBOLIC-HAWKS

This next section might seem confusing and even contradictory. Please keep going and all will be explained in the next chapter - *Being Wrong – Confessions.*

How much of an issue were false positives? The Data-Driven-Doves were pointing at all sorts of real world evidence showing that there was a false positive problem. The Hyperbolic-Hawks were saying that PCR was finding genuine cases and getting the diagnosis right. Broadly, the Hyperbolic-Hawks were right and we were wrong. When there was a covid wave the numbers of symptomatic people diagnosed on government testing by PCR was roughly equal to the number of people estimated to have developed antibodies from a sample of blood donors. Antibody testing is not perfect either. It is good for answering the question *"who has had covid?"* but there will be test errors as with any test. On a population level it is an effective measure of how many have had symptomatic disease. From all I have read, I have to assume that people who had a positive antibody test did have a symptomatic infection and would have been infectious. For example, when researchers at Imperial asked people who had developed antibodies when they had their symptoms, all of the people said they had had symptoms.[357]

Carl Heneghan worked with a team who gathered data on how weak the PCR positive test results were using the Freedom of Information Act. They found at least a third of diagnosed cases were not infectious at the time of testing.[358] The implication of this is that substantial numbers of people were only diagnosed by PCR before or after the period in which they would have been infectious. Either that or these people were overdiagnosed while a similar number of people were never diagnosed. That would be a lot of underdiagnosed covid.

Yet overall it seems that government testing did broadly work to diagnose the covid cases in the community. My concerns and those of others were exaggerated concerning the scale and true level of impact of false PCR positives. We were wrong on the big picture for community testing.

The impact of false positives was therefore mostly felt in the community between covid waves, in people who never developed symptoms and in the diagnosis of hospitalisations and deaths. During a covid wave 90 percent of positives were symptomatic, with periods where it reached 98 percent. The remainder were likely pre-symptomatic in this period. Between covid waves the number of positives who were symptomatic fell to as low as 0.5 percent.[359] This overdiagnosis of asymptomatic people was a major issue. It exaggerated the time for which covid was a threat, increased fear and for individuals caused unnecessary inconvenience and distress. The false positives and modelling that led the Office of National Statistics to produce an estimate for covid prevalence that was twice as high as the antibody measure was also hugely damaging in terms of inflating fear and reactionary policy making.

While the population measure of people who had covid may have been about right, the use of testing on individuals resulting in the removal of their freedoms would have been morally wrong even if it had been accurate. The law was clear that quarantine could be used to isolate the sick for the period for which they were sick but that law did not apply to healthy people.

Instead of arguing for the moral case that healthy people should not be quarantined, many people confused the argument by linking it to how diagnosis was carried out. For example, four tourists entered Portugal having tested negative in order to travel. Three days later one tested positive and all four were quarantined. A Portuguese court of appeal declared this *an illegal detention* in November 2020, because diagnosis can only be carried out by a doctor under the law and, *this test alone proves to be incapable of determining, beyond reasonable doubt, that such positivity does, in fact, correspond to the infection of a person with the SARS-CoV-2 virus.* [360]

The fact PCR testing of symptomatic people broadly worked to find community cases does not mean there was any benefit from testing. (I am referring to a benefit to the public. The huge amount spent on testing has been a massive financial benefit for certain individuals). Everyone I have met pre-Omicron who had a covid diagnosis knew they had it without testing. This was either because of characteristic symptoms or because contacts who were sick had those characteristic symptoms. A proof of benefit would have required two additional assumptions to be proved. Firstly, testing must have helped people to recognise it was covid earlier than they would have. Secondly, isolation must reduce transmission.

There were serious issues with PCR testing. I have no regrets in highlighting them because the problems they created from overdiagnosis for population estimates, asymptomatic cases, hospitalisations and deaths had a damaging impact. However, feeling unheard led to me shouting louder and exaggerating my concerns until I crossed a line in one tweet claiming, *"No-one is going to die of it (only with it)"* and that was wrong.

12

BEING WRONG – CONFESSIONS

The only man who never makes a mistake is the man who never does anything. **Theodore Roosevelt, 1900** [361]

Having weighed up all the evidence I was left confused. How could dripping trails of snot over other people's samples which are then amplified a trillion times *not* create a false positive problem in the community? If you were confused too at the end of the last chapter, that is what being wrong feels like. There is conflict, cognitive dissonance. The choice is to dig your heels in and ignore the conflicting evidence or to try to make sense of it. Science requires the latter but there are often loose ends when trying to fit facts to a new hypothesis. It is like permanently being one of the blind men feeling different parts of the elephant. There is evidence that it is like a tree trunk and evidence that it is like a snake and it is impossible to make sense of both. I have not made sense of all of it. Autumn 2020 still had many oddities.

I do think I over emphasised problems with the accuracy of tests finding the virus by suggesting that the false positive test results were from people who had encountered no virus. In retrospect, the key problem was misinterpreting the significance of small amounts of virus that the tests were correctly identifying. The test found the virus but the interpretation of what that meant was misplaced.

People who had sufficient immunity that they would never be infectious to others or develop symptoms were being described as positive cases.

When I was trying to make my point about overdiagnosis, I exaggerated the case. I made the mistake of ignoring the pockets of real covid deaths in an attempt to get across the message about the overdiagnosis that was happening in the majority of the country. I was thinking about London when I wrote a most regrettable tweet, since deleted, "*No-one is going to die of it (only with it). Flu diagnoses have been replaced by covid. This happens when you overtest people dying of respiratory failure until you get the result you are looking for.*"

Even for London it would have been wrong as there were enough covid deaths to match the number that would in the past have died from flu. It was sent on 18th October 2020, but it was shared by others repeatedly in December onwards after I deleted my tweets, when the Alpha surge returned and people in London were dying from covid. I had become emotionally engaged and exaggerated my case too strongly.

I have since seen a number of people behave similarly. When feeling unheard it is pretty human to want to shout louder, overemphasise your point and get emotional. For me, giving in to any of those emotions just backfires. For others, they can seem to make more impact tuning into their emotions and using them to amplify their conclusions.

In retrospect, the way I assessed the extent of overdiagnosis based on the illness being more deadly and less severe was an oversimplification. I was assuming that the virus behaved similarly throughout the year but subsequent data has suggested that there were seasonal changes in severity. It is as if the seasonal trigger creates not only susceptibility to infection but also to succumbing. That could account for some of the discrepancy.

I am human and have made other mistakes. I was wrong about the proportion of the population who remained susceptible after spring 2020 and the likelihood of a second wave, having not accounted for susceptibility to a new variant. I changed my views on masks and lockdown because my initial thoughts were wrong.

I made data errors. When calculating the death rates for non-covid hospital admissions in that September publication, *"Waiting for Zero,"* I mistakenly took the total number of people admitted to the NHS per year rather than the total number of admissions which is ten times as high.[362] I also made a data error in calculating the hospital death rate and those mistakes led to exaggeration of the level of overdiagnosis.

In an interview, on 20th November 2020, with Mike Graham on TalkRadio, I stated that the NHS would not be overwhelmed. I should have been clearer and defined what I meant by overwhelmed. There was no more pressure on normal hospital beds that winter than for any other winter but the NHS is overwhelmed every winter with the cancellation of surgery in winter being part of the NHS business model.

I was wrong about intensive care pressures that winter. Covid results in greater need for intensive care than influenza and so even when in other respects the changing course of seasonal respiratory infections was not significant the pressures on intensive care were.

There were other occasions when I failed to explain clearly what I meant. Sometimes this was unintentional and I could have been clearer but at other times I was being deliberately obtuse in order to protect myself.

I am sorry for the mistakes and any confusion and upset they may have caused. However, as I do not believe myself omniscient, I accept mistakes as an inevitability of speaking at all. Having tweeted over 15,000 times it would be very odd to claim that I never erred in any one tweet. A few tweets were used to try and discredit me in the hope that people would ignore the content of the others.

I had never used social media much before autumn 2020 and had to set up a personal Twitter profile in September 2020. It was a steep learning curve. Early on I made the mistake of adding in a link to a buy-me-a-coffee website and was given a total of £78 as a result, which I have since donated to HART. It was a lapse of judgement to have added the link. Since that time I have earned no money at all. All my work so far including publications, media appearances and preparing evidence for court cases has been unpaid.

SILENCING

Twitter has given me a platform I otherwise would not have had. Other social media sites have been more intolerant. I have barely bothered with other social media because their threshold for what could be said was so low. I have had YouTube videos removed and been given a strike such that I am reluctant to use that platform fearing a lifetime ban.

Disclosures in a Louisiana court case revealed a *"Censorship Enterprise"* which *"is extremely broad, including officials in the White House"* and numerous other federal departments. The result was *"intensive oversight and pressure to censor that senior federal officials placed on social-media platforms."* [363] Social media companies also had to negotiate EU law[364] which makes failure to remove content that the EU regards as disinformation an offence punishable with a fine of up to 6 percent of the company's global turnover.[365] Since January 2020, over 11 million Twitter accounts were challenged on covid content with over 11,000 accounts suspended and nearly 100,000 tweets removed.[366] The Twitter policy on what constituted false information included, *"claims… widely accepted by experts to be inaccurate or false."*[367] No-one could question the High Priests. From 23rd November 2022, Twitter announced they would no longer enforce their covid misinformation policy. While Twitter's new owner Elon Musk has reinstated numerous accounts, it remains unclear how he intends to comply with EU law. Issues around silencing blasphemers and witch-hunts will be explored further in *Spiked: A Shot in the Covid Dark*.

In her excellent book, *Being Wrong*, Kathryn Schulz explains how we view people with differing beliefs to us. They are assumed to be either ignorant, idiots or evil. There is no room for the option of having interpreted a complex web of data differently at a time when evidence is evolving. She describes the assumption of evil as *"the idea that people who disagree with us are not ignorant of the truth, and not unable to comprehend it, but have wilfully turned their backs on it."*[6]

I have been called a *"covidiot"*, *"charlatan,"* *"fraud, "grifter" "a truly diabolical human being,"* *"a malevolent force"* with *"vile behaviour"* and who *"kills people"*, but most commonly I was called *"dangerous."*[368] I was not dangerous except to these peoples' world view. Being labelled as a potential threat to people felt frighteningly close to me being, in their eyes, a person with evil beliefs whom it would be justified to destroy.

There is more than a little irony that catastrophic mistakes could be made and forgiven but only if they supported the government narrative. Anyone opposing that narrative is not allowed a single misstep. Likewise there is an irony that huge amounts of money have exchanged hands in order to buy influence to support the government narrative and yet anyone speaking against the government must sacrifice their current and potentially all future earnings in order to be allowed a voice. (Even having done that they will be incessantly called a *"grifter"* (an American word meaning a person who engages in petty or small-scale swindling – originally from a combination of grafter and drifter).

It is not always easy to admit you are wrong and there are certain people who never do and make excuses. The WHO described airborne spread as misinformation[369] saying that spread was through droplets that fell rapidly to the ground and labelling the results of a study showing aerosol transmission as incorrect. Those who tried to share information on aerosol spread were dismissed as spreading *"conspiracy theories."* [370] The word 'conspiracy' comes from the Latin 'conspirare' meaning to breathe together. They did indeed believe that we all share the same air ultimately and they quite literally were 'conspiracy theorists.' Perhaps we should call them *conspirationists* to make it crystal clear that the theory is based on rigorously tested empirical evidence.

There were several other occasions when those in authority were wrong. The lack of rebound in infections was attributed to covid not being able to spread outside despite subsequent spread at outside events. The autumn 2020 resurgence was blamed on students starting university even though that is the seasonal pattern and was repeated the next year when students had minimal covid. Children were described as vectors propelling the disease even though their trajectory followed the community with a rise in the Christmas holidays of 2020 and a fall when they returned in March.[371] The fall in deaths in that winter was attributed to the second lockdown even though it happened at the same time that deaths fall every year. Ineffective vaccines were blamed on the variants. When mistakes were admitted to they were blamed on the majority sharing in the misunderstanding as if that makes being wrong justifiable. I was guilty of that too in my believing the simplistic arguments presented for the initial lockdown. Both sorts of persistent error stem from the human desire to be right – *"I was right and I am still right."*

My mistake about lockdown being needed began with images of people dropping dead in the street creating my first impression of covid. They have stuck with me, despite being a total fabrication. I have often tried to remember what I was thinking in spring 2020. It was fear that made me buy the oxygenation machine. I took much of the Chinese propaganda at face value and didn't for a moment question the best intentions of those suggesting we enact similarly extreme measures. The moment schools were shut did cause me to hesitate as it seemed to be a huge step given children were at minimal risk. However, I fell for the *"just three weeks"* line and went along with it, for the Greater Good.

The fact I began sewing my own masks I think is more telling about my mindset at the time. I had been trained in how protective equipment worked and knew that aerosols would not be interrupted by a mask. At the time, the emphasis was all on droplet transmission and the idea that a cloth mask could reduce that made sense to me. However, I was not thinking very hard and did no research. The fact is that I was furiously busy at the time. I had been working full time and trying to fit home-schooling of four children into each day. I was starting work at 6am in order to be able to sandwich in three hours of helping the children with their work each day. Any interest I had in the media came from a desire to know the basics needed to engage in polite conversation. I had no capacity to question it or to read around the subject. In summary, I was too lazy.

All doctors outsource some thinking. It is not possible to keep up with every update in every aspect of medicine. When you have to outsource some thinking, it can be very easy to become reliant on outsourcing too much of your thinking.

The situation I describe is true for the majority of doctors. They work a long day, struggle to keep up with updates in their own area, have family and other commitments and just want to know enough to avoid any public faux pas. The BBC was a quick fix. Their health pages provided a quick summary that could be read with a sandwich during a lunch break. What the doctors reading the BBC may not have realised is that the journalists writing it were checking the veracity of what they wrote by asking doctors like them who were reading it each day. This created a positive feedback loop where no-one was actually providing critical feedback at all and everything that was written became a self-fulfilling prophecy.

Science is dependent on scientists being open to new hypotheses and willing to change their minds. One US Emergency doctor, Joe Fraiman, appeared on a broadcast round table held by the governor of Florida, Ron de Santis, in which he made a public confession of having been wrong about covid and apologised.[372] The mainstream media ignored this. But for those of us listening to a range of voices, he made a huge impression.

Democracy works because of the systems in place that allow for dissent and opposition and for politicians to change their minds. Ron de Santis, Governor of Florida, instigated a month of lockdown but later, having hired a red team to critique the advice he was given, he had the integrity to admit it was a *"huge, huge mistake."* By July 2022, he remained the only politician to have admitted this.

Sarah Waters is a psychotherapist specialising in childhood trauma and member of HART. She was part of a group of similar professionals, the Cornwall Trauma Informed Network. For two years she called out the harm that policy was causing to children and the responsibility the members of the group had to highlight this and protect children. She remained a lone voice while those in the group rolled their eyes and said she was wrong. Finally, when mainstream media took a view that aligned with Sarah, the remaining members of the group were faced with cognitive dissonance. They could no longer claim Sarah was wrong and ignore her. Rather than face up to their omissions they removed her from the group. The denial of any problem was followed by anger. Those with non-evidenced opinions directed their anger at those who had conflicting opinions which were confirmed as correct as more evidence emerged. After denial and anger, are the other stages of grieving: bargaining, depression and acceptance going to follow?

Given my experience of the last two years, it would be reasonable to predict that some extract of this book will be declared incorrect. It might be a number or a phrase. It might be something that is in fact correct and backed by evidence but goes against the majority belief. There are people who would use such examples to try and discredit me. I am not omniscient. I can only make an argument based on current evidence. All I can do is investigate as thoroughly as possible, own up to any mistakes and not let such attacks silence me. It is possible if you follow Twitter that you will know about what errors I have been accused of in this book as you read this. As I write, I know nothing about it. In time, as more evidence emerges, I may change my mind or would phrase certain aspects with more nuance. However, the evidence has to come first.

BELIEF SEVEN: 1 IN 3 PEOPLE WITH COVID SPREAD IT WHILE ASYMPTOMATIC

Covid 19 is very different from SARS and MERS, and the number one difference is that it has asymptomatic transmission. **Matt Hancock, June 2021**[373]

It is time to return to Dr Charles Chapin whose 1910 textbook, *The Sources and Modes of Infection*, founded the myth of all spread being through close contact. He can also lay claim to founding the idea of asymptomatic transmission. As well as evidence of airborne pathogens there were abundant examples of people contracting infectious diseases with no history of contact with an infectious source. If he was to prove his thesis that every infectious disease (bar tuberculosis) was *only* spread through close contact, then he needed to explain this anomaly.

In order to dismiss the evidence of spread seen in the absence of close contact he put huge weight behind the idea of asymptomatic spread. He leant heavily on the story of Typhoid Mary. She had a bacterial infection of the gut caused by *Salmonella Typhi* which does have a fairly unique ability to hide away in immune cells. He also referenced examples of bacteria being isolated from people who had recovered from the symptoms of their infection or people with, say scarlet fever, who had only mild symptoms. Finally, he referred to the fact that it is possible to find a bacteria that causes meningitis or one that causes pneumonia in the noses of healthy people.

He admitted that *"carriers often appear to be non-infectious"* and *"have been known to remain such for long periods of time without apparently infecting members of their families or others brought in close contact with them."*[374] He attributed this anomaly to four factors, intermittent excretion of bacteria, low numbers of bacteria, bacteria that have *"lost their virulence"* [375] and the exposed people being immune. He could prove pathogens could be found in bodies of the healthy but struggled to prove they were a source of symptomatic infection in others.

Influenza presented a particular problem. For well over a hundred years people had commented on the rapid global spread that could not be accounted for through chains of transmission. He agreed but decided, given his hypothesis that all spread was through close contact spread, that this must be evidence of asymptomatic spread, *"The rapidity with which epidemic influenza spreads, its sudden contemporaneous appearance at many distant points, and the difficulty of tracing the route of infection, render it almost certain that there must in this disease be many mild atypical cases, and many persons infected, but showing no symptoms."* [376] The evidence he quoted to support this conjecture was based on the culture of a bacterium now known not to be the cause of influenza. Nevertheless, that was the beginning of the belief that viruses were spread by asymptomatic people.

What does it mean to have no symptoms? In February 2020, the Chinese first coined the idea of an asymptomatic covid case. Something was lost in translation as it was clear that, for them, *'asymptomatic'* could include significant symptoms. The Chinese defined a mild instance as a patient sick enough to have had radiological investigations (carried out in a hospital) to demonstrate a pneumonia, *"A mild case was defined as a confirmed case with fever, respiratory symptoms and radiographic evidence of pneumonia, while a severe case was defined as a mild case with dyspnoea [shortness of breath] or respiratory failure."* [377]

Against this backdrop, *'asymptomatic'* has quite a different meaning, *"an asymptomatic case was defined as a confirmed case with normal body temperature or minor discomfort."*

In March 2020, Dr Maria Van Kerkhove, head of the emerging diseases and zoonoses unit at the World Health Organisation, said, *"Most of the people who were thought to be asymptomatic aren't truly asymptomatic. When we went back and interviewed them, most of them said, actually I didn't feel well but I didn't think it was an important thing to mention. I had a low-grade temperature, or aches, but I didn't think that counted."* [378]

For most of the public the word 'asymptomatic' means no symptoms whereas these definitions include mild symptoms. It is worth bearing that in mind when reviewing the literature on asymptomatic spread.

The assumption from public health authorities, in the medical literature and media was that an asymptomatic person would be perfectly healthy. Such a person could fall into one of two categories. They could be in the incubation period of the illness and about to develop symptoms i.e. presymptomatic. Alternatively, the assumption was that there were healthy people who never developed symptoms but were an infection risk to others. This latter group is what I call *'asymptomatic'* here.

Government advertisements which were ubiquitous from autumn 2020 said *"One in three people with COVID-19 don't have any symptoms, but can still pass it on."* Where did this myth come from? The DELVE report from the Royal Society went further saying, *"the majority of SARS-CoV-2 infected individuals (up to ~ 90%)*

remain asymptomatic throughout infection.[379] The myth of asymptomatic spread was really powerful as it was the basis of the justification for locking down the healthy, wearing masks and mass testing of the healthy. It created the ultimate generator of fear – the invisible enemy.

WHERE THE IDEA CAME FROM

The evidence that supports the belief in asymptomatic transmission, by those who never develop symptoms, appears to come from a combination of two factors:

> 1. Positive test results in the absence of symptoms (without evidence of transmission of disease)

> 2. The fact that people become infected without a traceable source

The former is explainable through poor testing and a failure to acknowledge immunity. PCR testing was positive with quantities of virus a thousand times too few to result in infectiousness. Where do you begin and the air ends? There is no cartoon-like invisible forcefield to keep infectious agents from entering your airways. Detecting virus in the air in someone's respiratory tract is of no consequence if they have immunity and will never become an infection risk, develop symptoms or acquire immunity themselves. The ability to detect even relatively large amounts of virus on testing is not proof of asymptomatic spread. Someone has to catch it to prove that. Only where transmission results in a person having more than fleeting, mild symptoms, would there be evidence of meaningful transmission and such evidence has not been found.

The fact that people become infected when there is no discernible source can be attributed to long-distance aerosol transmission.

Despite WHO officially acknowledging this mode of transmission, there seems to be continuing total denial of its implications by most authorities.[380] Remember how Delta in Australia, was spread without any chains of transmission and James Merlino, the acting premier of Victoria, said transmission had occurred due to people *"being in the same place, at the same time, for mere moments."*[381] Aerosol transmission could be denied by belief in asymptomatic transmission.

ANSWERING THE WRONG QUESTION

When testing did detect a virus what did that really mean?

The inventor of PCR testing, Kary Mullis, won a Nobel Prize for his work. He explained the limitations of PCR, *"It's just a process that is used to make a whole lot of something out of something. It doesn't tell you that you are sick and it doesn't tell you that the thing you ended up with was going to hurt you or anything like that."* [382]

In order to interpret any diagnostic test it is essential to know what question the test is meant to be answering. There are two questions in diagnostic testing for covid that require different strategies:

    1. Does this sick patient have covid?

    2 Could this apparently healthy patient have presymptomatic covid?

Virus could still be present when the answer to both those two questions was 'no'. Imagine that instead of diagnosing a disease we were trying to prevent a bank robbery. The detectives find people with the cash from the safe. They left DNA in the safe of the bank. But just to be sure they also arrest all the people whose DNA was only in the bank lobby – innocent bystanders. Air is filthy. All sorts of viruses and bacteria enter our airways but as long as they never pass the mucus barrier then they are innocent bystanders. There are numerous other microorganisms that could be found in the lobby of the bank in healthy people. Four in ten of us have a bacteria, *Neisseria meningitidis,* that causes meningitis in our noses, for example.[383]

Calling the bystanders innocent might be debatable. These viruses have records of causing harm even death. They also have the means. However, in a situation where the person is unaffected, there is no crime to convict them of and they therefore can be considered innocent in that context.

Testing had been set up to answer the question *"is there any virus in this sample?"* This meant that all the people with bystander virus were described as being cases, even when the virus may never have passed the mucus barrier or entered a cell. Even when the virus may have entered cells, if dealt with promptly by the immune system there could be no consequence in terms of sickness or infectivity to others. The test results, particularly weak positives, were clinically meaningless with results that did not relate to either sickness or infectiousness.

ASYMPTOMATIC POSITIVES

There is no doubt that there were plenty of people who tested positive and did not have symptoms although this was far more of a problem when testing the healthy than in outbreaks. For example, a hospital outbreak among 86 people, in September 2020, in Cambridge, was thoroughly investigated.[384] Every patient and all but two of the healthcare workers developed symptoms.

A group designing a nasal covid vaccine carried out a human challenge study.[385] They found zero asymptomatic infections where asymptomatic is defined as having no symptoms. Of the eighteen participants who tested positive the authors claimed only two had *"no reportable symptoms."*[386] Looking at their data seventeen had at least two days of symptoms according to the questionnaire and the remaining participant lost their sense of smell.[387] Infections were mild but the paper's claim that *"Many people infected with coronavirus have no symptoms (asymptomatic) and are a major cause of spreading the infection because they are unaware they are infected."*[388] is therefore utterly unsubstantiated.

Studies claimed asymptomatic positives accounted for between 4 percent[389] and 76 percent of people who tested positive.[390] If it was a feature of disease, like a cough, it would be a constant percentage wherever and whenever it was measured. It was not. It was a feature of a small percentage of tests giving meaningless positive test results. When there is plenty of covid around the meaningless positives are outnumbered by genuine positives. When there is very little covid around, the meaningless positives can become the vast majority of positive results reported.

Chris Whitty, Chief Medical Officer, hypothesised there could be *"a large iceberg of people who have asymptomatic infection"* adding, *"we will not know until we have a serological [antibody] test. Once we know what that number is in Wuhan and Hubei, that will change the way everywhere else in the world views this, in one direction or the other."* [391] At a press conference on 25th March 2020, Patrick Vallance said *"The other thing that is going to be important with the antibody tests is to be able to work out how many people have had the disease asymptomatically."* [392]

When the antibody tests became available it was clear that only those with symptomatic positive test results had had covid. Neither spoke out to say their concerns had been invalidated. The dread of the asymptomatic spreader was allowed to continue.

WAS THERE A REASON TO BE CONCERNED IT MIGHT BE A PROBLEM?

Occasionally, people have successfully demonstrated that repeat testing of someone who was apparently asymptomatic resulted in tests that were positive over several days with a rise then fall in the strength of the positive result.[393] Was this significant in the real world? When virus replication reaches high levels and there is considerable cell damage in the respiratory tract coughing begins. Viral transmission when coughing has been shown to be much higher. Logically, most viral spread will occur once coughing begins even if there was plenty of virus in the respiratory tract prior to that. Fauci commented on this in January 2020, saying, *"In all the history of respiratory borne viruses of any type, asymptomatic transmission has never been the driver of outbreaks. The driver of outbreaks is always a symptomatic person."* [394]

The WHO reported they had not found evidence of significant asymptomatic spread. As early as June 2020, a WHO representative said, *"Based on our data, it seems unlikely that an asymptomatic carrier will transmit the infection to someone else. We have a number of reports from other countries. They monitor asymptomatic carriers, their contacts, and do not detect further transmission."* [395]

SUPPOSED EVIDENCE TO SUPPORT ASYMPTOMATIC SPREAD

So why was there such concern about asymptomatic spread? In December 2020, I trawled through the massive body of work on asymptomatic covid (reports of positive tests in healthy people) to find the evidence for spread. The medical literature contained multiple attempts to analyse the totality of the evidence, called meta-analyses, but these recycled the same handful of poorly designed studies. Jonathan Engler and I published an article which claimed that the supposed body of evidence was an illusion.[396] The evidence to support asymptomatic transmission amounted almost entirely to a handful of people with positive test results who coincidentally had contact with each other but where neither of whom had symptoms. This is evidence of people breathing contaminated air or simply people with erroneous test results not evidence of the spread of a disease.

From all the studies at the time on asymptomatic transmission, we found only three instances globally of people becoming ill having allegedly been infected by someone who never developed symptoms. All three were reported together as part of an outbreak in Brunei when people returned from a religious festival in Malaysia. One had a runny nose and one a mild cough. The third was a 10 month old baby who had a *"mild cough"* for two days before testing positive but remained symptom free for all of a two week hospital stay until she tested negative.[397] In the grand scheme, given the numbers of people who were testing positive and asymptomatic, finding three people with otherwise common symptoms should be disregarded as coincidence. The possibility that any genuine covid may have been contracted through long distance aerosol transmission from a symptomatic person must also be considered. Even the most pessimistic take is that there was no evidence of anything other than the mildest of infections after asymptomatic spread.

We chose to omit what could have been a Chinese Communist Party propaganda piece in which they claimed to have tested ten million people in Wuhan in May 2020, resulting in a wholly unbelievable result of only 300 PCR positives.[398] Extensive swabbing may well have been carried out in an attempt to quell anxiety in Wuhan. Doing this would have allowed the authorities to announce that covid was over in Wuhan and encourage people to return to their normal activity. To put the ten million figure into perspective, a single sheet of paper to consent each person would have led to a pile of paper four miles high. Despite the huge focus on testing and the establishment of many industrial sized labs, the UK took until August 2020 before having carried out ten million tests, and yet the claim here was that this was done in one city within seventeen days. In my opinion, this was likely originally written up as internal Chinese Government propaganda. I would love to know more about how it ended up published in *Nature* as if it were entirely true.

In the last year there have been important reports which have increased knowledge in this area. These included a study of thoroughly tested marine recruits and a carefully reported outbreak in Japan from January 2020 but first published in February 2021.

MARINE STUDY

A study of trainee marines got a lot of traction with people claiming it demonstrated asymptomatic spread.[399] It is ultimately the story of five young people with mild cold symptoms and the ubiquitous nature of viruses. Before the advent of molecular biology this would hardly be newsworthy! The paper never reports on whether any asymptomatic person passed the virus to someone who then became symptomatic. Extensive genetic sequencing meant they had that information but chose not to include it. As a paper which set out to investigate the problem of asymptomatic positives, the failure to mention symptoms suggests either that asymptomatic spread resulting in a symptomatic case did not happen or that the scientists involved were using a very ill-constructed methodology. Details on symptoms have been repeatedly omitted from such studies of asymptomatic spread. A paper demonstrating that asymptomatic people may be contaminated with enough virus that others in their platoon test positive, asymptomatically, is of no clinical interest. If no-one developed symptoms due to exposure to someone who was asymptomatic then there was no meaningful asymptomatic spread.

JAPANESE STUDY

The Japanese study reported on an outbreak in Japan in early 2020.[400] The outbreak began in Tokyo at a party, on 18th January, attended by someone exposed by a tourist from Wuhan. Over six weeks, 36 people tested positive and 25 of them did not spread it to anyone. Despite this being the entirety of covid in Japan at the time they failed to trace one link in the chain who was the source of a small outbreak in a remote part of Kanagawa, a coastal town south of Tokyo. The authors assume that there must have been a person from the Tokyo outbreak who was responsible for this infection as they seem unable to think beyond person to person spread as the only possible mechanism of spread.

The article makes a big point of the fact that a woman in her eighties died in Kanagawa. The woman in question had pneumonia and had been ill from 22nd January. At that point only 2 out of the 36 cases had been reported in Tokyo. She died on 13th February. The diagnosis was made based on a positive test carried out after her death and the Ministry of Health tried to impress upon people that it was not clear that the virus had been the cause of her death.[401]

OTHER STUDIES

Other papers which reported on apparent asymptomatic transmission also failed to be clear on whether the source was symptomatic, pre-symptomatic or asymptomatic and whether those alleged to have caught covid from them developed symptoms.[402] Others failed to be clear on the timing of the asymptomatic positive compared to those who were said to have caught covid in their household.[403] But let's give them the benefit of the doubt. The possibility that there have been situations where spread has occurred between someone who was essentially a carrier of the virus and someone who then developed symptoms remains. Never say never in biology. What matters is not if it ever happened, but the overall impact. The evidence that it was a significant contributor to spread just is not there.

PRESYMPTOMATIC SPREAD

If we disregard all people who test positive but never develop symptoms there is still the issue of those who feel healthy but are in the incubation period of their illness, the presymptomatic phase.

On 30th January 2020, there was a report of a Chinese business woman spreading covid to two colleagues at a meeting. The authors, including Christian Drosten, the German Fauci equivalent, reported that, *"she had been well with no signs or symptoms of infection."* [404] An investigative journalist took until 3rd February to reveal in *Science Magazine* that the authors had only asked her German colleagues whether she seemed ill.[405] The businesswoman herself reported that *"she felt tired, suffered from muscle pain, and took paracetamol"* on the day of the meeting. The authors added supplementary information describing the symptoms but the main content of the paper has not been retracted or amended.

Despite this inauspicious start to the reporting of presymptomatic spread there is other evidence that it has occurred.[406] What matters is how much of an impact it has. One attempt at measuring the contribution these people made to an outbreak, in terms of close contact transmission, put it at 6.4 percent.[407] There is no reason to think that the proportion of long distance aerosol transmission from presymptomatic people would be different to the close contact estimate.

Our ability to test and detect minute, irrelevant quantities of virus has created an utterly distorted view of reality. The fallout from trying to minimise the already small risk of presymptomatic spread has resulted in significant harmful overdiagnosis of the healthy. The myth that healthy people are a significant threat to others needs to be quashed for good so that people can stop treating each other primarily as potential vectors of disease.

THE EVIDENCE MANIPULATION TRIAD	
EXTRAPOLATE	Positive test results in asymptomatic people were extrapolated to claim evidence of asymptomatic spread.
EXCUSE	Long distant aerosol transmission was ignored as a possibility with the explanation that asymptomatic spread was responsible despite a lack of evidence.
EXCLUDE	The fact no asymptomatic infections were confirmed with antibody testing was ignored.

TOP THREE MYTHS
1. One in three people with covid never develops symptoms
2. People who never developed symptoms have made others sick
3. An organism in the respiratory tract that never enters a cell is still an infection

13

HIGH PRIESTS

If I have a book to serve as my understanding, a pastor to serve as my conscience, a physician to determine my diet for me, and so on, I need not exert myself at all. I need not think, if only I can pay: others will readily undertake the irksome work for me. **Immanuel Kant, 1784**[408]

Since the Age of Enlightenment science has driven how we understand the world. Shaped by rational debate, measurements of the world around and experimental evidence our understanding of every aspect of the natural world has been brought into sharp relief. While the consensus among specialists in any particular area of science is born of years of work, questioning, debate and challenge, for those not at the cutting edge we take their consensus on faith. For hundreds of years this arrangement has served us well. No-one can be expected to go back to first principles on every subject in order to decide what to believe.

The breadth and depth of scientific evidence is so wide ranging that even scientists within a particular niche find themselves taking consensus on faith. How many people can say they have actually checked the measurements and arithmetic that led to Galileo's conclusion that the earth moves round the sun? Instead, the work of those who did and who changed their minds when they did is enough and we move on to other questions. To a non-specialist the difference between a widely held assumption in science and an evidenced and established fact can be hard to spot as both are treated similarly. Because we take this approach it is important to have trusted sources. The situation lends itself to appointing a position of high priest for any particular area of science.

The covid high priests were a combination of the public health officials, the politicians and the mainstream media. These trusted sources left very little room for doubt in their pronouncements – they were playing politics not helping people to understand science.

All of us rely on trusted sources as a shortcut to finding out about the world. For the public health officials deciding on the social distancing and perspex screen advice, they trusted the people who had written the guidelines. The mistake in the size of droplets vs aerosols (grapefruits and lentils from *Belief One: Covid only spreads through close contact)* was so long ago and repeated so often it had become an accepted truth. It is so much easier to be wrong when all your trusted sources are wrong. The people alerting the public health officials to the problem were outsiders – physicists – and were ignored. Bizarrely, there were still groups who identified as being primarily germ theorists and this resulted in an overly defensive rejection of anything remotely resembling miasma theory. That meant ideas were dismissed without proper consideration. Shortcuts to deciding who to trust are dangerous in science.

Prophets from the biblical era were considered to have an almost magical ability to see the future, a gift from God. Their believers were discouraged from questioning, checking or understanding the prophecies. Current experts are treated in a similar way and scientists are no exception. Science is treated in general media as being relevant to every aspect of life, knowing everything and having power over everything. In other words omnipresent, omniscient and omnipotent. Even scientists whose expertise lies in a different domain, can treat scientists from other specialties as expressing knowledge that is incomprehensible and must be taken on faith. In fact, a failure to do so can be taken as being disrespectful of the other's expertise.

Of course alongside our faith in scientists to know all that can be known, there is also a faith that scientists will always act in good faith and be omnibenevolent. Even well intentioned scientists will not always say and do things that can only result in benefit to mankind.

Where there is a conflict of interest be that money, career progression or an offer of power, then what is said may also be distorted. What made the Age of Enlightenment a time of such amazing progress was at least in part because of the independence of the researchers. Funding came from benefactors with the occasional government funded prize worth the equivalent of millions in today's money. The boundaries of what was known were constantly expanding, with openness to plentiful new ideas alongside rigorous debate and challenge.

WHO WERE THE PROPHETS AND HIGH PRIESTS?

Neil Ferguson, lead of the doom modelling, had a track record of failure. In 2001, his models for foot and mouth disease led to the unnecessary slaughter of over 6 million farm animals.[409] In 2002, he predicted up to 50,000 people would die from 'mad cow disease' and there were 177 deaths but 4 million cattle were slaughtered. In 2005, he estimated bird flu could kill 200 million people and that failure to act could be catastrophic for the UK. It killed 74 people worldwide. In 2009, he estimated swine flu would kill between 19 and 65 thousand people and it killed 457. In February 2020, he said he would *"much prefer to be accused of overreacting than underreacting."* [410] Somehow he made a career of overreacting.

Anthony Fauci, Chief Medical Advisor to the US President, had been one of the most powerful scientists in the USA and the world for many years before covid. He has been head of the National Institute of Allergy and Infectious Diseases since 1984 and has advised every president since Ronald Reagan. The power seemed to go to his head and in June 2021 he said, *"A lot of what you're seeing as attacks on me, quite frankly, are attacks on science."* [411]

By November 2021, he was warming to his theme saying, *"If they get up and really aim their bullets at Tony Fauci… they're really criticising science, because I represent science… And if you damage science, you are doing something very detrimental to society long after I leave. And that's what I worry about."* [412]

Fauci pushed lockdowns in the USA saying, in October 2020, *"I recommended to the president that we shut the country down."* [413] However, by July 2022 he was distancing himself from the policy saying, *"I didn't recommend locking anything down."* [414] Israelis awarded him a $1m[415] prize for *"speaking truth to power"* in February 2021. The implied bravery of *"speaking truth to power"* should surely require that those doing so have little power themselves. However, Fauci was a high priest not a heretic.

Christian Drosten is a German virologist and became the go-to advisor for the German media. He also ran a commercial PCR testing laboratory in Berlin. On 21st January 2020, when there had been only six covid deaths worldwide, he co-authored a paper describing a new laboratory protocol to be used for covid PCR testing. Other similar papers took 180 days to be published before 2020. No doubt through great effort, that delay had been reduced to an average of 90 days in 2020. Scientist, Dr Simon Goddek[416] and data analyst Wouter Aukema[417] exposed how the PCR test paper was published at breakneck speed, within one day. The fact two authors, including Drosten, were editors at the journal may have helped them speed through all normal peer review processes.

Twenty two scientists identified numerous serious flaws in the design of the PCR test described in the paper largely to do with errors that could result in erroneous results. No calibration was carried out, even in retrospect, to ensure the test was measuring clinically meaningful levels of virus. His evangelical attitude to mass testing seems to be unique to *SARS-CoV-2*. He had a very different attitude in the MERS outbreak saying:

"The fact is that there has been a clear case definition so far, i.e. a strict scheme that stipulates which patient was reported as a MERS case. This included, for example, that the patient has pneumonia that affects both lungs. When a whole series of MERS cases suddenly appeared in Jeddah at the end of March this year, the doctors there decided to test all patients and the entire hospital staff for the pathogen. And to do this, they chose a highly sensitive method, the polymerase chain reaction (PCR).

…The method is so sensitive that it can detect a single genetic molecule of this virus. If, for example, such a pathogen flits over the nasal mucous membrane of a nurse for a day without becoming ill or noticing anything, then it is suddenly a MERS case. Where previously terminally ill were reported, now suddenly mild cases and people who are actually very healthy are included in the reporting statistics. This could also explain the explosion in the number of cases in Saudi Arabia. In addition, the local media boiled the matter up incredibly high."[418]

By April 2020 he was saying that testing and tracing of all contacts was the only way out. *"You enter a symptom in a checkbox on your phone and the app will order a diagnostic test for you and identify the people you've had contacts with in the last five days. Then you and all your contacts lock yourselves away immediately, while everyone else who is not tarred by your transmission chain has the freedom to travel and work."* [419]

It seems he was happy to change his views with the political tide. Changing your mind is an attribute that we should praise but only if the change is based on new evidence. There is an extreme contrast between his insistence on making an accurate diagnosis for MERS and this remote proxy measure of a case in order to isolate the healthy. Despite his self-contradictions his views were taken as gospel truth.

The UK press conferences saw the prime minister or the Secretary of State for Health flanked by Chris Whitty, Chief Medical Officer, Patrick Vallance, Chief Scientist and others. The pinnacle of the arrogance of these high priests came in autumn 2020, when Patrick Vallance presented a prediction of an imminent rocketing number of cases. The result was a claim that we would reach 4,000 covid deaths a day, four and a half times higher than the April peak. Mark Woolhouse, professor of epidemiology on SAGE tells how the shortest doubling time and most pessimistic projection of all the models had been used for this presentation. Worried about a loss of scientific credibility, he published his views online. Thereafter, he was reprimanded and told to *"correct"* his comments. [420] The High Priests graph led directly to the second lockdown. In due course Woolhouse was proved right.

There followed a rebuke from the Office of Statistics Regulation concerned about the *"potential to confuse the public and undermine confidence in the statistics."* [421] A Select Committee session was also called where Chris Whitty and Patrick Vallance were held to account for the graph.[422] Jeremy Hunt, who was health minister until 2018, called them out saying *"if it was not important or reliable enough to present to Ministers I am surprised you both decided it was important or reliable enough to present to the public."*

Greg Clark, the chair, pointed out at the same select committee, that Ministers were not given much choice about policy, *"if the advice from advisers to the Prime Minister is that the capacity of the NHS is likely to be overrun within weeks, that is quite difficult advice to gainsay, is it not?"* Chris Whitty replied, *"Ministers... have to [consider] multiple other things that have big social and economic impacts. Ministers have to take them into account. It is right that elected Ministers make those decisions."* Greg Clark hammered the point home, *"It is much more difficult to make choices and decisions if the bottom line is that people are going to be dying in hospital car parks."* [423] By anointing these high priests and failing to have other scientific advisers to hold them to account the government transferred all power to them.

Three favourite high priests of the BBC were Dr Deepti Gurdasani, Professor Christina Pagel and Professor Devi Sridhar. Between them only one had qualified as a doctor and none claimed to have ever been employed as a doctor. They were entitled to their opinions as much as anyone but they were often presented as omniscient by the BBC and their views were rarely if ever opposed or debated.

The elevated status of the High Priests was emphasised with the use of giant screens. The UK Secretary of State for Health and Social Care, Matt Hancock's giant face was being projected into a Nightingale Hospital at its opening while the staff stood in socially distanced rows like a science fiction drone army. Similarly, Bill Gates was interviewed by the BBC beneath the skeleton of the blue whale in the natural history museum, from a screen the size of the blue whale's skull while the interviewer stood some distance away looking utterly diminutive.[424] He declared, *"the pandemic will come to an end because these amazing vaccines were invented in a year."* Bill Gates has honorary degrees but no PhD and dropped out of his undergraduate degree.[425] Qualifications were not a necessity to be a covid High Priest.

Politicians avoided all complexity parroting their hypnotic triplet word chants: *"Stay Home, Protect the NHS, Save Lives"*; *"Stay Alert, Control the Virus, Save Lives"*; *"Hands, Face, Space"* and *"Get boosted Now"*. Evidence for decision making always amounted to 'we know best': *"we are following The Science"*, *"The Science is clear"* or *"We know this works"*.

HOLDING THE PRIESTS TO ACCOUNT

The media failed to hold the high priests to account for their contradictions, their lack of evidence and for the harm they caused. This is likely in large part thanks to threatening Ofcom guidance which said, *"We remind all broadcasters of the significant potential harm that can be caused by material relating to the Coronavirus... Ofcom will consider any breach arising from harmful Coronavirus-related programming to be potentially serious and will consider taking appropriate regulatory action, which could include the imposition of a statutory sanction."* [426]

There would have been ambiguity in deciding whether or not something causes harm but who decides what is a *"potential harm"*? Given how unpredictable live broadcasting can be, why would broadcasters risk inviting guests that might say something which could be a *"potential harm"* and which could risk their broadcasting licence?

The same restrictions were not in place for print media, but broadcast media set the tone. The Government spent over £180m on advertising,[427] including in print and broadcast media in 2020 and committed to spend another £320m by spring 2022.[428] Newspapers saw a decline in sales with reduced commuting making them even less keen to bite the hand that feeds them. On top of that incentive, uncritical, covid fear propaganda sold newspapers.

Although people were largely oblivious to the Ofcom guidance, there was an awareness that something was not right. Seventy percent of the public complained, in a survey in April 2020, that journalists were failing to hold politicians to account.[429] In September 2021, Julia Hartley-Brewer, TalkRadio presenter, put ITV's main reporter, Robert Peston, on the spot asking why he had never once asked for the evidence for restrictions and lockdowns. He replied, *"I'm not 100 percent sure, you know, if you say genuinely I didn't ever ask for sort of empirical evidence then I'm slightly surprised."* [430]

An attempt to keep a grip on the official narrative saw Jacinda Ardern, Prime Minister of New Zealand, saying, *"You can trust us as a source of that information… to clarify any rumour you may hear… dismiss anything else. We will continue to be your single source of truth… Take everything else you see with a grain of salt."*[431]

The USA went further still. In April 2022, The Department of Homeland Security set up a Disinformation Governance Board (colloquially known as the Ministry of Truth). By 18th May the board was *"paused pending review"* after enormous backlash from civil rights groups.[432] The UK had its own secretive counter disinformation unit or 'cell' about which few details have emerged.[433]

QUALIFICATIONS

While journalists were failing to ask questions, questions could still be asked on social media. A common refrain in 2020 was to reject questions or comments from interested parties because they were *"not an epidemiologist or a virologist."* Of course people with training should have a better handle on the details but, in an epidemic, they also have a responsibility to help educate the public. These dismissive comments were usually said because there was no good answer to the question or comment being put. I think experts should be listened to but shouldn't be able to hide behind their expertise or avoid answering questions.

The idea that only certain parties were entitled to comment on the situation reminded me of the way the church tried to protect its own scriptures by making it illegal to translate the bible. I was also reminded of those scientists who transformed their field without having had the privilege of being channelled through the official system. Scientists like Michael Faraday, who was working as an assistant when he made his breakthroughs in electro-magnetism or John Harrison, the carpenter who cracked the longitude problem or even Einstein who wrote his four groundbreaking papers while working in a patent office. There was a certain irony in how the advisory bodies like SAGE were hugely diverse from a disciplinary perspective including experts in astronomy, architecture and adolescent sexual health while failing, by May 2020, to include anyone with expertise in respiratory disease, intensive care or covid testing. The right to a voice was not really to do with expertise. It came from being part of the inner sanctum.

Instead of listening to a diverse range of voices with their own specialist skill sets only those who had conformed to a traditional scientific career path were given an audience. Unfortunately, many of these scientists appeared to think their job was to lend their credentials to official lines without seemingly any scrutiny of the underlying claims, reading of the current evidence or nuance of any sort. Having heretical views as a scientist, even if right, makes for a challenging life. Unfortunately, there's a long history of scientists who would rather conform to societal beliefs than let the evidence shape their beliefs.

TRUSTED SOURCES

Many doctors were relying on the BBC or other news sources signed up to *The Trusted News Initiative* to summarise the science on their behalf.[434] Fact-checkers were another easily digestible resource. There is nothing independent about many of the prominent fact-checkers. Organisations who choose to spend their resources on fact-checkers, more likely than not have an agenda or there are likely conflicts of interests. For example, Jim Smith, the former CEO and president of Reuters, responsible for Reuters Fact Checker, was also a Pfizer board member.[435] Investigative journalist Sharyl Attkisson also described the problem of a *"circular feedback loop of verification"* whereby *"like-minded journalists or often liberal Silicon Valley gatekeepers... frequently rely on partisan news sources and political activists to control narratives."* [436] The problems have led to fact checkers having to retrospectively alter their articles or even retract them.[437]

Scientific journals are a key route to scientists' voices being heard. The editors have always protected the journal's reputation by publishing papers by sources they trust and through a system of peer review where papers were critiqued by experts before publication. In the avalanche of covid publications, the peer reviewers, who are unpaid, repeatedly failed to critique papers adequately. Raw data was often not presented before adjustments and modelling were carried out and conclusions drawn that were not supported by the evidence presented.

Their publishing of untruths is only half of the story. Others were silenced or findings only published in papers where incongruous lines that read like declarations of faith to the official narrative were included even where they contradicted the findings of the paper. (I have written more about the stories of witch-hunts and blasphemers in *Spiked: A Shot in the Covid Dark*).

The high priests were in a position where anything they said would be believed. This applied even when they contradicted either themselves or opposed self-evident real world facts. *The New York Times* epitomised this contradiction when they ran with a headline in June 2022 *"Do covid precautions work?"* [438] A question which they answered with a subheading *"Yes, but they haven't made a big difference."* What better way to praise the high priests as always being right than to say they are right even when no effect is discernible?

Knowingly or not, these High Priests created all the ingredients needed for a new faith to be born. There was fear, the unknown, the unseeable and the unpredictable. There were a number of unshakable beliefs not grounded in reality and the promise of simple solutions including the vaccine (our saviour).

14

FIRST DO NO HARM

The welfare of the people in particular has always been the alibi of tyrants, and it provides the further advantage of giving the servants of tyranny a good conscience. **Albert Camus, 1995**[439]

The concept of not causing harm, forms the basis of all ethics. It is universally accepted that harming people is morally wrong. There will be perpetual arguments when opposing parties both emphasise what they believe is a greater harm. The abortion and gun debates will rage on as each side of both arguments believe their way prevents more harm. The covid conflict was similarly emotive because people who believed interventions prevented harm from covid were opposed by people who believed the harm from interventions, which were often denied altogether, far outweighed any reduction in harm from covid that they achieved.

'First do no harm' has formed the backbone of medical ethics since the seventeenth century and harks back to phrasing in the Hippocratic Oath from ancient Greece. My medical training rightly placed great emphasis on this. As a junior doctor we used abbreviations in the notes when writing a plan for a patient. A favourite of one boss was CLOMI, which stood for, cat-like observation, masterful inactivity. There was nothing disrespectful about it. It was an admission that what the patient needed most at that time, was nursing care not medical care. It was also a recognition of a phenomenon that most patients are familiar with. The moment you step foot into the doctor's surgery or a hospital, the worst of the problem has often already passed. That is not true for everyone, which is where the cat-like observation comes in. The nurses would keep a very close eye, measuring and assessing and at any sign of a deterioration we would return and reassess if the time was right to intervene.

It was also an acknowledgement that medical interventions all have a risk. Waiting to intervene was in the patient's best interest. Every treatment and even every test a doctor carries out, has the potential to cause harm. Even a chest x-ray has a low, trivial radiation dosage associated with it. It also comes with a risk of finding a shadow of no consequence which would be best ignored. However, having found it, the only way to prove it is not a cancer is to biopsy it. A lung biopsy is a serious intervention. For benign conditions there is a mortality rate of 4 percent, largely from accidental puncturing of the lung.[440] It is the knock on effects of testing that can be the most harmful. That is why smokers are not routinely screened for lung cancer using chest x-rays. It would cause more harm than good despite the high incidence of cancer among older smokers and the relative ease with which x-ray screening could be done.

The reason the concept of 'first do no harm' is at the heart of medical education is because it is not about intentional harm. The harm doctors do, (apart from a very occasional criminal doctor), is done with the best of intentions. Junior doctors would always be ready to do something, anything. It was senior doctors, with their wisdom and experience, who taught the juniors not to be hasty.

The simplest way to do no harm is to do nothing, but there are occasions when doing nothing is wrong. If a child has an acute appendicitis, a surgeon minimises harm to that child by removing the appendix. However, if they misdiagnose an acute appendicitis then removing the appendix will have caused only harm. With a sick person, where doing nothing might cause harm the ethical decision becomes a balance between the potential harm caused by either doing nothing or intervening. The balance is different for the healthy. Doctors should not risk harming a healthy person but a sick person can accept a greater risk in order to benefit.

A dying person can accept a huge risk. One of the most dangerous operations is fixing a burst aorta in the abdomen. Half of those operated on die within a month.[441] However, a person with a ruptured aorta will die without an intervention. Ethically the dangerous operation is an utterly justifiable procedure.

Many people appear to conflate interventions to prevent harm in the sick with interventions in the healthy. For someone who is in immediate danger, attempts to minimise harm can come with a risk. For someone who is not in danger, great care must be taken not to cause unnecessary harm. For example, rugby tackling a granny to the ground could cause terrible harm but would be justified if it prevented her stepping into the path of an oncoming car she had not seen.

There is also confusion when these ethical principles based on balancing risks and harms in an individual are extrapolated to a population. Minimising the risk of harm in an individual is not the same as harming a minimum number of individuals for the benefit of the many.

Would you sacrifice a minority for the sake of the majority? For example, should minorities who are more likely to be convicted of offending be incarcerated preventatively? The very concept of sacrificing the innocent is an anathema to our culture which also teaches us, rightly, that the moral thing to do is to defend minorities, even when they represent views, cultures and opinions that are different to our own.

The greater good' is a phrase that sounds positive but would not exist without the concept of sacrifice. By definition the sacrifice must be of a minority or else the *greater* number would not benefit. The phrase 'the greater good' has been used to justify all sorts of harmful covid policy. The sound of that phrase sends shivers down my spine.

Although everyone agrees that a minority group should not be sacrificed on a population level, our ethics on sacrificing a minority apparently fall apart when scaled down to smaller numbers. The Trolley Problem is a hypothetical ethical dilemma devised by Phillipa Foot in 1967, where a train can be diverted away from a track on which there are five people, saving their lives. However, the track to which the train would need to be diverted has one person on it such that the person diverting the train would be directly responsible for the death of that person. Do you choose the option with the least total death as the outcome or stick to a moral code that prohibits causing someone's death? Nine out of ten people say they would pull the switch and divert the train, saving five lives but killing one. They take the utilitarian view that the most ethical choice is the one that will produce the greatest good for the greatest number.

The framing of this dilemma is critical. When rephrased, such that a person must be pushed onto the track from a bridge to save the five lives, only one in ten would push that person to their death. The outcome in terms of causing a death and saving the most lives is the same. Something about proximity, about touching the flesh brings the killing into sharper focus and people can see it is wrong. Perhaps it also highlights that you yourself could jump in front of the trolley in an attempt to save those five lives. If you would not sacrifice your own life, how could you justify sacrificing someone else's? (Frighteningly, some have argued for training artificial intelligence algorithms with the trolley dilemma. The idea that a car could be trained to ever, under any circumstance, aim directly at a person is monumentally stupid).

The trolley problem was recreated in real life by my son's favourite YouTuber, Vsauce, who makes science, maths and psychology videos.[442] The scenario was enabled with the help of several actors, a dummy signal station and video streaming. Carefully screened participants who thought they were taking part in a transport survey were persuaded to wait their turn in a signal box with an actor. They each did a dry run pulling the lever causing a train to switch tracks before he left them alone while workmen, seen on the video stream, distributed themselves unevenly on the two tracks and a second train approached. At the end of the experiment, moments before the crash, the screens cut to a message saying *"End of test. Everyone is safe."* A debrief followed immediately. The majority did not pull the lever. Those that did not act described feeling terror and anxiety as the train approached. They froze in a shocked state, but in the end decided it was not their responsibility to act. The two that did pull the lever, seemed to think only of the lives that were at stake and the families concerned and not of what others would think of them for their actions. But they were emotionally distraught afterwards, unlike those who didn't act, suggesting that pulling the lever had conflicted with their inner moral compass. They had acted to cause a particular individual to lose their life.

Covid policy has been like a runaway train. At the outset it was on one track and there was a pandemic plan in place for managing the problem. Everyone agreed that those who were susceptible would be infected at some point regardless of interventions and a proportion would die. No claims were made that deaths could be prevented in the absence of effective vaccination or treatment. They could not. The only claim was that it was possible to intervene to slow the rate at which the virus spread to flatten the curve so that the peak pressures on hospitals would be more manageable.

Neil Ferguson et al. printed cardboard cutouts from a computer model and placed them on the tracks the covid train was heading towards. Politicians panicked at the last minute and pulled the lever diverting the covid train to a different track. They knew there were real people on that other track.

Sunetra Gupta was right when she said, *"The one thing we did know was that… lockdowns and other restrictions would have enormous cost. That was the one thing we were certain about and yet that is what we went in and did. So we inverted the precautionary principle…"* [443]

The covid train will keep careering out of control down the other track killing people who were out of sight in the original count for decades to come. There is no other lever to pull that will save these lives. Politicians have tried to blame the harm caused by their runaway covid train on covid itself but it was their policies that are to blame. Worse, attempts have been made to claim the harm was justified. This is reminiscent of the masochistic concept of medical quacks persuading patients that when medicine causes pain and damage you know it's working.

It would be hard to imagine any ancient religious text suggesting that sacrificing a minority for the majority, as in the trolley problem, was ethical. Every religion recognises the sanctity of all human life. What would such a parable be called? Even if you are not a religious person, these texts, as the bases for societal morality, have stood the test of time. If a concept currently accepted as normal and ethical would be unthinkable in that context then perhaps that should cause us to pause and rethink.

BELIEF EIGHT: LOCKDOWN SAVED LIVES

We don't even talk about containment for seasonal flu – it's just not possible. But it is possible for COVID-19. We don't do contact tracing for seasonal flu – but countries should do it for COVID-19, because it will prevent infections and save lives. Containment is possible. **WHO, March 2020**[444]

In Cloud-Covid-Land lockdown logic goes like this: *SARS-CoV-2* is spread through close contact transmission. Lockdowns reduce close contacts. Therefore lockdowns reduce transmission. It is a very simple, intuitive theory but, for various reasons, notably long distance aerosol transmission, it is flawed.

THE ARGUMENT FOR LOCKDOWNS WORKING

Spread from symptomatic people was already minimised by asking sick people to stay at home. The only way lockdown could minimise close contact transmission further would be by reducing asymptomatic spread by keeping those people at home too. If spread from people who never developed symptoms was negligible and presymptomatic spread only contributed to a tiny fraction of an outbreak then lockdowns would do little to change the speed of spread. The fact that the first UK lockdown was timed neatly to fit with when infections fell is often cited as evidence that lockdowns did work. The bigger picture tells us there is more to this story when looking across the world or comparing regions within the UK.

There are undoubtedly huge numbers of instances where *SARS-CoV-2* did spread due to close contact. However, we have already encountered the possibility of spread occurring through the air over long distances because of aerosol transmission. There were therefore two competing hypotheses for what was the prime driver of each wave. The first is that aerosol transmission seeded infections throughout a community and these led to close contact transmission as a secondary event. The second is that close contact transmission was the key driver and the numbers were topped up by aerosol transmission.

Lockdowns were a highly unethical way of showing which was the key driver of each wave. Lockdowns reduced close contact transmission but they could not alter long distance aerosol spread.

WHAT WOULD HAPPEN IF WE DID NOT LOCKDOWN?

Given that there was a population that was susceptible to the virus, the virus would pass through that population. That would involve exponential spread as the numbers infected repeatedly doubled until the virus began to run out of susceptible people to infect. No-one would expect a trajectory to correlate with mobility data in such a situation.

Team In Theory told us that close contact transmission was slow, so even with seeding in multiple cities close contact models predicted it would take 10 to 14 weeks for the wave to peak from when sustained community transmission began as the virus progressed through almost the entire population.[445]

On 14th February 2020, Warwick University modellers made a range of predictions[446] all of which resulted in a peak after about four months in June or July 2020, before starting to decline. This held even when they accounted for seasonal impacts.[447] They anticipated 82 percent of the population being infected. Imperial models presented at SAGE on 16th March 2020, also had a June peak in the UK and a July peak in the USA with 81 percent infected.[448] Ferguson's team had their model published by the WHO on 16th March 2020. It said, *"The epidemic is predicted to be broader in the US than in GB and to peak slightly later. This is due to the larger geographic scale of the US."* [449]

The idea was that close contact chains of transmission would keep going until the most isolated people, either geographically or socially had been reached.[450] SAGE minutes from 2nd March 2020, say that the peak in rural areas would be later, *"Peak timings in different parts of the UK would be expected to vary by around 4–6 weeks in an unmitigated reasonable worst case scenario."* [451]

What was predicted was not a covid wave but a covid tsunami. It would take around 8 months to see the back of it if nothing was done. A seasonal low in September or October could be followed by a winter resurgence before it would finish once 80 percent had been infected.[452]

WHAT WOULD HAPPEN IF WE DID LOCKDOWN?

The aim of the lockdown was to slow the spread. No claims were made that spread could be stopped. Slowing the spread would lead to a gentler gradient and a later peak. Chris Whitty said, *"I think it is important to stress that if you pull down a peak that is spreading through the population, generally, you do not reduce the number of people who get infected, but you reduce the number who get infected in that central period."* [453] The predicted four months to reach a peak in June with close contact transmission would therefore be even later with lockdown.

Chris Whitty spoke in the house of commons of how a balance had to be struck between the risk to the NHS of a short period of intense disruption compared to the risk of a much longer period of disruption which might cause more public health harm in total. He said: *"some of the things that pull the peak down extend the peak. They basically put it further out. You might have something that made it easier to cope over the peak period, but it might counterintuitively have a bigger effect on waiting times, for example."* [454]

The difference in timing of urban and rural peaks, instead of being 4-6 weeks, according to SAGE would, thanks to lockdown, *"vary by a greater amount."* [455] We will return to that number shortly.

WHAT WOULD HAPPEN IF THE CLOSE CONTACT MODEL WAS WRONG AND IT WAS LONG DISTANCE AEROSOL TRANSMISSION THAT WAS THE KEY DRIVER OF SPREAD?

If the key driver of spread was long distance aerosol transmission there would be rapid spread. The rate at which spread occurred would not be abated by a reduction in close contacts. The peak would arrive regardless.

In theory, with either close contact or long distance aerosol spread the peak would be earlier in cities because of there being either more chances for close-contact or there being a higher dosage in the air, respectively.

The main difference between the two models would be in how long it would take to reach every remote region. Aerosol transmission would reach many more remote regions in just a few weeks whereas droplet, close contact transmission would take much longer to reach remote places.

A critical aspect to the SAGE modeller's hypothesis is that more remote regions would have been saved from covid by lockdowns. That is why Professor Keeling, one of the SAGE modellers, said in June 2020, about a second wave *"We are not going to see a uniform spread of infection across the country. It is going to be very isolated in small pockets."* To be fair, he was assuming that interventions would be repeatedly applied to prevent further spread.[456]

WHAT HAPPENED IN THE REAL WORLD?

There is now a huge body of peer reviewed scientific work demonstrating that the trajectory of covid cases was not interrupted by lockdown.[457] Naturally there are also papers that claim the opposite. The most prominent one being the Flaxman paper.[458] These authors make various assumptions around how lockdowns work; carry out modelling based on those assumptions; draw some colourful graphs to represent their results and then conclude that lockdowns work. The papers based on real world evidence all point in the same direction showing that lockdowns had no impact. The lag between lockdowns starting and the peak in deaths varied hugely from country to country as well as regionally within countries.

In the real world, with lockdown, rural and remote regions did have death peaks after more densely populated and well-connected regions, with Brighton and many parts of London peaking first. The virus did reach remote areas like the Isle of Wight and Isle of Anglesey, less well-connected cities like Lincoln, Hull, Merthyr Tydfil and rural areas like West Norfolk, the East Riding of Yorkshire and the Highlands of Scotland. Deaths peaked in these areas a month after the first urban peaks. These places with peak deaths in early May clearly had a case curve that had both risen and peaked, well after lockdown.

SAGE predicted a slower spread would lead to a longer than six week gap between the urban and rural peak. Based on that prediction, lockdown did not slow the spread. The timing of the peak in different places was what they predicted it would be *without* lockdown.

A much wider lower curve can be seen quite starkly by comparing deaths in the thirty hospitals with the earliest date for their maximum daily deaths with those thirty hospitals that had the latest maximum daily deaths. Had these places successfully squashed the sombrero? Did covid arrive in these regions late enough that lockdowns did slow the spread? The places that had a flattened curve were all more rural and the steepest curves and earliest peaks were densely populated urban areas. It seems more likely that the key differences between them was simply population density and that the virus took longer to reach people with a meaningful dose in less densely populated parts of the country. The virus kept spreading despite lockdown, with slower spread in more rural areas then peaked at a time that would be expected if there had been no effective interventions.

There has been confusion about the role that population density plays in covid spread. Very often the researchers asked the wrong question. The key question is whether spread was faster. However, most waited until the end of a wave and then demonstrated that over the course of the whole wave the same proportion of the population were susceptible to a particular variant, which tells us nothing about the speed of spread. Unsurprisingly, places with higher population density do have faster spread[459] and earlier peaks.[460] This has been attributed to people having more close contacts each day depending on how many people live near them. However, it would also be true that the dosage would climb more rapidly in the air in a more densely populated area. Only the latter could explain the dramatic surges of infections seen despite lockdowns in e.g. the Czech Republic in the first covid winter and in Australia and China in 2022.

The USA incidentally demonstrated that changes in behaviour had no impact. There was a marked difference in the timing of behavioural changes across the USA. In Seattle one research laboratory decided to break the rules for the sake of public health. Rather than wait for a CDC approved covid test, they adapted an existing influenza test and started testing. Consequently, Seattle became the first major hotspot for covid positive test results in the USA. According to mobile phone data, the people of Seattle started staying at home from the beginning of March,[461] a week earlier than any change in New York[462] and well before the Seattle lockdown on 23rd March 2020.[463] However, the deaths in Seattle peaked along with every other northern state around 10th April 2020. The implication is that Seattle was successfully measuring what was going undetected elsewhere but that the change in human behaviour had no impact on the timing of the peak.

To an untrained eye a wave takes a dramatic turn from increasing to decreasing at a particular point in time. However, a mathematician can plot, not the number of cases, but the way cases grow day to day and how that changes over time. Michael Levitt, a Nobel prize winning Stanford professor of biophysics demonstrated that instead of a dramatic rise and fall, when using the Gompertz formula the change in cases day to day can be presented as a dead straight line showing that the brakes were being applied at a predictable rate. At the beginning growth was rapid, as it slowed it reached a point where case numbers were no longer increasing and then, with continued application of the brakes, the growth (i.e. the increase from day to day) went into reverse and the case numbers started to fall. Any impact from changes in human behaviour should have caused that straight line to deviate but it did not falter.

WAVES SURGE AND FALL INDEPENDENT OF LOCKDOWNS

Waves have begun and surged during lockdowns. The most extreme example for this was the strict lockdowns in Australia that failed to prevent a surge of Delta infections and, later, Omicron.

As different states in America had different policies there was plenty of real world evidence that lockdowns did not work. South Dakota had minimal restrictions and yet its trajectory was identical to North Dakota's which closed schools, restaurants, cinemas and gyms. The lifting of restrictions in Texas, Florida and other states did not stop them tracking the same trajectory as neighbouring states.

The fact that lockdowns have failed to prevent waves and have not been needed to bring waves to an end does not prove that they did nothing at all. After all, the initial aim was to slow the spread, to, in the words of Boris Johnson, *"squash the sombrero."* However, the fact that surges occurred despite minimal close contact does imply that aerosol transmission over long distances was a prime driver of each wave. Lockdowns therefore could not prevent a wave nor stop it in its tracks, pausing it until the lockdown was released. Despite this admission early on, the emphasis moved from slowing the spread and avoiding an overwhelming peak to *"stay home, protect the NHS, save lives."* This gave the false impression that a change of behaviour could stop the virus and prevent deaths rather than briefly postpone them.

FREEDOM DAY

Waves have also plummeted when restrictions were released. On the 19th July 2021, 'Freedom Day', restrictions were released. There would be no limit on numbers who could attend events and no further rules on masking or social distancing. The British press had a hysterical frenzy about the inevitable huge numbers of cases (up to 200,000 per day) that would follow the release of restrictions.[464] I was invited onto the GBNews breakfast programme to comment. I was taken aback by so much of the coverage which amounted to interviewing random members of the public and asking whether they thought having 100,000-200,000 cases a day was a good idea. GBNews had, on the whole, managed balanced coverage and grown their audience hugely as a result, but this did not seem balanced.

I pointed out that the forecast was from Neil Ferguson and his team of Imperial modellers whose track record spoke for itself, especially recently, and asked why people still gave them any credence. I was then asked to make a better prediction. My answer was to say that it's a fool's errand to predict the future. I then pointed out that positive test results had peaked a few days earlier in Scotland and had already peaked in the unvaccinated cohort measured by the King's College London ZoeApp team who asked millions of volunteers to input data about their symptoms and test results. Given that the English are not so very different to the Scottish and the unvaccinated are not so very different to the vaccinated, the peak was probably imminent.

Two surprising things followed. Firstly, they brought on the next interviewee, Dr Simon Clarke, a microbiologist from Reading and asked him what he thought of what I'd said. Caustically, he replied that people should *"check her funding."* (No-one was paying me). He then tweeted *"Did a hit with @GBNews this morning (happy to chat to anyone except Philip Schofield, before anyone starts) but to have to listen to Cla*r Cra*g's arrant bullshit beforehand made me want to vomit and I'll be making sure that doesn't happen again."*

Next, the ZoeApp team decided to change their methodology[465] such that the trajectory they had published for the unvaccinated population was retrospectively increased.[466] There is no big scandal about cases in the unvaccinated peaking before the vaccinated. It was well established that the young were infected earlier in each wave than the old and it should therefore be expected that the unvaccinated would peak before the vaccinated. The ZoeApp team said that they changed their methods to make their results better fit the Office of National Statistics estimates of how many were infected.[467] The result was that the plateau they had estimated around Freedom Day disappeared and instead supposed case numbers soared with the unvaccinated figures cranked up to peak at the same time as the vaccinated. Whereas the original peak had had a similar proportion of unvaccinated infected as the vaccinated, the adjusted graph doubled the estimate of the proportion of the unvaccinated who were infected. The Office of National Statistics estimate turned out to be far too high.[468] They had estimated 15 percent more cases in the week after Freedom Day compared to the week before whereas actual diagnosed cases fell to two thirds of the week before.

The number of positive tests peaked on that very day, Freedom Day. The models were yet again shown to have flawed assumptions and, yet again, none of those core assumptions were adjusted about how spread occurred and how many were susceptible. The defining story that showed the erroneous nature of those assumptions was the course of events in Sweden.

SWEDEN

Sweden had a constitution which prohibited lockdown and a working class prime minister who had been a welder and did not have his head in the clouds.[469] They did not lockdown. People did adjust their behaviour to an extent but shops, restaurants and other small businesses continued to open with people able to gather and socialise if they wished. What is critically important about Sweden is what did not happen. The predicted 66,000 to 90,000 deaths that the modellers claimed would result in the absence of lockdown did not occur.[470] The 16,000 to 34,000 deaths that were predicted if only social distancing and enhanced protection of the elderly were implemented also did not happen. Sweden had fewer than 6,000 deaths by August 2020. At that time, the USA had had 500 deaths per million, the UK 600 and Sweden lay between them with 570 but having had no lockdown.

The importance of the discrepancy between the predicted number of deaths and reality, which turned out to be 73 times lower, cannot be overstated. The modellers had wildly overestimated what would happen had the advice from years of pre-covid pandemic planning been followed. In fact, although Sweden experienced more total deaths in 2020 than 2019, the mortality rate was lower than in 2012 and every year prior to that.

The Imperial team had to come up with a reason why the Swedish trajectory peaked and fell without lockdown. They published yet more modelling in *Nature* in June 2020.[471] In it they claimed that all the measures introduced pre-lockdown in countries other than Sweden were insufficient to stop growth and yet lockdown had still worked. In Sweden, they claimed that stopping mass gatherings had the same impact as full lockdown elsewhere. They made no allowance for non-mandated behavioural changes nor for the fact that not everyone was susceptible to that variant.

Neil Ferguson defended his dire prediction, in February 2021, by comparing Sweden to its neighbours, *"Sweden made one set of choices, Denmark and Norway another. The result is that Sweden has had fewer restrictions overall, but has had 3–4x the per capita death toll of its neighbours. [The UK's] death toll is higher still not because we over-reacted, but because we introduced measures too late last March, and then repeated the mistake last autumn. And because of factors which were just bad luck – the level of seeding last February and the new variant last November."* [472]

It was claimed that within Sweden the space available per person is much higher than in the UK, making it harder to spread disease. Looking at the country as a whole this was true. However, the large swathes of empty northern Sweden are irrelevant. What matters is the population density in the places where people do live. Sweden is in fact very comparable to the UK with Stockholm and London having almost identical levels of population density and 88 percent of Swedes living in urban areas compared to 84 percent in the UK. However, points made about differences in urban density, ethnicity and care home size were all missing the key point. Sweden was the only country to test the Ferguson hypothesis and it demonstrated how extremely inaccurate the predictions were.

Sweden's head epidemiologist, Dr Andres Tegnell's response to this accusation was to say that countries engaging in the novel[473] experiment of lockdowns should not be judged prematurely.[474] Where lockdowns were thought to have postponed deaths the populations would remain susceptible to having a later epidemic. Deaths caused by lockdown, for example from restricted access to healthcare for cancer patients, will take years to measure.

Tegnell was right about waiting. Overall Denmark, Norway, Finland, Latvia, Lithuania and Estonia all have higher total cumulative case numbers by the end of 2022 and the three Baltic states all have higher cumulative covid deaths per million too, compared to Sweden. The three Baltic states also all had higher excess mortality at the end of 2022.

The measure of deaths is imperfect and highly dependent on the culture of death certification in a country (see *Belief Four: Death certificates are never wrong*). According to the official data, of all the people who tested positive they were four times more likely to die in Sweden than in Denmark and Norway. It is highly likely that differences in the strictness with which covid deaths and cases are labelled is causing the discrepancy in mortality estimates. A strict criteria for calling a covid death is bound to result in a lower cumulative death total, even if everything else were identical.

For both deaths and cases, Sweden is not an outlier compared with neighbouring states when including the Baltic states. The real outliers are Norway and Finland. To understand why this might be, we need to look earlier to the winter that preceded covid. Deaths in Norway, Finland and Denmark had been higher than expected for the age of the population whereas Sweden had had a quiet year. When a country has a quiet year for winter deaths the following year is often a bad one because the frail, who were lucky to survive the previous winter, are more likely to succumb.

REVISITING THE LOCKDOWN MODELS IN LIGHT OF REAL WORLD EVIDENCE

The SAGE prediction of a June peak was clearly a mistake. May and June have seen a low in the Northern hemisphere for covid infections in 2020, 2021 and 2022. With hindsight it is clear that prediction was wrong and levels would have been low by June regardless of interventions. Modellers were very slow to admit a strong seasonal impact but were right about a second wave.

The belief in the covid tsunami did not falter despite a much smaller fraction of the population being infected than predicted. Instead, the modellers persuaded themselves that lockdown had pressed pause and, as noted in a Ferguson (et al.) paper first published in March 2020, continued believing that *"returning to pre-pandemic social contact patterns leads to rapid resurgence of the virus."* [475]

Ferguson's proposal at the time, to manage this was to repeatedly introduce measures and shut down education. He said, *"Given suppression policies may need to be maintained for many months, we examined the impact of an adaptive policy in which social distancing (plus school and university closure, if used) is only initiated after [the number of] ICU patients... exceeds a certain 'on' threshold, and is relaxed when ICU case incidence falls below a certain 'off' threshold."* [476]

In June 2020, the modellers still believed that lockdowns saved us from a tsunami sized peak, claiming that models showed lockdowns had saved 470,000 lives. [477] (Another way to reach that number would be to carry on believing that 510,000 deaths would have occurred without lockdown and just subtract the real life death curve). No major modification of their model had been made even after its rude introduction to the real world.

Bizarrely, on 5th March 2020, Chris Whitty admitted that they had already known for a couple of weeks that under 20 percent had been infected in Hubei, the province in China where Wuhan is situated. [478] He attributed this to two things, *"some of the people who are counted currently as not having had it actually have had it with no symptoms, and some of it is to do with the remarkable efforts of the Chinese state and people"* warning that if it was the latter, *"when they take their foot off the brake the epidemic will surge back again."* [479] Within a week or two most of Hubei had ended lockdown with no such rebound. The Wuhan lockdown ended on 8th April, before the UK was due to end its *"three weeks to flatten the curve"*. There was still no rebound. Despite this the official narrative remained that over 80 percent would be infected if nothing was done and that there would be a rebound if life returned rapidly to normal. How did this information not result in an urgent adjustment of the model?

The refusal to adjust the models led to the claim by Ferguson that *"had we introduced lockdown measures a week earlier, we would have reduced the final death toll by at least a half."* [480] This utterly unverifiable claim was seized on by a frightened public and irrational scientists to berate the government and demand zero covid policies moving forward. The final nail in the coffin of lockdowns came with the first Omicron wave. [481] The Netherlands entered a third lockdown. Other European countries such as Germany, France, Denmark, Scotland and Wales reintroduced strict measures including masking, working from home, closing businesses and restricting family gatherings at Christmas. There was no positive impact from these restrictions.

There has been a refusal to admit that covid has occurred with a seasonal trajectory, with one wave in each season. Similar to other seasonal respiratory viruses, such as the flu, the highest rates of hospitalizations and deaths occur in the winter months. Furthermore, only a fraction of the population is susceptible to each variant of the virus. In contrast the official narrative was a pandemic where a tsunami of infections has been repeatedly interrupted thanks to interventions or seasons. For early waves every seasonal rise was blamed on a new variant and every fall on public health interventions. The idea of a pandemic that can last perhaps as long as several years is a pretty novel concept. The whole reason to be concerned in a pandemic is because of the short period during which a disease passes through the population. A disease that takes years to pass through, like a new strain of influenza, requires a very different response.

DID LOCKDOWNS ACHIEVE ANYTHING?

A beautiful feature of the mathematics of an epidemic outbreak is that, because of the constant slowing of growth, the entire trajectory of the outbreak can be plotted from data gleaned from just the first few weeks. The slowing in growth is constant because a virus, uninhibited by human behaviour, will be restricted only by the number of remaining susceptible people in the population. It becomes increasingly difficult for it to find new hosts to spread to.

By extrapolating how the growth slowed using data measured at the beginning of the wave, prior to lockdown, Joel Smalley, a data analyst and member of HART, along with others, demonstrated what would have happened if growth had continued unabated and compared it to what did happen. What we see when we do this is that, at the very most, two thousand deaths were postponed by the spring 2020 lockdown. Regions where an argument could be made that lives were saved saw higher rates of lives lost only a few months later at the beginning of the next wave. These findings were confirmed by researchers at John Hopkins University who estimated a reduction in deaths of 0.2 percent.[482] Even then it is hard to argue that this was due to a change in behaviour as the timing does not fit. It seems more likely this was due to seasonal shifts bringing the first wave to an end.

In autumn, places that had been least affected in spring saw a resurgence. Covid reappeared all over but in many places death levels were the same as for previous years with other respiratory viruses. What was notable about the places where excess deaths occurred in autumn 2020 was their geographical remoteness. Joel Smalley, data analyst, noted that they included regions that were separated by geographical features. For example, North East Lincolnshire is far from any urban area, Havering in London is separated from the rest of the city by a forest and Gosport is separated from the city of Portsmouth by its harbour etc. For the latter two, it is hard to believe that people were not continuing to commute into the city from these regions in the lead up to lockdown. All these locations did have a spring wave of deaths. When the seasons changed and the conditions were ripe again, these regions were worse affected. Their geographical remoteness may well have played a part in sparing them from the higher dose of aerosol transmission affecting regions close to them in spring 2020.

Were we ever really doing anything to affect the transmission of a virus? Those with the means changed their routines, hid at home, avoided public transport. Those without, worked as hard as ever and kept the country running. If we look at the beginning of the epidemic, before lockdown, those in the most deprived cohorts were at greater risk of dying. The most deprived cohorts saw deaths in the first two weeks that were 73 percent higher than expected levels based on averages for previous years.[483] In contrast, for the least deprived, deaths were 56 percent higher. If hiding at home was having an impact then we would expect this gap to widen after lockdown had had an effect. Instead it fell. In the two weeks of deaths from 18th April, when lockdown should have had an effect, the most deprived had deaths 85 percent above expected levels compared to 81 percent in the least deprived.

WHAT ABOUT STOPPING OTHER INFECTIONS?

There are other illnesses that could act as a control for this particular unethical experiment. Gastroenteritis is primarily caused by infectious spread through the faeco-oral route (where poor toilet hygiene leads to passing virus or bacteria via direct touch or via a surface to someone's hand and then their mouth). Close contact is a prerequisite for spread. With lockdown, the attendances at the emergency department for cases of gastroenteritis plummeted. (It was not eliminated, but reduced to a half of normal levels). Attendances for gastroenteritis correlated perfectly with mobility data from mobile phone networks with the fall occurring before the official lockdown. The more contacts, the more spread. Lockdowns worked at reducing gastroenteritis. Or at least they seemed to. Unfortunately, trying to find an example of a disease to use as a control just creates confusion.

On closer inspection, the dip then recovery in the mobile phone mobility data also correlated perfectly with the total number of people attending the emergency department. The proportion of all those attending that had gastroenteritis remained the same. Was the apparent fall in gastroenteritis attendances just down to people not seeking care? Notifications to public health authorities of food poisoning cases also halved.[484] Can we be sure that everyone with food poisoning sought medical care in the way they usually would? Notifications for acute meningitis also halved and it would be hard to claim that people with acute meningitis could just man-up and ignore it.[485] Can we use acute meningitis as a control example to say lockdowns reduced close contact spread? Even that is challenging because the fall in acute meningitis has not recovered since, despite interactions returning to near normal levels. Could this be because people are still not accessing healthcare normally? Have levels really reduced or has something changed about how doctors are classifying meningitis? When it is so hard to find an example of lockdowns having reduced the incidence of an infectious disease, the question arises as to what lockdowns could ever achieve even for diseases where only close contact transmission is thought to be responsible for spread.

Claims have been made that lockdowns or masking were the reason influenza disappeared. This is demonstrably wrong. Influenza disappeared globally in spring 2020 including in places that did not lockdown like Japan, Sweden and Belarus. Similarly, influenza disappeared long before masking was introduced in most countries and disappeared from everywhere. When influenza returned at the end of 2021 it had developed a reciprocal relationship with covid waves. *SARS-CoV-2* waves would be split in two by an influenza wave and vice versa as if the two viruses were competing for dominance. The same was true for other viruses, the most notable of which is respiratory syncytial virus (RSV), a respiratory virus which can be lethal in children. In the past RSV had been a winter disease but in outcompeting *SARS-CoV-2* it shifted its seasonality and waves began much earlier in the year.

COULD LOCKDOWN EVER SAVE LIVES?

There are strong ethical principles which mean we should never have allowed the government overreach that led to lockdowns. There is separately the question of whether lockdown could ever, even hypothetically, be of any benefit. There is no rationale for lockdown for an airborne virus. What about for a very deadly disease spread through close contact transmission, like Ebola? First of all, Ebola spreads through exchange of bodily fluid with someone who is sick. However, for a hypothetical more deadly illness with a presymptomatic infectious period, there is still no rationale as people would adjust their behaviour for their own self-preservation. A government could (and did) easily scare people by advising them to stay at home without needing to legally enforce restrictions.

What about for a moderately deadly disease spread by close contact? Lockdown may have slowed the spread. Remember the SAGE and BBC pandemic models which predicted it would take 14 weeks for asymptomatic transmission to reach the furthest rural areas? If spread was already this slow, how much would we need to slow it and what would be gained by that? Remember, at best, as we saw for gastroenteritis, it could only reduce spread to about half of normal levels. It did not eliminate disease. In reality, such spread would be so slow that any change in behaviour would have to be sustained for many months to have any impact.

Every country demonstrated a massive reduction in close contacts before lockdowns demonstrating that people adapted without a legally enforced house arrest. People behave in a way that is appropriate to the perceived threat. Any hypothetical benefit from a reduction in close contact for any future disease would have to be balanced against the ethics of outlawing earning a living or holding the hands of the dying or even sitting on a park bench. The fact that our governments implemented lockdowns has meant they have crossed a Rubicon in terms of the social contract we have with the state and it may take many years to repair that damage and put in place laws that prevent such overreach in future.

ALTERNATIVE APPROACHES

Lockdowns were presented as inevitable but they were not. Dr Donald Henderson, who is widely credited with eradicating smallpox said in a co-authored article, in 2006, *"Experience has shown that communities faced with epidemics or other adverse events respond best and with the least anxiety when the normal social functioning of the community is least disrupted."*[486] Even the WHO said in 2006, *"forced isolation and quarantine are ineffective and impractical."* [487]

Mark Woolhouse, a Professor of Epidemiology on SAGE, reported that SAGE was never asked to model alternatives to lockdown and that ministers *"opined on minutiae such as the maximum queue for takeaways* while *"strategic questions were left unasked and unanswered."* [488]

Three eminent public health experts did propose an alternative strategy, Professors Sunetra Gupta from Oxford, Jay Bhattacharya from Stanford and Martin Kulldorff from Harvard. They carefully worded a moderate appeal in October 2020, stating that foundational principles of public health should be remembered and a strategy adopted that would minimise overall harm. At its heart was a plan to ensure the elderly and vulnerable were given some protection and they referred to this as *"focused protection."* [489] The three of them met in person at the American Institute for Economic Research, located in Great Barrington, Massachusetts. Their proposal was published as The Great Barrington Declaration and has approached one million signatories including 45 thousand doctors. [490] At the top of their website was a photograph of the three authors under a blue sky in front of the ninety-year old stone institute built to resemble a Cotswold manor and fittingly reminiscent of the distinguished institutions that they represented.

Not only would their proposal of focused protection have prevented much of the harm from lockdowns but it may have also prevented harm from covid. When society is functioning normally, the youngest people have the most social interactions and the elderly have fewer contacts. Had close contact transmission been the driver, then allowing viral spread amongst those with the strongest immune systems who were least at risk of complications would have been safer. The population as a whole would reach the peak at a point when enough people were immune to protect the whole community, while the elderly had had minimal exposure to that variant. Attempts were made to compare different levels of contact for different age groups assuming close contact transmission. These showed that the least exposure for those over sixty years old was seen when those under sixty years of age were free to mix such they they the elderly were not standing alongside the young on the frontline achieving immunity. [491]

Lockdown meant that everyone behaved like the elderly and hid. If the close contact model had been correct, this would have meant the elderly and vulnerable were on a par with the young and strong on the journey towards community immunity for that variant. If close contact had been the driver then the theory that spread could have been slowed would have been true. However, it would have been slowed while the most vulnerable were as exposed as the strongest.

The authors of the Great Barrington Declaration were attacked through a concerted smear campaign following its publication. Jeremy Farrar, head of the Wellcome Trust and a prominent member of SAGE said that Dominic Cummings, advisor to Boris Johnson *"wanted to run an aggressive press campaign against those behind the Great Barrington Declaration and others opposed to blanket Covid-19 restrictions."* [492] The campaign included articles that were arguably defamatory calling the authors *"COVID-19 deniers, conspiracy theorists, and grifters."* [493] The authors of the declaration responded to the accusation that they were proposing a *"herd immunity strategy"* saying, *"To characterise the Great Barrington Declaration as a 'herd-immunity strategy' is like describing a pilot's plan to land a plane as a 'gravity strategy'. The goal of a pilot is to land the plane safely while managing the force of gravity. The goal of any Covid pandemic plan should be to minimise disease mortality and the collateral harms from the plan itself, while managing the build-up of immunity in the population."* [494]

Given what we now know about aerosol transmission it is questionable whether a focused protection policy would have any benefit. For example, take care homes. Mass testing, restrictions on visiting and mini-lockdowns with every positive test result did not prevent waves in care homes. [495] Winter waves of covid cases and deaths in both 2020/21 and 2021/22 were of the same magnitude as the original spring 2020 wave. This policy failure has not been acknowledged and thus far there seems to be no end in sight to the policies to separate the frail elderly from their loved ones. Even if aerosol spread meant that focused protection might not have worked, it would have been an excellent compromise. It would have enabled politicians to have a policy while acknowledging people's fear of the disease and their belief that close contact transmission was key, all while minimising the harm from lockdown.

March is one of the times when we should expect seasonal cases to peak. In reality, the first peak of cases occurred in March 2020 in Western Europe. Cases peaked again in March 2021 in Eastern Europe. Cases peaked again, across Europe in March in 2022. For influenza, in England, there were March peaks in 2013, 2014 and 2016. The lockdowns coincided with this peak and fall. They did not cause it.

There are two reasons so many people continue to believe lockdowns save lives. The first is because of the false belief that covid is only spread through close contact. The second is that those in authority have failed to admit that the hugely destructive policy of lockdown had no benefit.

At the end of *Belief 1: Covid only spreads through close contact* we were left with the question of whether aerosol transmission drove a wave which was topped up by close contact spread or whether close contact spread was the main driver with cases topped up by aerosol transmission. The fact that changes in human behaviour failed to impact on the trajectory is evidence that aerosol transmission was indeed the driver of waves. Exposure was widespread and almost inevitable but not everyone was susceptible.

If close contact transmission through asymptomatic spread was the driver as we were told, lockdowns would have worked. The fact they did not is further evidence that aerosol transmission combined with meaningless positive results created the illusion of asymptomatic spread.

THE EVIDENCE MANIPULATION TRIAD	
EXTRAPOLATE	Modelling was used to extrapolate wild estimates of potential death tolls based on false assumptions. Evidence of an apparent success of lockdown policies in Eastern Europe or South East Asia was extrapolated to claim that lockdowns could all but extinguish covid if carried out early and fast enough. This was despite this clearly being a geographical effect.
EXCUSE	The pro-lockdown hypothesis that without lockdown infections would rebound was disproved in summer 2020. However, the excuse used was that covid could not spread outdoors so the marches and packed beaches did not count as a risk for rebound. Swedish successes were excused on the basis that the population underwent a voluntary behavioural change (while ignoring the fact that other countries would have had voluntary behaviour change in the absence of lockdown).
EXCLUDE	The body of work demonstrating that lockdowns have not impacted on Covid's viral trajectory has been ignored by public health officials and the media.

TOP THREE MYTHS
1. Reducing contacts would reduce spread
2. Viral waves only peak with a change in human behaviour
3. Predicted death estimates modelled on extreme assumptions were realistic

15

SUNK COST FALLACY

No matter how far you've gone down the wrong road, turn back. **Turkish proverb**

When we have invested money in something it is easy to 'throw good money after bad'. Say you had an unreliable car and had spent hundreds or thousands on car repairs. Should you stay loyal to your disappointing wreck or cut your losses and buy something else? How much would another car cost? Did spending so much on the old car earn it some loyalty? Economists argue that, to make a good decision, the money already spent should not be considered part of the equation because the spent money cannot be recovered. Rationally, the car has shown itself to be unreliable and any money that would otherwise be spent on future repairs should be spent instead on something new. The refusal to do so because of all that has been spent is a sunk cost fallacy. A sunk cost fallacy comes from an inability to admit to error.

The costs that cannot be recovered are not necessarily financial. The same can be true of investments of our time, our commitment and our emotions. Just because we have devoted time, sacrificed our fun, our experiences, even our relationships and spent hours feeling distressed about the risks does not mean that our assessment of the situation should be influenced by what went before. The economists argue that only future costs and benefits should feature in how we decide what to do next. It is easier said than done. The more we have invested the harder it is to face reality. If we were sold a duff deal then we need to give up on it and get a new one.

Accepting that all that we've been through has been for nothing is painful. It requires admitting we were wrong, our trusted sources were wrong and that all our sacrifices were in vain. To do this requires humility. Stubbornness and loyalty to the group make it harder. Accepting that lockdown failed also removes hope, for those still afraid, that there is a strategy we can rely on in future. It is so much easier to just keep believing lockdowns worked and thousands of lives were saved as a result making them worthwhile. Unfortunately, no amount of sunk cost can make a wrong belief right.

Imagine a world in which Boris Johnson left hospital after three weeks of lockdown and promptly ended the lockdown as promised. *"We tried our best, it did not make much difference. What is important now is that we minimise the harm caused and make sure it never happens again."* Instead, the extended lockdown and the second and third lockdowns meant that policy makers were more and more committed to the idea. Massive sacrifices were made for lockdowns. The thought that it may all have been for nothing prevents us opening our minds to the truth that that is the case. It is really important that we do that though because otherwise we justify doing it again and causing more harm.

Imagine a world where £15 billion was *not* spent on protective equipment.[496] It was spent at prices inflated to six times the 2019 rates, often with contracts that benefited friends of the powerful.

It was a wasteful error and in 2022, £9 billion of it was written off and £700,000 a day was being spent on storing the rest.[497] How much easier would it have been to admit that none of the equipment made any difference to infection rates? Gowns, gloves, visors and masks do not stop an airborne virus. Antibody levels among healthcare workers were in line with young city dwellers and rose for each variant. Those who were susceptible were infected whether at work or in the community.

Imagine a world where the sunk cost of £13.5 billion spent on test and trace in its first year did not influence plans for spending a further £14 billion the following year. The UK Public Accounts Committee described the programme as *"muddled, overstated, eye-wateringly expensive."*[498] Sir Nicholas Macpherson, former Permanent Secretary to the Treasury said it was *"the most wasteful and inept public spending programme of all time."*[499] Mass testing was an expensive error which made no impact on the trajectory of infections.

Imagine a world where over purchasing of vaccines did not lead to pressure to find any arm, no matter how young, to inject. By spring 2022, only 80 percent of the UK stockpile had been used with an expectation that £4 billion would be written off.[500]

In total the UK spent £400 billion[501] on furlough payments, £30 billion on test and trace, £26 billion[502] thought not to be recoverable from business loans and £15 billion[503] on protective equipment. It is hard to see how they could have spent more money with the public's consent. Having had this spending binge in the first year it was hard not to continue it into successive years rather than admit how wasteful it had been.

Imagine a world where covid did not happen until Omicron arrived. For a variety of reasons Omicron was around half as severe and deadly as the variants that had gone before. On that basis it was no different to a mild influenza and there were fewer deaths over the winter than would have been expected in an average season. It was the sunk cost of all that had been sacrificed in the name of covid that meant that Omicron was not simply ignored. Parts of Europe and some states did ignore it with public health advice to stay home if you were sick and not to test. While in other parts of the USA they continued to mask toddlers and school children. The federal government prevented the unvaccinated entering the country, with continued quarantine on arrival and the military and students faced continuing vaccine mandates.

Imagine a world where fear had not been used to increase compliance. The fear created was, at best, like a sunk cost, a harmful mistake. For those in power to try and reverse the fear with reassurance meant not only admitting to the error and the harm caused by the error but also letting go of the power and compliance which the fear provided. There has been a telling lack of reassurance from those in positions of authority, even after years.

Sunk cost fallacy has led to mistakes being repeated and magnified. There may yet be more harm due to sunk cost fallacy if there is a refusal to admit to these errors.

BELIEF NINE: LOCKDOWNS ARE NOT HARMFUL

It's a communist one party state, we said. We couldn't get away with it in Europe, we thought… And then Italy did it. And we realised we could. **Neil Ferguson, December 2020**[504]

What could have been done to spend more money, ruin more businesses, damage more health, disrupt more education? It's hard to think of a strategy that could have been more harmful and still have had the cooperation of the population required to make it happen.

There was total denial of the harm lockdown caused. The lifting of restrictions was described, as late as March 2021, as *"neanderthal thinking"* by President Biden[505] and the removal of restrictions was referred to as a *"dangerous and unethical experiment"* in the *Lancet*.[506] Yet, the destructive and truly experimental policy of lockdown was treated as a default. Lockdowns were not an inevitability – far from it – the pre-covid pandemic plans made no mention of it.

The harmful impact of lockdown was easier to predict than the impact it would have on the virus. Could infections be slowed by behavioural changes? Would any impact prevent or merely delay infections? If so, by how much? Even if lockdowns could prevent harm from the virus, it was outweighed by the predictable certain harm that lockdown would cause. Sunetra Gupta, Professor of Epidemiology at Oxford and one of the three authors of the Great Barrington Declaration, said, *"We inverted the precautionary*

*principle of trying to minimise harm by doing the one thing that we
knew would cause harm: lockdowns."* [507]

Government estimates for the damage caused by the economic and
healthcare impacts of lockdown were published in April 2020. [508]
It was estimated that a 75 percent reduction in non-emergency
healthcare alone would result in 185,000 deaths. That means the
government were prepared to enact policies that could result in the
death of 185,000 people because this number was lower than the
hypothetical hundreds of thousands of lives that Neil Ferguson
claimed lockdown could save. The alternative presented to them
was that no lockdown would result in all non-emergency care
being cancelled and, as well as the 520,000 deaths from covid, a
further one million lives lost to covid due entirely to lack of
treatment. By the time of the second lockdown the errors in these
assumptions were clear and yet more lockdowns followed.

A second report in July 2020 revised this 185,000 figure down on
the optimistic assumption that treatment would be postponed
rather than cancelled because the NHS would find capacity to
work through the backlog post-lockdown and catch up. [509] This
further report estimated a *reduction* in deaths from dementia,
cardiovascular disease, alcohol and drug misuse. It was a bizarre
conclusion. Even at the time the report was written there had
already been a significant increase of such excess deaths (above the
average for preceding years) in all these categories. Deaths from
cardiovascular disease in the absence of covid rose in lockdown
because patients were scared to go to hospital and emergency staff
were isolating. The NHS estimates there will be 25,000 excess
deaths just from alcohol related to lockdown. [510] They gave no
reason to justify their predicted reduction in dementia deaths.
Oddly, they did quote social care staff commenting on isolation
leading to deterioration, *"We are seeing them going downhill without*

visits from their families". This deterioration may have contributed to these deaths.

POLICY INDUCED DEATHS

There is no doubt that policies will have caused death. There was a group belief that disaster from an overwhelmed healthcare service was imminent and that difficult decisions had to be taken to prepare for and prevent that. Those decisions were difficult because they were going to result in harm to others. To prepare for the expected pressures, patients were discharged from hospital to care homes or even hotels. The criteria used to decide who would be offered an intensive care bed were made stricter. Intensive care beds and other beds where monitoring was possible were 'freed up'.[511] Hospital staff, including ambulance and emergency department staff were made to isolate for two weeks if they had a fever or a cough. General Practitioners wrote to patients with certain conditions saying they would not be taken by an ambulance or admitted to hospital during the crisis.[512] Do not resuscitate orders were expanded and a communication error led to otherwise healthy people with learning difficulties being included.[513] In King's College London Hospital 85 percent of covid deaths up to November 2020 had a do not resuscitate order compared to only 32 percent in Barts hospital in London.

The investigative team of *The Sunday Times, Insight*, carried out an investigation into deaths in the elderly during spring 2020.[514] In it they reported that Mark Griffiths, a Professor of Critical Care Medicine at Imperial College London created a document *"Covid-19 triage score: Sum of 3 domains."*[515] It suggested patients would be scored based on age, frailty and underlying conditions. Anyone over the age of 80 would score too high to be treated and anyone aged over 75 years with any frailty score would too. Even people over the age of 60 would be denied care if they were frail and had underlying health conditions. After discussion the document was rewritten with adjustments to the scoring, given an NHS logo and presented to ministers on 28th March 2020. A version of the document was uploaded onto NHS Highland's internal internet. The intention was that such triage should be used only when capacity was at risk of being breached but the chair of the group who wrote the original document said it had been distributed to doctors and hospitals as part of the consultation process and *"we were aware that some of them were looking at that tool... Some of them were using it."* [516]

A survey of care home staff included sixteen respondents who reported blanket do not resuscitate orders and that these led to hospital admission being refused. The report by the Queen's Nursing Institute said, there were *"blanket DNACPR [do not attempt cardio-pulmonary resuscitation] instructions from the GP or the CCG [Clinical Commissioning Groups – who contract for care] or hospitals putting DNACPR in place without discussion with the resident, family or care home. Also, hospitals refusing to admit patients who had DNACPR or blanket "no admission' policy."* [517] It quoted one respondent who said, *"We were advised to have them in place for all residents. We acted in accordance with medical advice and resident wishes, not as advised by a directive to put in place for all by a CCG representative. We challenged this as unethical."*[518] Another respondent said the decisions were, *"Put in place without family consent by Trust staff, no consultation with staff in home."* [519]

On 3rd April, NHS England had to write a letter to GPs to clarify that *"treatment decisions should not be made on the basis of the presence of learning disability and/or autism alone."* [520] A further letter was written by NHS England on 7th April emphasising that *"each person is an individual whose needs and preferences must be taken account of individually. By contrast blanket policies are inappropriate whether due to medical condition, disability, or age."* [521] As late as 30th April, Matt Hancock said at a press briefing, *"And we're making crystal-clear that it is unacceptable for advanced care plans, including 'do not attempt to resuscitate' orders, to be applied in a blanket fashion to any group of people. This must always be a personalised process, as it always has been."* [522]

Of those who died of covid before 4th April 2020, 90 percent had not had access to intensive care and those that did were far healthier than the usual cohort of patients admitted to intensive care with viral pneumonias. It was during these weeks that half of excess deaths in care homes were from non-covid causes.

Ventilators were overrused, partly to protect staff and other patients from the risk of infection (for the greater good), and in retrospect we know these covid patients had a higher mortality. [523] Stress has a negative impact on health, including through suppression of immunity and the high levels of fear induced will have contributed to stress. Each of these factors will have raised mortality. The uncertainty is in how many deaths were caused by policy.

There is a notable mismatch in the timing of people experiencing symptoms when covid first reached Europe, in Lombardy. Cases rose from the beginning of February and were falling in Lombardy by 20th February. [524] However, the excess death spike did not peak until the end of March and the excess was very localised geographically. There were 24,000 excess deaths in Lombardy but only 14,000 were thought to be due to covid. [525] What proportion of excess deaths were non-covid or covid is likely different for different locations.

I can still recall the emotional response I had to each ambulance siren during the initial lockdown. Only, now when I look back, I realise that not all of those patients were covid patients. People were not accessing healthcare when they needed it. Before lockdown, there was a rise in both ambulance calls for chest pain and for cardiac arrests. These were likely predominantly covid patients. However, after lockdown the chest pain calls plummeted to two thirds of normal levels while the arrest calls climbed higher still.[526] It is hard not to believe that some of those sirens were for people with cardiac disease who could have been saved if they had only called with chest pain before they experienced an arrest. The mortality rate for cardiac arrest calls is around 90 percent or more. How many non-covid patients died having not accessed clinical care early enough?

Patients themselves did not attend as usual even with life threatening conditions. In England, attendances at the emergency department fell from the beginning of March dropping to half of normal levels in lockdown.[527] Some may have been afraid of attending. Others were told they should isolate rather than attend, because they had a cough or a fever. Pathologists in Oxford pointed out this resulted in deaths because no treatment was given for life threatening diabetic complications or heart attacks.[528] A study from Milan showed that even for life threatening conditions that needed an urgent operation the numbers fell to less than half of 2019 levels.[529]

Looking at those deaths in excess of normal levels in spring 2020, more than a third were attributed to causes other than covid in England and also in the USA overall. In Texas and California more than half of deaths above normal levels were attributed to causes other than covid.[530]

New York City had dramatically higher numbers of deaths from all causes, than other parts of the USA. Neighbouring states including Vermont, New Hampshire, and Maine had deaths that were close to or at expected levels. What was different about New York City? As well as the footage of mass burials on Hart Island, there was a hospital ship with 1,000 beds moored next to Manhattan to treat the sick and fourteen tents erected in Central Park as a 'field hospital'. A reduction in the working hours of funeral directors had contributed to a backlog of corpses and the need for 45 mobile morgues to be set up.[531] The impact in terms of fear was demonstrable. Whereas there was a doubling in deaths from certain non-covid causes in the USA, there was a nearly four-fold increase in deaths without covid involving diabetes and heart disease in New York City.

One community hospital, Elmhurst in Queens, bore the brunt.[532] A quarter of the city's covid hospital admissions in spring 2020 were in this one hospital. Elmhurst served a community living in densely packed, often crowded accommodation with a high proportion of Hispanic people who were at greater risk. In the meantime, certain hospitals in Manhattan were relatively quiet. The public were staying home to flatten a curve while no-one was flattening the discrepancy in patient numbers between hospitals. City officials did intervene to move patients to some of the 3,500 empty beds elsewhere but not until the end of March 2020.

New York had an additional problem. In April 2020, it was reported that 97 percent of over 65 year olds who had been put on ventilators had died.[533] For all ages the percentage was nearer 80 percent which is around twice the expected death rate for people who have respiratory distress from other conditions. Doctors began to question whether the ventilation itself may be causing more harm than good and tried using other strategies such as nursing the patient face down and using high pressure oxygen. These techniques saved lives and after months of doctors persistently trying to share this message, they were widely adopted. In retrospect people questioned whether some of those who died from ventilation should ever have been started on it. There were two main perverse incentives. Hospitals in the USA were given extra money if their covid patients were ventilated.[534] This was justified in terms of covering the cost of intensive care for these patients but the amount of money provided far exceeded that given for patients ventilated for other diagnoses. Secondly, there was the belief that having the patient breathe through an enclosed system was keeping the staff safe. One doctor told the Wall Street Journal, *"We were intubating sick patients very early. Not for the patients' benefit, but in order to control the epidemic… That felt awful."* [535]

New York was bad but not as bad as Peru which had one of the strictest lockdowns in the world. Four in ten people in Peru have no refrigerator let alone internet access and three quarters have only informal employment. Leaving home for fresh food and work was essential for these people. During the lockdown, the military patrolled the streets and only essential workers who had completed an online form were allowed out. The police even used tear gas to stop people leaving the cities. Peru had the highest excess deaths globally. The reasons for this will be complex but Mateo Prochazka, a Peruvian epidemiologist attributed this to people suffering from other illnesses who could not be treated, *"If you add additional layers of stress to a health system that was already on the brink of collapse, you'll have collapse."* [536]

FRAGILITY OF HEALTH AND CARE SECTORS

There have been previous instances where patients have made detrimental decisions leading to them not attending hospital when they needed to and where a fragile care and health service led to high deaths. A severe heatwave in France in 2003 resulted in a sudden spike of more than fifteen thousand deaths. The sequence of events has some parallels with lockdown. Doctors declared that hospitals, which had reduced their bed numbers for the summer, were overwhelmed on 8th August.[537] Three days later, *Le Parisien* had a headline saying, *"Heat wave now a national tragedy."* The majority of deaths were due to heat related factors including dehydration, heat stroke and hyperthermia. However, there were indicators that other factors may have contributed to the excess. There was a significant rise in deaths among the middle aged[538] as well as the elderly.[539] Even areas that only had excess heat for two days saw a 50 percent increase in mortality. Causes of death unrelated to heat including tumours, traffic accidents and suicide also rose. A primary problem was that *"many victims did not call"* the emergency services.[540]

The elderly were particularly at risk during the heatwave with one fifth of deaths being in care homes. One retirement home demonstrated that deaths from heat had not been inevitable.[541] With minimal interventions they had only one death in the home despite the region where it was situated experiencing a 171 percent rise in deaths. A failure of care can lead to excess mortality especially among our frail populations.

With covid, care home staff were put into impossibly hard situations. Firstly, they were forced to admit new residents from the hospital, some of whom had covid. It takes more work to settle a new patient into care than to care for established residents, even at the best of times. Secondly, understaffing was a serious issue with colleagues having to isolate and visitors, who normally help, absent. Thirdly, staff had minimal protective equipment but also had a belief that the virus was a serious threat to them personally and that the equipment would protect them. The result for some residents was that contact was massively reduced while equipment was eked out. Critically, elderly patients with dementia need encouragement to drink little and often. A large part of a carer's role is to help residents stay hydrated and that requires popping in for brief visits as often as possible. A combination of equipment issues and fear meant that the frequency with which residents could be offered drinks was much diminished and may well have led to the demise of a number of care home residents. Prior to June 2020, there were as many extra deaths of care home residents attributed to non-covid causes on death certificates as to covid. It is hard not to conclude that understaffing, scared staff with consequent lack of hydration, a lack of in-person care from doctors, blanket do not resuscitate orders and a lack of visitors able to identify issues before they escalated all contributed to a genuine rise in non-covid mortality for care home residents.

The impact of patients not accessing healthcare; the interventions in hospitals and the problems in care homes will have inevitably resulted in extra deaths. Not all excess deaths were directly due to a virus and not all deaths due to the virus were inevitable (this is explored further in *Spiked: A shot in the covid dark*).

People denying aerosol transmission and a seasonal trigger have a dilemma. If you believe the Imperial model then there needed to be an explanation for there having been twice as many deaths in spring 2020 as was predicted to happen after implementing lockdown. The reason given by Neil Ferguson was, *"The higher level of seeding of infection in the country meant that, effectively, at the end of March... there were a lot more infected people around than we had anticipated."* [542] He was claiming that his model was right but the rise in cases had begun earlier than he had accounted for. If this was the case where were the excess deaths associated with covid prior to lockdown? Prior to lockdown deaths were at levels below what would be expected based on previous year's averages.

His model would imply that the excess seen on lockdown was almost entirely due to the lockdown policy not the disease.

However, if the long distance aerosol transmission model is used then a remarkably sudden surge is possible because spread can occur at a faster rate than close contact spread would allow. Also, a separate team from Imperial collected blood samples and showed when people who developed antibodies by February 2021 had had their symptoms.[543] A fraction had covid from autumn 2019 but there was a clear and dramatic surge in symptoms that occurred in February. Influenza had remained the dominant virus before then based on random testing. These facts support a model of aerosol transmission with a seasonal trigger causing a surge in susceptibility in February 2020. With this model no further explanation is needed for the lack of excess deaths prior to March 2020. Yes, policy-induced deaths did occur and they may have been significant. However, by adopting this model a reasonable proportion of the sudden surge in excess deaths in March can be attributed to covid.

WAS THE CURE WORSE THAN THE DISEASE?

The talk of balancing risk and benefit on a population level has been incessant. Some claim that so long as the lives lost to lockdown balance the modelled estimate of lives saved from covid then lockdown was, for them, morally justifiable. Setting aside the point that lives should not be traded, even if they were real lives saved, the balance would not have been even. When measuring life lost these people fail to account for years of life lost or quality of life.

There is a totally different way of calculating the effect of lockdown on quality of life.[544] Ask the question, how many months of life would you trade to avoid lockdown? Given the choice would you choose to live 12 months like the ones we had including lockdowns or would you prefer a shorter period of normal life? Some people may have enjoyed aspects of lockdown and would not have given it up. Others have had a horrendous time and would give up a year or more of their life expectancy rather than have to live through their experience of it. For most of us the figure would be measured in months. It might be that asking this question now would result in a different response as people have forgotten the pain they went through. Anecdotally most people would trade at least two months of life. The result would be living only 10 months of that year but without lockdown. Adding all these months up gives 13.4 million years of life that the UK population would be prepared to trade to avoid lockdown. That would be equivalent to saving 1.34 million lives of people who had an average of 10 years of life left to live. That number far exceeds even the most fantastical claims of Neil Ferguson for the number who were at risk.

Two analyses, using different methodology, both concluded that in the first year alone, lockdowns had led to the deaths of around 100,000 US citizens.[545] If later restrictions and the longer term impacts were included, the death toll would be higher still. It is hard to see evidence of deaths having been prevented, when the USA saw deaths peak at similar levels regardless of restrictions each winter at the end of January and a second peak at a lower level in August or September.

IMPACTS

The immediate impact on access to healthcare was only the beginning of the negative consequences of lockdown. Parents with cancer were told to shield from covid while their chemotherapy was stopped. A number of young children lost mothers or fathers after they had been seemingly abandoned in this way by the NHS.

Cancer services from screening for diagnosis, to referrals from GPs and surgery were all postponed. Deaths of people on NHS waiting lists doubled.[546] Globally, deaths from malaria, TB and AIDS all also rose significantly.

The World Bank estimates that the children impacted by school closures will miss out on a total of $17 trillion worth of lifetime earnings, equivalent to an entire year's worth of EU GDP. Mark Woolhouse, epidemiologist on SAGE points out, *"The arguments for closing schools were that the children were at some risk, the staff were at high risk and that schools would drive community transmission... None of these concerns were supported by the epidemiological data."* [547]

As well as school closures children were impacted in many other ways. There were delays in children coming forward for care particularly cancer and child protection referrals.[548] Declines were seen in uptake for existing childhood vaccination programmes. Children saw a marked increase in short-sightedness because of e-learning and lack of sunlight exposure.[549] Chris Whitty commented, *"During the last two years obesity, particularly in children, has got significantly worse, and the reasons for this, I think, are complex"* [550] going on to highlight the fact that the trend affected every socioeconomic group. His policy to lock children in their homes was perhaps too simple an explanation for him.

UNICEF estimated an increase in preschool aged deaths of over a million in just 6 months.[551] Certain countries saw a more than 50 percent rise in teenage pregnancies, which according to a US think tank was *"due to reduced access to contraceptives, increased poverty, and a spike in sexual abuse."* [552]

Child mental health services were inundated by an epidemic of Tourette-like tic conditions especially in teenage girls. Dr Alasdair

Parker, the president of the British Paediatric Neurology Association said, *"The most severe tic disorders I have seen over the last 20 years have all presented in the last five months to my practice."* [553] There were unusual features to this epidemic which caused psychiatrists to search for a new cause and, somehow, they managed to all point the finger at TikTok videos.[554] However, as one review of animal models for tic conditions said, *"motor and behavioural sequences that are repeated purposelessly… are typically exhibited by most captive animals kept in spatial restriction (which interferes with the expression of behavioural needs)."* [555] If you want to give an animal tics you can put it into lockdown. This fact was overlooked. In 2022, John Stone, neurology professor at Edinburgh University attributed the problem to lockdown saying, *"teenagers have had a horrendously abnormal time… stuck in their bedrooms, away from their friends."* [556]

Loneliness, anxiety, depression all increased. There were excess deaths in the young which the FDA attributed[557] to *"despair"* while the Canadian Stats authority said they were due to *"indirect consequences of the pandemic, which could include increases in mortality due to overdoses."* [558]

ECONOMICS

A blinkered strategy of trying to minimise covid deaths led to a disregard for deaths and harm from non-covid causes. Much of the harm including premature death will be from the longer term economic impact of lockdown. For example, being homeless knocks 30 years off a person's life expectancy.

Globally, in mid-2022 the number[559] of people on the brink of starvation was five times higher than the number[560] when Band Aid begged us to *"feed the world,"* in the 1980s.[561]

India saw the largest internal migration in human history in spring 2020,[562] as tens of millions of people who had migrated to cities and were employed on a day to day basis lost their work and often their workplace accommodation. They walked for days, hundreds and sometimes thousands of miles under constant threat of police brutality for not obeying lockdown rules to stay at a home they did not have. Some were still stranded and starving over a year later.[563]

Support packages in the UK may have limited the short term economic harm but the inflationary impact of massive spending was only just starting to be felt from autumn 2022.

HUMANITY

The disregard for harm caused by policy was not just from a blinkered approach. The impacts above were predictable but numerous people lost their humanity.

Children who were infected were put into solitary confinement by their own families and this was recommended in Canadian government guidance.[564] I was told of a child put into solitary confinement at boarding school shortly after learning that their mother was dying of cancer and of children forced to isolate for two weeks after being removed from their families into foster care. Those children in care were not granted the usual access to their biological parents and siblings. Some of those whose very role was to care for children seemed to have lost some of their humanity to a culture of rule following. Mary Bousted, co-leader of the National Education Union described young children who were *"mucky, who spread germs, who touch everything, who cry, who wipe their snot on your trousers or on your dress."* [565]

Schools were repeatedly referred to in the media as petri dishes. Children were forced to mask without any consideration of the effect that would have on their ability to communicate and learn, with some making no exception for children with special needs or the very young. Children were considered an amorphous threat to adults. Adults disregarded their duty to create a society where children can trust authorities to act rationally and in their best interest, ensuring their basic needs are met. Respect for children as individuals with unique needs was missing entirely. This lack of respect extended beyond children to young adults.

Students at Leeds and Nottingham University discovered one evening their fire doors had been locked with cable ties.[566] The University of York advised that, in the event of fire, students who were isolating should wait behind until all those not isolating had exited the building.[567] Firefighters were outraged after the Scottish first minister, Nicola Sturgeon said cutting the bottom off some school doors to improve ventilation was *"common sense."* [568]

Finn Kitson, just 19 years of age, committed suicide three weeks into his degree at Manchester University, while 12 days into a two week isolation period. He did not have covid but had been a contact. He was not the only one. A survey in June 2020 revealed that a massive 25 percent of 18 to 24 year olds in USA had contemplated suicide in the preceding thirty days.[569] Despite this tragic failing, in November 2020, students from Manchester University awoke to find they had been barricaded into their halls of residence. The students tore down the barriers and the university relented. These stories get forgotten when people relate how the Chinese welded people into an apartment block in Wuhan in spring 2020 and how people died in Xinjiang in November 2022, after being locked in their apartments when a fire broke out, saying that could never happen in the West.

Instructions were given to 'protect' the terminally ill by letting them die alone. The elderly were deprived of days out, sunlight, fresh air, family, touch and love – all that makes life worth living. The impact on loneliness and the rapid deterioration in dementia was catastrophic.[570] Two sons were even stopped from offering comfort to their elderly mother at their father's funeral[571] and, as late as April 2022, the queen sat alone at her husband's funeral.

A meta-analysis from John Hopkins University of the effects of covid lockdowns concluded, *"Lockdowns during the initial phase of the COVID-19 pandemic have had devastating effects. They have contributed to reducing economic activity, raising unemployment, reducing schooling, causing political unrest, contributing to domestic violence, loss of life quality, and the undermining of liberal democracy. These costs to society must be compared to the benefits of lockdowns, which our meta-analysis has shown are little to none."* [572]

The harm caused extended into many aspects of life from broader health issues and economic impacts to devastating impacts on children. As if there was not enough damage, it has also altered how people treat each other, and the relationship between people and the state. Not only did the cure not treat the disease but it was worse than the disease. After all this was known, and the evidence of harm was unequivocal, governments still locked down a second time in the first covid winter. Worse, even with Omicron and even while their neighbours remained free, people complied with a further lockdown in the Netherlands in winter 2021/2022. Even now many people in positions of power believe that lockdowns worked and might be needed again.

TOP THREE MYTHS
1. All excess deaths were due to covid
2. Lockdown harms were not predictable
3. Lockdowns were worth it

16

THE OPPOSITE OF KEEPING YOU SAFE

Those who would give up essential liberty, to purchase a little temporary safety, deserve neither liberty nor safety. **Benjamin Franklin, 1755**[573]

Biology has primed us to avoid danger. Our emotional inner beings flee it automatically. We have an inner homing device that seeks safety. There is an attractive notion of security that comes from living a peaceful, protected, predictable life. While it's fine to yearn for safety, we have to be honest about what that means. I too want to feel security for both me and my children. While I want to protect my children from any harm, I realise that they must have adventure, challenge and risk in order to live a full life. My job is to prepare them to take personal responsibility as they enter the adult world, not to protect them from it. Restrictions in the name of keeping you safe risk preventing people making their own decisions and taking their own risks. The opposite of keeping you safe might be freedom and personal responsibility.

The cultivation of safety culture by the government has been long in the making. It is fair to say that many in the general public also welcome this shift and to a certain extent the government is responding to that sentiment. The laws and regulations to improve on road safety and safety in the workplace and in the home have been building for decades. There was always a hint of public health type prevention about how these laws were presented. Indeed, Public Health England listed *"keeping people safe"* before *"preventing poor health"* when describing their purpose.[574] Taken to an extreme we might have speed-restricted cars, semi-contact or no contact sports, knives with blunted tips, or even restrictions on how hot kitchen appliances can become. Why not make the elderly wear inflatable suits to prevent hip fractures, even if it makes them less agile and causes them to stop going out? How much of safety-ism is a result of the slide towards ever older more cautious politicians, an increasing proportion of whom are over sixty, leaving the voices of the young unheard?

Every authoritarian regime of the past has promised increased safety as it has increased its power. Covid meant that governments kept people in their homes, prevented people earning a living or sitting on a park bench and even stopped the Welsh buying birthday cards. People were not made safer as a result but even if they had been, this was massive overreach by governments. There is no justification for such a power grab. The argument that legally enforced restrictions were necessary while we did not know about the severity of covid falls down on two counts. Firstly, we did know the severity and even the worst estimates were not nearly as bad as the viruses for which pandemic plans recommended minimal intervention. Secondly, had the virus been a severe threat, like airborne Novichok, people would have adjusted their behaviour of their own accord. In any case the risk of spread from apparently healthy people would always be tiny.

Professor Frank Furedi, an emeritus professor of sociology at the University of Kent and author of *How Fear Works*, speaks passionately about what happens when freedom is reduced in the name of safety, *"Rather than making us safe or making us feel safe, our loss of freedom, which is an indirect way of losing our agency or power as human beings, actually has made us far more insecure and anxious. And I think what human history has shown us over the centuries is that every time we give up a bit of our freedom for the illusion of safety, we don't become safer. But if anything, we become even more obsessed about our safety and it almost becomes a self-fulfilling prophecy that the more you're thinking about your safety, the more you organise your life around being safe – the less safe you are."* [575]

Furedi, rightly points out the danger in criticising the harms of lockdowns in terms of deleterious health outcomes. Arguing that lockdowns should not have happened because they impacted on healthcare for non-covid conditions or because of the impact on mental health implies that our end goal should always be optimum health. The core of the problem with lockdown was that it was unnecessary and handed enormous power to a select few. The core issue was the loss of freedom.

Gregory Keating, professor of law and philosophy, wrote in 2018 about how hard it can be to balance harms against benefits in the real world, or as was often the case with covid one set of harms against another set. *"Precautions may be fair or unfair as well as efficient or inefficient; they may respect or disrespect people's rights; they may enable or disable desirable forms of choice; they may be sensitive or insensitive to the distinctive values realized by some activity."* [576]

Keating gives the example of soldiers risking their lives on the battlefield to rescue the bodies of their dead comrades. Their motivation for doing so lies in their values. They decide to honour and show solidarity to the fallen. There may also be a desire to save the families from the uncertainty of having no body to bury. How can a monetary value be attached to those things? Keeping the soldiers safe would mean undermining their values and their freedom. There are no numbers that can be put into a spreadsheet to represent these decisions.

Cost benefit equations that are used to justify restrictions in the name of safety end up being in tension with our values as a society. Each harm, so long as it is considered, can be assigned a monetary value to be entered into a spreadsheet but no such number exists for human rights or societal values – what price should be put on restricting funerals for example?

As a way of deciding what harm is permissible, cost benefit equations necessarily erase the distinction between individuals. As we only have one life, it is fair to object to other peoples' lives being traded against our own. Similarly, trading harm to one group to prevent harm to another is not morally justified even if the numbers in a spreadsheet are balanced.

Trying to weight the importance of different factors and arrive at an answer that is rational and based on evidence is laudable. However, this method will always fail when unmeasurable outcomes, incommensurable values or unexpected outcomes are underestimated. It is why the precautionary principle is so important (see *The precautionary principle*). Pretending that everything of value in our society can be reduced to numbers in a spreadsheet is to deny our worth as more than parts of a biological entity. It is to deny our individuality, our values and our humanity. The numbers assigned to the benefits of safety have thus far risked trumping proper consideration of the resultant loss of personal responsibility and freedom.

Safety-ism results in a blurring of the lines between it being wrong to cause others direct harm, which we can agree with and it being wrong to not prevent indirect harm to others, which is a different issue. Almost everything we do can be presented as having an indirectly harmful effect from driving a car, speaking our minds and finding that someone somewhere found our words offensive, to buying food and thus adding to the demand for food that keeps it unaffordable for some. While causing direct harm is wrong, life cannot be lived without an acceptance that our actions might affect others negatively.

My short stay in New York demonstrated that the Land of the Free had become the Land of the Safe. The children were kept particularly safe, a lovely policeman was the first face we would see on entering the school, sun cream was essential but could not be applied by a teacher, Kinder eggs were banned and random people would scream at my children if they were on a bike in the park without a helmet on. To comply with Californian law almost everything was labelled as potentially carcinogenic from mugs to caravans. On one trip to the beach with friends in Rhode Island an announcement came over the tannoy banning everything you could possibly have fun with from frisbees to inflatables. I didn't catch the full list and asked our host who quipped that everything was banned but it was nevertheless legal to carry an AK47. I had never understood the gun lobby before but this helped me understand a little despite being against guns. As well as the usual arguments, part of the gun lobby's success is that they are one of only a few voices representing those in America who are against all the safety rules.

The Land of the Free lost its bearings with covid. The ultimate irony came in a bulk email from then New York City Mayor, Bill de Blasio, in February 2021. The banner across the top said "*All in favor of a Covid-19 vaccine, raise your arm*" and next to it was an image of the Statue of Liberty, with a plaster on her obediently raised upper arm and a mask over her face, leaving her, like the tired, poor, huddled masses that she welcomed, *"yearning to breathe free."* [577]

BELIEF TEN: MASKS REDUCE TRANSMISSION

> *Surgical masks are not intended to provide protection against infectious aerosols. There is a common misperception amongst workers and employers that surgical masks will protect against aerosols.* **Health and Safety Executive, 2008**[578]

If social distancing was sufficient to handle the larger than a grapefruit sized droplets, could mask wearing deal with those that were smaller? When there is political pressure because *something must be done*, then mask wearing has the feel of something that might work and the public wanted something that could work. If the infected could be persuaded to wear something to filter out droplets then could the virus be controlled? The answer, according to public health officials across the world in spring 2020 was a resounding 'no'.

On 11th March 2020, Professor Chris Whitty, Chief medical Officer said, *"In terms of wearing a mask, our advice is clear: that wearing a mask if you don't have an infection reduces the risk almost not at all. So we do not advise that."* [579] Dr Jenny Harries, England's Deputy Chief Medical Officer summarised the position in an interview with the Prime Minister saying, *"it's really not a good idea and doesn't help,"* and *"in some ways you may actually risk catching the disease rather than preventing it."* [580] Fauci said, *"There's no reason to be walking around with a mask. When you're in the middle of an outbreak, wearing a mask might make people feel a little bit better and it might even block a droplet, but it's not providing the perfect protection that people think that it is. And, often, there are unintended consequences – people keep fiddling with the mask and they keep touching their face."* [581]

A review of all the medical literature on the subject, with an international authorship, was published in 2020. It showed no clear reduction from mask wearing in seasonal influenza, which was the most similar virus to use as a measure.[582] The Cochrane Collaboration, who are a well-respected international authority on collating medical evidence stated that surgical and medical grade masks made no difference as to how many people caught a flu-like viral illness.[583] That was the starting point. However, from summer 2020, every authority did a lockstep u-turn and masks were promoted as an essential, sometimes mandatory intervention that would save lives.

HOW WAS MASK WEARING JUSTIFIED?

The evidence that was presented by SAGE at the time of the U-turn were two reports from groups both convened by the Royal Society. The first report from Data Evaluation and Learning for Viral Epidemics or DELVE, reasoned thus, *"40%-80% of infections occur from individuals without symptoms… Droplets from infected individuals are a major mode of transmission… Face masks reduce droplet dispersal."* [584]

A single paper from an international group of scientists in January 2021 presented a good summary of the justifications used to bring in masking.[585] Firstly, they claimed there were no randomised controlled trials of masking to prevent covid transmission but did cite a study showing reduced community transmission. The Czech Republic was cited as a country where masking was introduced early and kept covid at bay. As there were no other population studies they heavily relied on laboratory mechanical studies and modelling. Each of these studies were scientific analyses that appeared to show a benefit from masking. We will come back to the problems with the studies but first let's look at the broader picture.

HOLES IN MASKS

Many scientists spoke out when masks were introduced saying that the holes in the fabric were too large to stop the spread of a virus.[586] Individual virus aerosols are a similar size to tobacco smoke particles and, similarly, fill the air. If someone sent you into a smoke-filled room wearing a cloth mask and you could smell tobacco, then you could also breathe in the virus (and exhale it back into the room). The holes in the fabric were so large it would be like trying to protect a double decker bus from grapefruit and lentils being thrown at it by using badminton or even tennis rackets with no strings in.

Dr Colin Axon, a former adviser to SAGE, said masks were just *"comfort blankets… an imperfect analogy would be to imagine marbles fired at builders' scaffolding, some might hit a pole and rebound but obviously most will fly through."* [587]

Mask proponents objected to this view saying that the cloth holes were more like tunnels. The holes were large but were deep and a proportion of aerosols would hit the sides and not progress further. The very largest droplets would be stopped anyway and that, alone, might have reduced spread. Maria Van Kerkhove, technical lead of covid response at WHO said, *"We have evidence now that if this is done properly it can provide a barrier… for potentially infectious droplets."* [588]

Here were two conflicting hypothetical positions. When there is more than one prediction of what may happen, the way to find which is right is to test it with an experiment. There was no shortage of real world evidence of neighbouring regions with and without mandates to compare their covid rates.

Ian Miller, author of *Unmasked: The Global Failure of COVID Mask Mandates,* carried out a comprehensive analysis of the USA.[589] He plotted the cases over time in areas with mask mandates against neighbouring areas where no mandate was introduced. In every case the covid trajectory in the masked state perfectly tracked the one in the neighbouring state. Without labels no-one could tell from his graphs which state wore masks. Where there was a slight difference it was the masked state which fared worse. There was not a single example where the masked region fared better, (even thanks only to chance), that the pro-maskers could use to push masking. This whole population measure is the one that matters. The results were in, from the real world, and the masks did not work. No amount of hypothetical ideas about how they could work or even evidence that they do reduce droplets is of any relevance when in the real world there was no impact on spread. However, because the authorities ignored this data and continued pushing masks, the majority continued to wrongly believe they did work.

BIOLOGICAL MASKS

Let's return to the land of giants and see what happens to an inhaled aerosol. The infectious source has left behind a cloud of droplets, or perhaps they wafted out of the window of someone who was sick in bed. Those larger than a grapefruit (100 microns) would have fallen to the ground but for every one the size of a grapefruit or larger there are thousands that were smaller.[590] The droplets smaller than a grapefruit have evaporated and shrunk in size and are suspended in the air. The majority of virus was in those droplets smaller than a lentil (five microns) so remained in the air.[591] A second giant now walks by, hoovering up air into his bus sized nostrils. Those droplets that remain larger than a lentil are deposited in the nose and throat.[592] The remainder flows down through the trachea, into the branching bronchi and down into the lung.

Any that land in the airway will hit a layer of mucus. Zooming in to the mucus reveals tiny holes the size of the virus, acting like a biological sieve.[593] The final layer of mucus, protecting the cells, has no holes big enough for the virus to fit through and the virus becomes entrapped.[594] The mucus is then swept along by hair cells on the surface of the airway, which waft the mucus upwards creating a slow moving stream that reaches the throat and is swallowed into the acid of the stomach killing the viruses and bacteria. In order to start an infection the virus must break through the mucus to reach the cells.[595] In order to do that it must, ultimately, leave the aerosol it travelled in to be small enough. Even then, the rate at which a virus can cross the mucus barrier is slower than the rate at which mucus is swept away. Therefore, other factors must contribute to allow entry. One theory is that characteristic bacterial infections, which often cause infections alongside viral infections such as influenza and *SARS-CoV-2*, are not coincidental. These bacteria have mechanisms for crossing the mucus barrier and virus can hitchhike on these bacteria in order to penetrate the mucus.[596]

The mucus lining of the respiratory tract creates a natural barrier far superior to any mask. Wearing a mask is much the same as wearing a hearing aid in your ear designed to alert you to any dangerous noises. There is already a finely tuned system for dealing with that problem which is far superior. All the focus on cloth masks has distracted from the much more important, unanswered question of why some people's mucus layer fails and what can be done about that.

Because the aerosols trapped in mask fabric will keep evaporating and shrinking, they may be closer in size to sand than lentils, in the giant analogy, when they reach the mask. Most aerosols will be channelled with the flow of the air. Those that become stuck on the fabric will not just disappear. Instead, the person wearing the mask continues to breathe through the mask. The virus can be absorbed into aerosols in the exhaled breath which the wearer or others may inhale later. On contact with the mask, the liquid surrounding the virus may be absorbed, freeing the virus to travel alone into the respiratory tract.[597] The idea that viruses would sit obediently on the fabric waiting for laundry day was a fantasy.

Everyone can agree that a mask would reduce the volume of what was expelled in a sneeze or cough in a similar way to a tissue or handkerchief. How many people would honestly keep their mask on to sneeze into and then continue wearing it? Even if they did, would they create new aerosols as they breathed over the trapped droplets? Either way, with every breath smaller aerosols would penetrate the mask and enter the air. The major problem though was that the vast majority of the virus in the air would have originated from maskless sick people in their homes who could have been some distance away.

WHY DO WE SEE NO IMPACT FROM MASKING?

As anyone who has worn a mask will know, the premise above about holes and tunnels is ridiculous anyway. The vast majority of air taken in and exhaled enters and leaves by gaping holes at the sides and top of the mask. In theory this could have helped reduce person to person transmission, as larger droplets would be directed away when two people were facing each other, reducing spread. It would not have helped hairdressers however, who stood behind the vents of their client's mandatory masks or school children who sat alongside their neighbours' vents. Because air was mostly redirected the only real impact of masking was to reduce the spread of large droplets during close face to face contact. In theory such a reduction could have reduced spread to some degree but in practice it did not. There was no impact on real world data on infection trajectories. In fact the similarity of speed of transmission between areas where droplets are stopped by masks and those where they are not is a further indication that aerosols drive spread.

As the evidence on the size of aerosols that contained the virus came to light, there should have been a change of position on masks. As John Maynard Keynes said, *"When my information changes, I change my mind."* The open letter to the CDC about aerosol transmission by the physicists and other scientists we met in *Belief 1: Covid only spreads through close contact*, was sent in July 2020. The consequence of this information could have been to acknowledge that the virus was airborne, that close contact was not the prime mode of transmission and that there was little that could be done to prevent aerosols from the sick filling the air. It could, however, have led to a change in hospital policy of stopping masking, returning bed capacity to normal and installing air filters. An announcement to that effect at that time would have been politically difficult and it has only become harder since.

Instead, an utterly illogical position was taken. There seemed to be a strong desire to 'do something' about aerosol transmission and the only idea out there was masking. Trish Greenhalgh, Professor of Primary Care at Oxford University, was a key proponent of masking on the basis that, *"Even limited protection could prevent some transmission of covid-19 and save lives."*[598] Taking this approach meant that proof of a meaningful difference was no longer the bar for intervention. A hypothetical idea about what might work coupled with an aim of preventing even a single death meant that the entire world could be asked or forced to mask.

Mask proponents eventually cottoned on to the literal gaping hole in their reasoning and started to advocate for tighter fitting masks. In February 2021, New York City's *Mask to the Max* campaign included recommendations for everyone over the age of two to wear snug masks by wearing a cloth mask over a disposable mask; a nose wire; a mask fitter or brace or to *"knot the ear loops and fold or tuck extra material,"* in an attempt to create a tighter seal.[599] Trish Greenhalgh had been a vocal advocate of masking with some extreme suggestions. She went from suggesting people wear panty liners[600] in their masks (which have an unbreathable waterproof plastic layer) to advocating for wearing a pair of nylon tights over the head to improve the fit.[601]

However, by December 2021, she was insisting only high quality medical grade masks would work, not cloth masks saying, *"Breathing, coughing, talking generate aerosols. So EVERYONE needs high-grade masks"* [602] and *"that droplet theatre won't cut the mustard"* [603] before begging Elon Musk to buy enough to supply the world, *"If you did that, you'd be very popular and it could end the pandemic."* [604] Leana Wen, one of CNN's favourite high priests said, in a most patronising way, *"Cloth masks are not appropriate for this pandemic. It's not appropriate for Omicron. It was not appropriate for Delta, Alpha or any of the previous variants either because we're dealing with something that's airborne."* [605]

Calls to wear close fitting medical grade masks have increased since that time and the US administration had 750 million such masks in storage by January 2022.[606] These masks are meant to have a sealed fit and have very small holes (20 microns in diameter or around an inch in the giant analogy). However, they do not work through mechanical filtering but through static electricity which traps small particles in the fabric. Good quality masks filter 95 percent of particles three times the size of the virus. No-one knows the exact percentage for particles the size of the virus and there is disagreement about the true filtration ability for larger particles with estimates of between 54[607] and 85 percent[608] rather than 95 percent.

Again, there was a hypothesis that medical grade masks could make a difference where regular masks failed. The experiment to test that hypothesis was carried out by Germany and Austria which both mandated medical grade masks. Despite these mandates there was no difference in case numbers compared to neighbouring countries. Perhaps this is surprising given that medical grade masks are clearly superior to cloth masks. However, in the real world they do not make an impact. There was no need to rerun this experiment after Germany[609] and Austria did that for us in 2021. However, since then Japan, South Korea and Hong Kong have confirmed their findings. By 29th November 2022, a randomised trial showed that healthcare workers given N95 masks had similar infection rates to those wearing surgical masks i.e. they did not work.[610]

There are at least five reasons which could explain why medical grade mask mandates did not work. The issues are where the air enters, what happens to aerosols either trapped in the fabric or later when the mask is removed, how *SARS-CoV-2* spreads and who is emitting the virus.

WHERE THE AIR ENTERS

Air enters and leaves through sometimes barely perceptible gaps in the sides. These vary in size with movement of the face. Pushing a medical grade mask onto the face to improve the seal results in a clear increase in difficulty breathing demonstrating that the air normally enters through gaps in an inadequate seal. Think about how hard it is to maintain even a large, tightly fitting, rubber seal on a snorkel mask. Given that aerosols containing virus are so small perhaps the recommendation should have been for Hazmat suits and personal oxygen tanks. Such bulky equipment probably could have protected from the virus, but only when worn.

WHAT HAPPENS TO AEROSOLS TRAPPED IN THE FABRIC

An aerosol that is trapped in the filter will continue to evaporate as it is breathed over. The liquid portion will shrink as this happens until virus particles themselves could be released. Ultimately, it is the smallest aerosols that cause infection as larger ones cannot penetrate the respiratory mucus layers.

WHEN THE MASK IS REMOVED

One study showed that the highest levels of virus in a hospital could be found in the room where protective equipment was removed.[611] Having carefully collected virus in the filtering mask, on removing the mask, healthcare workers would unwittingly fill the air around them with virus, while imagining this air away from their patients was safe. It is a bit like thinking that wearing a snorkel to swim in the sea would stop you getting salty.

HOW *SARS-COV-2* SPREADS

If you are still left wondering if medical grade masks are worth wearing, remember that *SARS-CoV-2* can also spread via the surface of the eye anyway.[612] If it turns out that touching contaminated objects led to significant spread, masks would achieve nothing.

WHO IS EMITTING THE VIRUS.

The most crucial point is that no-one can wear medical grade masks all the time. Remember that each infected person emits 72 million infectious particles overnight when there is no UV light to cleanse the air. Even if a mask could protect you for a few hours a day, what about the rest of the time?

HOLES IN THE EVIDENCE

Let's take another look at the single paper that listed the justifications for the introduction of masks. Their first claim of there being no randomised controlled trials at the time for covid was invalid as they totally ignored a Danish randomised controlled trial showing no statistically significant difference in infection rates between the masked and unmasked.[613]

The one study they cited which overclaimed about a reduction in community transmission was a Chinese study. The Chinese produced huge volumes of academic papers on *SARS-CoV-2* in early 2020. Of all the papers on respiratory virus infections in March 2020 produced by the UK, USA and China, two thirds were Chinese. In the twelve months prior to that only four in ten had been Chinese and it was a full year before that ratio returned to normal. From April 2020, the Chinese Communist Party mandated that all scientific literature on covid must be approved by the Chinese Ministry of Science and Technology or Ministry of Education before publication.[614] Microbiology professor, Sarah Cobey said, *"it would be very problematic if results from China were being filtered or suppressed for reasons other than quality."* [615] This political interference means that all Chinese scientific publications should be regarded sceptically and with consideration around the motivations of the Chinese Communist Party. The fact that China reported fewer than 250 covid deaths over the two year period from 17th April 2020 should also be treated with the scepticism it deserves. In fact, no reliance should be placed on official Chinese Communist Party data.

Those justifying masking cited the Czech Republic as an example of masking working. The Czech Republic had made masks compulsory in public from March 2020 and, as with all of Eastern Europe, were relatively spared from covid in spring 2020. The hypothesis that they had hit on the solution that prevented spread was proved erroneous in autumn 2020 when Eastern Europe as a whole, and the Czech Republic in particular, were hit hard with covid. Since February 2021, the Czech Republic became and has remained one of the hardest hit countries in Europe for covid deaths.

The use of hypothetical modelling studies to support masking were based on assumptions at the outset about the impact that mask wearing would have at a population level.[616] The results of that impact were graphed and then the paper concluded that mask wearing would be worthwhile. These modelling studies were entirely works of fiction using circular logic.

For a short time in October 2020, the real world evidence appeared to support the modelling studies. A paper was published showing that masked states in the US had lower case rates.[617] The paper was withdrawn in November 2020 because the rates had rocketed in those same masked states.[618] Looking back after a few more months it was clear that masks 'worked' in the summer and stopped 'working' in autumn and winter.[619]

The justification for mask wearing also heavily relied on laboratory studies in which aerosols were synthesised using tubing to demonstrate protection from masks. The unrealistic base assumption was that two people would be in close proximity, face to face while one coughed or sneezed directly at the other.

Measurements of mask effectiveness in such a scenario were taken with a machine called the Gesundheit, named after the German phrase for wishing someone good health after they sneeze.

Hector Drummond wrote a superb summary of the evidence around mask wearing and its manipulation in *The Face Mask Cult*.[620] He points out that only the front of a person's face enters the collection cone. Any breath that escapes at the sides of the mask is not measured. Even allowing for this misfiring, the claim from a key experimental study showed that, when unmasked, nine out of ten people emitted fine particles and this fell to only eight out of ten with masking.[621] There was no evidence that such a reduction, even if genuine, would have a real world impact. He also points out that one in six of the participants had more measurable virus in the cone with the mask on than without it, a point the authors failed to mention.

SHODDY EVIDENCE

Hector Drummond goes on to expose how the UK authorities have created their own perpetual motion engine for evidence on masks. An original paper from SAGE stated that *"Face coverings are likely to reduce transmission through all routes by partially reducing emission of and or exposure to the full range of aerosol and droplets that carry the virus."* [622] Note the lack of claim about the extent of any reduction. This paper referenced another government paper which referenced another government paper and so on in a chain. The fourth paper had become more bold saying *"Promoting high levels of wearing face coverings or face masks can potentially reduce transmission through all transmission routes especially via close range and long range airborne transmission (high confidence)."* [623] Where was this high confidence from? The reference used for that claim was the original paper that made the much lesser claim. The claims had spiralled into more and more confident assertions without the addition of any further evidence.

Hector Drummond also did a deep dive into the references used for a face mask report from The Royal Society, from June 2020, '*Face masks and coverings for the general public: Behavioural knowledge, effectiveness of cloth coverings and public messaging*'.[624] Their entire meta-analysis was based on four Chinese reports having excluded a randomised controlled trial which showed no benefit. Of these, two showed a minor effect of masking, one showed a negative effect and the other showed a dramatic difference. The latter study was of healthcare workers treating *SARS-1* patients and claimed that eight out of ten unmasked healthcare workers were infected compared to only two out of ten who were masked. The original paper is only available in Chinese, only one author is listed along with an affiliation and he appears to be a state employee, there was no randomisation and participants reported their mask use in retrospect (i.e. those who became infected might then report that they had not always worn masks much after all). The fact that these results have never been replicated anywhere else did not seem to cause the Royal Society any concern.

The second Royal Society report in May 2020 was the DELVE report which proposed masking to stop droplets.[625] Dr Antonio Lazzarino from the Department of Epidemiology and Public Health at UCL said about the DELVE report, *"That is a non-systematic review of anecdotal and non-clinical studies. The evidence we need before we implement public interventions involving billions of people, must come ideally from randomised controlled trials at population level or at least from observational follow-up studies with comparison groups. This will allow us to quantify the positive and negative effects of wearing masks. Based on what we now know about the dynamics of transmission and the pathophysiology of covid-19, the negative effects of wearing masks outweigh the positive."* [626]

As well as the justifications used by the UK above, the CDC in the USA chose an odd paper as their second most important piece of evidence on their website when introducing masking.[627] It was a report of a covid outbreak in a Missouri hairdressers, Great Clips.

It starts, as most papers do with an introduction on background scientific knowledge of the area. This said, *"Consistent and correct use of cloth face coverings is recommended to reduce the spread of SARS-CoV-2."* That does not sound anything like a scientific review of the evidence base.

The paper reports on one hairdresser who had what were described as *'covid symptoms'*, tested positive and spread covid to four close contacts. A second hairdresser had *'respiratory symptoms'* and tested positive but none of her close contacts developed covid. No further details were given as to what the symptoms were but it is worth noting that this was in the second week of May 2020, only a week after their five week lockdown ended. No-one would have been keen to cough in public. If someone started coughing at work they would surely be sent home, especially in a job where masks had been thought necessary to reassure the clients. There is no elaboration on what the *'covid symptoms'* were but it is odd that a cough was not mentioned. Coughing or not, the air in the hairdressers may have contained *SARS-CoV-2*.

There were 139 clients who were exposed while the first hairdresser was symptomatic. (They ignored clients from her presymptomatic phase). The public health authorities tracked them all down, interviewed them and made them quarantine for a fortnight. A quarter of the clients had had respiratory symptoms in the preceding three months unrelated to covid. With that background rate, we would expect five of them to have these symptoms, unrelated to covid, in the following two week period, even if there was no covid. Robin Trotman, a local infectious disease doctor predicted five to ten covid cases among the clients in addition.[628]

Public health officials contacted the clients during quarantine to ask if they had developed symptoms which would have meant a longer quarantine period. Not one of them had a single symptom, which either proves that masks are a panacea or that people do not like extensions to their house arrest. It is worth appreciating what the atmosphere was like at the time. Only a couple of weeks earlier headlines said, *"900 Missouri residents who 'snitched' on lockdown rule-breakers fear retaliation after details leaked online."* [629] These were people whose names and addresses had been released in public, alongside reports they made of their neighbours breaching regulations, thanks to local freedom of information law. Even with no intention of leaving the house when unwell, with a heightened awareness of snitching neighbours, it might be best not to draw attention to any issues. The final indication of how keen the clients were to cooperate with public health authorities comes from the fact that half of them refused covid testing when offered it. The ones that were tested had negative results.

What is particularly egregious is that the CDC used this as prime evidence when there was no control group. The City Journal article, *Do masks work?"* demolished this study concluding *"Nobody has any idea how many people, if any, would have been infected had no masks been worn in the salon. Late last year, at a gym in Virginia in which people apparently did not wear masks most of the time, a trainer tested positive for the coronavirus. As CNN reported, the gym contacted everyone whom the trainer had coached before getting sick – 50 members in all – "but not one member developed symptoms." Clearly, this doesn't prove that* not *wearing masks prevents transmission."* [630]

One final point, is that the study took place in Springfield, Green County, Missouri. Like the other states in the central band of the USA there may have simply not been a seasonal trigger. Green County was relatively spared in spring 2020 with only thirteen patients in hospital at the spring 2020 peak compared to over 230 in the subsequent peaks in December 2020, July 2021 and January 2022.[631] Missouri was part of a central band of US states from Montana and Minnesota down to Oklahoma and Arkansas which were relatively spared in spring 2020 but not spared thereafter.

Why on earth was the CDC using this paper as the second reference for the effectiveness of masks?

Proving something works which does work is relatively easy. Proving that something does not work is much harder. When something has no impact, it may still be possible to measure an effect by chance some of the time. Only by including the times when it had a negative effect, can you see that averaging all the evidence out shows it achieved nothing. Singling out individual studies of apparent benefit is not an adequate way to prove the case because there will be other studies where there was no benefit. Yet, the CDC did just that, with a study that was effectively an anecdote of an outbreak at Great Clips, to justify the introduction of masks.

Another time woeful evidence was used to support masking policy happened in England. Public Health England carried out a study of covid rates in schools with or without masks.[632] They showed that schools with masks had higher covid rates. Their excuse was that those schools had introduced them *because* they had more covid. That should have led to a better designed experiment but instead led to modelling of the data to show what they believed the results would have been had covid rates been equal. This was not just a slight adjustment but an inversion of the results. Even their fictional claim was only that masked schools had 0.6 percent fewer covid absences than unmasked schools. That data was used to push the policy to introduce masking in UK schools.

Interestingly, Jenny Harries England's Deputy Chief Medical Officer, who has said masking was not a good idea was not particularly persuaded by the evidence. She said in Aug 2021, *"The evidence on face coverings is not very strong in either direction… but it can be very reassuring in those enclosed environments."* [633] However, by July 2022 she was totally contradicting her position in March 2020 saying, *"we want people to take sensible precautionary advice like… wearing a face covering if you're going into enclosed, poorly ventilated spaces. If I've got any respiratory infection it's a good thing to do, and I think it's a new lesson for the country."* [634] The damage such flip-flopping does to public confidence in health messaging is the real negative here.

In February 2022, the Cochrane Collaboration, the world centre for collating evidence on medical matters, finally published Thomas Jefferson's updated summary of the evidence on masking showing *"little to no difference."* [635] The first review of the evidence was delayed by seven months until November 2020 due to *"unexplained editorial decisions"* such that mask mandates were already in place.[636] Public health officials and others in positions of authority continued to undermine themselves by dismissing this comprehensive analysis and continuing to insist that masking worked and in USA continuing with mandates even for toddlers.

In March 2023, there was a further twist. Karla Soares-Weiser, the Editor-in-Chief of the Cochrane Collaboration published an apology regarding the review without having consulted the authors.[637] The lead author, Thomas Jefferson complained that *"Cochrane has thrown its own researchers under the bus again,"* adding, *"I think Soares-Weiser has made a colossal mistake. It sends the message that Cochrane can be pressured by reporters to change their reviews. People might think, if they don't like what they read in a Cochrane review because it contradicts their dogma, then they can compel Cochrane to change the review. It has set a dangerous precedent."* [638]

Masking has been supported by government policy documents, Royal Society reports and laboratory and modelling studies. However, even a relatively superficial dig into the scientific real world evidence shows these documents have a foundation of dust. It was hope and politics which drove masking not scientific evidence.

	THE EVIDENCE MANIPULATION TRIAD
EXTRAPOLATE	Laboratory and modelling studies were extrapolated as the basis to justify the introduction of masks and anecdotal evidence, like the Great Clips story, was used as if it was solid evidence.
EXCUSE	Public Health England claimed the reduced covid rates in schools without masks were because they were at lower underlying risk of covid. When masks did not prove effective, it was said not enough people were wearing them or they were not wearing them properly or for long enough. Another common excuse was that people had let their guard down because of overconfidence around safety provided by the masks. (Even though evidence shows people give mask wearers a wider berth than the unmasked).[639]
EXCLUDE	Real world evidence was ignored and experts trying to explain how masking could not work were silenced on social media.

TOP THREE MYTHS
1. Stopping droplet transmission at close distance would reduce spread
2. A small decrease in transmission in a fraction of interactions would have a measurable impact overall
3. Medical grade masks work where cloth masks do not

17

THE POLITICS OF MASK RULES

But not in the classroom because that's clearly nonsensical. **Boris Johnson, August 2020**[640]

If the evidence didn't support it, why were masks really introduced? In December 2022, the ex-UK Health Secretary, Matt Hancock's *book Pandemic Diaries*[641] revealed that the decision to introduce masking in schools was entirely political.[642] The first minister of Scotland had introduced them in, as he described, *"one of her most egregious attempts at one-upmanship to date."* Despite having written guidance that specifically excluded schools from masking, they did a U-turn, to avoid *"a big spat with the Scots."* The harm to children from this policy did not seem to enter the discussion.

Matt Hancock, had told parliament, in July 2020, *"face coverings increase confidence in people to shop."* [643] Was the original intention for masking to give people confidence to start shopping again and to reboot the economy? That sounds to me as if they were intended as a national security blanket. But the result was not security. The result was a world bereft of smiles where signals from strangers' faces that we use for reassurance that there is no threat were hidden and we were constantly reminded of covid.

Laura Dodsworth, author of *A State of Fear*, interviewed members of the Government's advisory group on using psychological influences to alter behaviour, SPI-B, and said, *"I am sure from talking to the psychologists on SPI-B for my book* A State of Fear, *that they wanted masks to provide a signal of danger as well as symbolise solidarity. But they do the opposite. They divide us."* [644]

Laura Dodsworth also interviewed Robert Dingwall, a Professor of Sociology and a government advisor, who said, *"They create a sense of threat and danger and that social interaction might be something to be anxious about. So mandating masks can feed the fear."* In his view, they were designed, *"to make people compliant."* [645] Masks amplified the fear, they did not reduce it. [646]

ABSURDITIES

Because they were introduced for psychological reasons not for virus prevention there was never any advice on how masks should be handled and cleaned and instead the rules around them were nonsensical. Politicians seemed happy to follow illogical rules and ignore their cognitive dissonance. They wore masks when seated in the debating chamber of the Houses of Parliament but removed them when standing to speak; then wore them when standing in the parliamentary restaurant or bar, but removed them when sitting. Those same politicians who felt the need to be unmasked to communicate, mandated masks in classrooms. People playing wind and brass instruments in bands and orchestras wore masks with holes cut in so they could blow hard through their instruments. People attended church services and sat masked but sang whole heartedly once they were standing with their mask removed.

There was considerable hypocrisy with politicians and public health officials parading in front of the camera with them on but removing them when the cameras were off. The rich and powerful went mask free while being waited on by masked staff and in the USA powerful adults were frequently pictured unmasked alongside masked, young children.

There seemed to be an absurd paradox such that masks could never fail. If the cases did reduce it was because masks were working. If the cases rose it was because it would have been worse without masks. If cases fell just as much in a place with a mask mandate as one without then it was because people were more careful with their behaviour when not wearing a mask. There was no evidence that could be presented that would shift the belief. One doctor, Robert Redfield, director of the CDC went as far as saying, in September 2020, *"We have clear scientific evidence they work. I might even go so far as to say that this face mask is more guaranteed to protect me against COVID than when I take a covid vaccine, because the immunogenicity may be 70 percent and if I don't get an immune response, the vaccine's not going to protect me, this face mask will. Masks are the most important, powerful public health tool we have."* [647]

The mask mandates are a paradigm example of political theatre along with virtue signalling trumping evidence based policy making.

18

THE PRECAUTIONARY PRINCIPLE

He's suffering from politician's logic.
Something must be done! This is something! Therefore we must do it!
But doing the wrong thing is worse than doing nothing.
Doing anything is worse than doing nothing.
Yes Minister[648]

Mask wearing was often justified on the basis of the precautionary principle, which can be summarised as *"better safe than sorry."* The phrase was first used widely by the EU during debates around genetic modification. The basis of the precautionary principle used to be the admission that there were consequences of any action that we did not know. Any action should therefore only be taken with great care. The precautionary principle has been turned on its head. Instead of being cautious when changing anything, in case of causing unanticipated harm, the term has been used to push policies with unproven but hypothetical benefits even when there are known harms. Trish Greenhalgh (the vocal mask advocate we met earlier)[649] and colleagues twisted its meaning in a paper in April 2020, saying, *"The precautionary principle states we should sometimes act without definitive evidence, just in case. Whether masks*

will reduce transmission of covid-19 in the general public is contested.
Even limited protection could prevent some transmission of covid-19
and save lives. Because covid-19 is such a serious threat, wearing masks
in public should be advised." [650]

The pushing of masks as a way to improve safety is just one
example of how the precautionary principle was inverted. The key
to being better safe than sorry comes down to careful weighing up
of benefits against the known and unknown risks. There are
parallels with the medical principle of first do no harm. Evidence
of benefit needs to be strong before taking on risks of harm from
an intervention which may be either unknown or unmeasurable.

When deciding if the benefits of an intervention outweigh the
risks, all the risks must be considered. However, with mask
wearing all risks were brushed aside or never measured. The Danish
randomised controlled trial of mask wearing included several
outcome measurements of harm caused by the masks but the results
from these were never included in the publication. [651] Instead,
justification for mask wearing, even if benefits were minimal, were
made on the basis that they were a harmless intervention. They were
not harmless. A particular harm may seem relatively trivial.
However, when considering the number of people affected and the
long time period they were affected for, the total harm to the
population was significant. The harms stretched from harming
child development to destroying communication for the deaf and
from extending the fear of covid to stripping society of reassuring,
bonding smiles (see below). There was never a sound justification
(because masks do not work). Even if the evidence supported greater
benefit than risk there were reasons not to mandate masks based on
ethical principles (see 22: *Finding a Moral Compass)*. Even if those
ethical principles were discarded there is the argument that there
is always danger in interfering with complex systems which cannot
be understood in advance.

Robert Conquest, pointed out that *"Everyone is conservative about what he knows best."* [652] Once the detail of a complex system is understood, people defend it against change because they understand the enormous potential for unexpected consequences. Mask mandates – an untested innovation – were carried out as if the public were a simple homogeneous mass of disease vectors with no consideration of what a complex system human interaction is. Lockdowns were imposed as if there was only one infectious disease and, setting aside the many harms caused; even the consequences to immunity of reduced exposure to other infectious diseases did not appear to be a consideration.

FIRST DO NO HARM

Being reminded of covid, made to feel subordinate and having facial expressions stripped from our interactions were not insignificant in themselves, especially once scaled up to millions of people affected. However, the harms from masks extended beyond these issues. Face coverings are a barrier to all person-to-person interactions in hospitality, retail, schools and the workplace.

When in Legoland for my son's birthday, I witnessed one set of parents desperately trying to get their two year old to smile for a photo, making all the usual high-pitched enthusiastic noises in their attempt. The little boy just stared blankly back at their expressionless masks. It reminded me of a video I had seen of a one year old whose young brother tries desperately to get a reaction from him through a mask. Despite the brother's huge enthusiasm the baby remained disengaged but when he removed his mask the baby immediately lit up and vocalised. Masks make it harder to recognise people (even their voices) and make us misjudge emotions.[653] An inability to read faces is detrimental to social confidence and quality of life[654] and masks prevent accurate reading of faces for children and adults.[655]

But more than any of that they have impacted on child development. Children learn how to interact by watching faces and responding to them, yet masks were mandatory in some nursery settings where staff cared for babies. A team of scientists at Brown University showed that children who were at least 15 months old by March 2020 had normal development.[656] However, children, who were born after that time had significantly lower verbal, non-verbal, and overall cognitive scores, suggesting a policy impact on child development. The lead author, paediatrician Dr Sean Deoni, said, *the closest thing we've seen in other research – and this is horrible, not a good comparison to be making – is the studies that were done of orphans in Romania. The effects of institutionalisation and lack of interaction on them were profound, but what we're seeing here is on par with that.*[657]

As well as the interpretation of facial expressions, children's language development was damaged. CDC guidance used to say that at 24 months of age a child should have a vocabulary of more than 50 words but that goal has now shifted to 30 months of age. As well as hearing clearly, seeing mouth movements is fundamental to learning language.

Seeing mouths is also fundamental to communication for the deaf. The Guardian reported that the National Deaf Children's Society surveyed 500 parents and found a quarter of deaf children were being taught by teachers in masks and that for sixth formers half of them were.[658] Masks have also hindered communication for the one in five UK adults who are hard of hearing.[659]

One wise friend of mine told me how she had entered a shop with a mask and, on seeing the owner was maskless, offered to remove hers. He was relieved, explaining that he was hard of hearing and needed to lip read to be able to communicate properly. This conversation was not overheard by the masked couple who made loud derogatory remarks on her maskless status as she left the shop. For some people masking was embraced and policed with huge enthusiasm.

Alex Gutentag, a teacher for special needs and underprivileged students described the cultural shift of masking in education, *"mask mandates have given rise to a wide array of new educational materials, such as dystopian singalongs and call-and-response routines that teach young children to cover their faces in order to keep their friends safe. In some schools, students are rewarded with 'mask breaks' and can be suspended for mask noncompliance. Back-to-school activities that once had names like "All About Me" now have names like 'Me Behind the Mask'."*[660]

The broader impact of masking on children was brought to life by a conversation between a six year old American girl and her mother who had just had a parent's evening meeting with her teacher over zoom.

Daughter: *"Did my teacher wear a mask?"*

Mother: *"Of course not. We were on zoom."*

Daughter: *"What does she look like?"* [661]

At the age of six a child's relationship with their teacher is one of love. How was this child's ability to connect to her teacher affected by having spent every school day only able to imagine what she might look like?

CREATING DIVISION

The Masks4all campaign claimed the risk of catching covid

remained high for the mask wearer and the risk was only lowered if others were wearing them.[662] They even went as far as saying *"Anyone without a mask puts you and your family at risk,"* and *"Spreading COVID-19 is not a right."*

A *Telegraph* article entitled *"Why face masks became the symbol of a divided Britain"* by Harry de Quetteville described how masks had become *"personal political broadcasts on behalf of the wearer"* and quoted Professor David Halpern, Chief Executive of the UK government's Nudge Unit as saying in November 2021 *"Most of the heavy lifting is done when we look at each other and think 'Why aren't you wearing a mask?' and frown."* [663]

The CDC permitted the fully vaccinated to remove their masks in May 2021. This created a dilemma for the compliant. If they removed their mask they could no longer be easily identified as being someone who obeyed the rules. For a short time there was a market in badges and T-shirts and even masks that said *"Not a Republican. Just vaccinated."* I saw a woman wearing a chain around her neck meant for reading glasses with a mask attached like a pendant to a necklace. She could declare her faith without the inconvenience of wearing the mask. If we want to declare a belief then the most obvious way is to wear something to show it. People have publicly declared their beliefs by wearing Christian crosses, Jewish skullcaps, Muslim hijabs and Leninist red ties. Mask wearing had become a marker of belief in the official covid narrative.

BELIEF ELEVEN: CHILDREN ARE RESILIENT

Children themselves show remarkable resilience, creativity and adaptability. ***UNICEF, 2020***[664]

The claim that children were resilient was made repeatedly. The UNICEF report quoted above listed the ways in which policy had harmed children including domestic abuse, lost education, food insecurity, reduced healthcare and raised anxiety and stress. All of these problems were caused by adults and resulted in direct harm to children. It is for adults to address the causes of the harm. Focusing on the resilience of the child is a way to abscond from responsibility.

In 2020, academic and parenting websites published on a theme of how to help build resilience in children. Resilience was discussed as if a child with enough resilience should be expected to cope with any eventuality. Not only are they supposed to remain unharmed but they are supposed to adapt positively. Any horror is justified as character building. Of course children tolerate difficult times and adapt so that they can continue to grow into adults. These adaptations are not necessarily a good thing. In the words of an eight year old, *"just because we're resilient doesn't mean you should treat us like this."*[665]

Children that demonstrate resilience tend to have supportive loving relationships with their parents and peers providing them with the ability to regulate their emotions. Rather than policy makers considering how to maximise or substitute for these protective environmental factors, the emphasis has been on the characteristics of the child. Are they 'hard' enough or 'tough' enough? The result is that responsibility to be resilient has been treated as an issue for the children themselves rather than the adults and support networks around them.[666] Children who are disadvantaged or exposed to abuse, bullying or pressures to perform are least able to be resilient because these factors result in stress that stops them learning how to regulate their emotions. The children most likely to be negatively impacted by policy were least likely to be resilient.

Molly Kingsley and Liz Cole formed the campaign group UsForThem and campaigned tirelessly in the interests of children since 2020.[667] They have written up a comprehensive book, *The Children's Inquiry*, on the failings of government and others whose role was to protect children.[668] It opens with the line, *"They say it takes a village to raise a child, but in March 2020, that village slammed shut its gates."* In it they interview Dr Sunil Bohpal, a paediatrician who persuaded the Royal College of Paediatrics and Child Health to support an open letter to the Prime Minister. The letter was signed by 1,500 specialists and summarised the issues saying:

"It seemed quite acceptable in some circles that children and childhood ceased to be an issue in public policy, and they were simply expected to stay at home with parents — whatever circumstances, whatever hardship they were facing at home. There was very little understanding that what happens to children in those early years influences their whole life, health and well-being. We heard decision makers and the public claiming that children are 'resilient' and 'will be fine' in the face of any hardship or degradation. I always found this extremely troubling, based on decades of scientific understanding showing that what happens to you in childhood matters lifelong – not just for individuals but for whole societies too." [669]

CHILDREN WHO WERE NOT RESILIENT

Educational disadvantage has lifelong implications. It is not something character building that children can just brush off. Jay Bhattacharya, Professor of Medicine summed up the issues thus, *"You get learning loss that then echoes throughout a kid's entire life. They live shorter, poorer, less healthy lives."* [670]

The impact of school closures fell hardest on the most disadvantaged. The Education Endowment Foundation expressed concern that the gains for disadvantaged children seen since 2011 were reversed by school closures. The Chair, Sir Peter Lampl said, *"The repercussions of these months of lost learning are devastating and will be felt for a lifetime, especially by those from low-income backgrounds."* [671]

Schools provide much more than just education. For many children they are a chance to be heard and respected, for others they are a refuge from harm. Professor Lucy Easthope, an expert in disaster planning told the authors of *The Children's Inquiry*, *"The idea that school was first and foremost a place of education was quite an elitist view – the framing of the school as the educational environment is the last thing it becomes in a disaster: it's a place of safety where children are fed, kept warm, given a respite from abuse and safeguarded."* [672]

Anne Longfield, The Children's Commissioner was charged by the government with representing the interests of children in policy making. She tried her best to call out the issues and prevent harm. In April 2020, she reported on children *"living in dangerous circumstances – who are experiencing neglect, abuse and serious harm. For these children, as for adult victims of abuse, they are now forced constantly to stay at home in places where they are likely to be scared and in danger, with no way out. These are the children for whom school can be a reprieve, where trusting relationships need to be built up with social workers, and where close monitoring is needed to understand if they can stay at home or need to be in care."* [673] Only 5 percent of such children attended school in spring 2020. [674]

In March 2020, legal protections that were designed to protect vulnerable children were relaxed at the same time as services with a role in detecting and preventing harm were suspended. Longfield protested saying, *"I would like to see all the regulations revoked... If anything, I would expect to see increased protections to ensure their needs are met during this period."* [675] Longfield also fought tirelessly to keep schools open pointing out that the scientific evidence around covid was limited but there was overwhelming evidence of the harm from shutting schools. Despite her role being to call out policy problems, especially those driven by the Department of Education, her boss and funding came from the Secretary of State for Education. She nevertheless spoke out but even with her important role, most of her pleas were ignored.

Social workers moved most of their services online where it was impossible to know who was listening in on the conversation. [676] A report on policing expressed the police's *"dissatisfaction at the constraints the pandemic imposed on their ability to effectively safeguard young people."* [677] It continues by saying how lockdown led to, *"isolation, a need for belonging and emotional neglect,"* among the young. Drug gangs exploited this to recruit children living in towns, through social media. A report from the National Youth Agency said they targeted *"girls, who are less likely to be picked up by police"* and went on, *"The number of potential child victims of criminal exploitation has overtaken adults for the first time driven, in part, by an increase in the identification of county lines."* [678]

CHILDREN WHO BECAME LESS RESILIENT

Experts in child safeguarding emphasise that it is not possible to identify those children who are at risk at any point in time because there is a spectrum of risk and the situation changes over time. Children who would have thrived normally were locked into sometimes cramped accommodation away from their all-important network of friends and with adults who were themselves starved of healthy human interactions. These adults were also afraid of covid and, perhaps, their financial situation and had had their autonomy stripped away by the state.

If you had wanted to create a toxic environment for children, what more could possibly have been done? How about cancelling exams so that there was nothing to strive for and children were left aimless and their feelings of worthlessness confirmed? Perhaps, shutting the only outside space available as happened to Victoria Park in London which provided the only escape outside for tens of thousands of people who lived in cramped accommodation in that area. What about cordoning off the playgrounds or even welding the metal gates to children's playgrounds shut and blocking off the basketball hoops as occurred in New York or filling the skate parks with sand as happened in California. How about repeating all those mistakes again the coming winter?

The key to building resilience is for children to feel supported, for their voices to be heard and for them to learn to articulate their emotions. That way, instead of being consumed by difficult emotions they can learn to better put situations into context and to seek help and advice when struggling. How can children feel supported after their supports have been stripped away? How likely are they to feel heard when the only people left to listen to them are stressed out? How can they articulate the unfairness of what happened to them when it was happening to every other child around the country to a degree?

One example of a child whose resilience was destroyed was related by Dr. Zenobia Storah, a child psychologist who had numerous stories of the harm she had witnessed to children caused by covid policy.[679] She treated a child who was forced to strip naked on arrival at home each day and head straight to the shower. The pathological level of fear among the parents of that household meant they were unable to recognise the harm they were causing their own child.

One nine year old child said, *"I killed my mommy."* Her mother died two weeks after the child tested positive for covid. Feelings of guilt are not uncommon in bereaved children and I feel for those who cared for this child. It must have been exceptionally hard to persuade the child that they were not to blame in a world where blame for covid was being daily attributed to the unmasked or the unvaccinated. The wider environment of blame would have been responsible for yet more children losing their ability to be resilient.

Teachers commented on how changed children were after lockdown. Some had been showered with attention and became demanding of it. Others had been ignored and were in desperate need of attention. A debate in the commons that did not happen until February 2022, discussed how, *"more than one headteacher has used the word feral, in terms of behaviour."* [680] Children had forgotten how to play and there was such severe separation anxiety that they had *"whole classes who are crying for hours."* By 2022, the press reported on children who had fallen behind on crawling, toileting, understanding facial expressions, language development as well as academically. [681]

There were a whole series of children who were navigating difficult teenage transitions who may have had a bit of a wobble in the absence of lockdown. Their chances of skirting that disaster were utterly undermined with lockdowns and huge numbers of children who would have never required mental health services in the past, overwhelmed the provision available. Ellen Townsend, Professor of Psychology, advocated powerfully for young people and wrote in June 2020 how, *"Half of young people aged 16-25 report deteriorating mental health, with 1 in 4 feeling 'unable to cope'."* [682]

The numbers of children who lost their resilience was not just hypothetical but it was apparent in the data. For some children whose mental health was damaged, parents decided that school was no longer in their best interests and they never returned to school. Others may have had different motivations but in the UK alone 100,000 children did not return to school after lockdown.

WHAT ABOUT THOSE WHO KEPT THEIR RESILIENCE?

Children were recast as pariahs. Supermarkets in Ireland had signs up saying, *"No children at any time as they are vectors for Covid-19."* Old people, made pathologically anxious, saw them as a threat, keeping their distance, shooing them as they passed and even shouting at them to keep clear. In June 2020, Wimbledon and Wembley were packed and pubs, shops, hotels, libraries, places of worship, restaurants and hairdressers were all open while schools remained shut. Children were asked to sacrifice their education, their friendships, their interests, their rights of passage, their ability to share facial expressions and, with vaccination, were asked to undergo a medical procedure for the alleged benefit of those around them. Resilient children may have avoided letting this rejection affect them but it is bound to have clouded their opinion of the adult world.

For other resilient children the effects had a lasting impact on their lives. In *The Children's Inquiry*, Cole and Kingsley, described how the suspension of services for disabled children was *"nothing short of barbaric"* and how the parents felt *"abandoned"* and some feared that cancelled appointments had *"resulted in deteriorations in their children's conditions that may be irreversible."* [683]

UNICEF estimated that 'pandemic' policy could lead to 10 million more child marriages and 8.5 million more child labourers before the end of the decade. Henrietta Fore, the UNICEF Executive Director, said, *"The COVID-19 pandemic has been the biggest threat to progress for children in our 75-year history."* A large proportion of these children will have been resilient but that is of little consolation to them now. [684]

CHILDREN'S ALTERED RELATIONSHIP WITH AUTHORITY

Children witnessed first-hand as their needs became subordinated to covid. The UN convention on the Rights of the Child makes clear that in all public decision making affecting children, *"the best interests of the child shall be a primary consideration."* [685] For decades, this central principle had primacy in the decision making of policy makers and the courts but for covid it was discarded.

The extent to which voiceless children bore the brunt of public health policy was exemplified in the USA in summer 2022. Parts of the USA were continuing to mandate masking in children even under the age of five while at the same time being careful *not* to advise against unprotected gay orgies, to reduce the spread of Monkeypox, for fear of offending the gay community.

Children have been subject to coercion from the state with access to their bodies being requested in exchange for ice cream, pizza or gift cards.[686] Nudging was targeted at ever younger audiences with Big Bird, Elmo and Peppa Pig all playing their part in pushing vaccination. The head of the largest secondary school union gushed about how *"peer pressure"* could be used to increase uptake.[687] The South African government struggled to get to their target percentage of vaccinations and decided to focus on vaccinating children without their parents' consent in order to reach that target. Campaigners started a *"just say no"* campaign reminiscent, for my generation, of the 1980s campaign against recreational drug use, only this time the pusher was the government.

Children were asked to endure pain. My own child was one of many primary children who were asked to repeatedly apply hand sanitiser until their skin dried and cracked leaving open sores which burned with the next applications. In our case the problem was quickly resolved with a sympathetic teacher and our relationship with her was preserved but not everyone was as lucky.

Children were asked to lie. Del Bigtree, founder of the Informed Consent Action Network, won a court case which stopped a new law from passing in the District of Columbia in the USA which would have allowed the vaccination of children without their parents' knowledge.[688] Prior to that win, authorities in D.C. could force schools and doctors to lie to parents with fake immunisation records and even forced medical insurers to lie with bills sent incognito to the medical insurer hidden from the parents.

On a personal level, I will never forgive those in power for putting me in a position of having to tell my children there are some laws that it is right to ignore. Teenagers need boundaries set by society as well as family. When their own parents are unable to justify intrusive societal rules then they will make the first steps towards adulthood on the path of a cracked and crumbling social contract.

Trust in medicine has also been damaged. Some doctors decided that the risk benefit balance was not in favour of vaccinating their own children but they continued to vaccinate other children. My children had learnt how I had trusted their doctors with decisions about their physical health and I had started to pass that baton of responsibility on to them. The pressure from authority supported by the medical profession for my children to have an unnecessary and potentially harmful 'vaccination' meant that I had to snatch that baton back and with that they suffered a further assault to their trust in authority.

In the long term, cynicism of authority and the ability to take personal responsibility may be a good thing for them as adults. However, having a distrust of authority from pre-adulthood is in a different league to a healthy and growing level of cynicism that comes with age. What the impact will be of a generation whose trust in authority has been so abused will be evident in time.

Policy decisions were taken to remove services for children in the full knowledge that harm would result but with doubt about whether there would be any personal benefit with respect to covid. Having justified causing harm by saying children were resilient, the extent of the harm done was worse than feared. After all that, no lessons were learned and lockdown was repeated.

TOP THREE MYTHS
1. Children's lives and development can be paused and switched back on
2. The social and societal role of schools can be temporarily ignored
3. Sacrifices have to be made

19

FUELLING THE FEAR

The whole aim of practical politics is to keep the populace alarmed (and hence clamorous to be led to safety) by an endless series of hobgoblins, most of them imaginary. **Henry Louis Mencken, 1922**[689]

Can the government affect your behaviour and even your identity? The government published a discussion document called *MindSpace* in which they claimed they could, *"but if government is seen as using powerful, pre-conscious effects to subtly change behaviour, people may feel the relationship has changed: now the state is affecting 'them' – their very personality."* [690] Regardless of affecting identity they certainly made every effort to influence behaviour using their team of behavioural psychologists. It started quietly in 2012 by changing the wording of messages to reduce the rate of late or unpaid taxes. Subsequent projects included reducing the size of glasses in pubs to decrease alcohol consumption and trying to increase organ donation through messaging. The latter ended with us having an opt out rather than opt in system. The opt out systems counterintuitively reduced donations in Wales. (It is easier for relatives to agree when they know their loved one actively consented. It is a good example of the law of unintended consequences). The same group produced effective covid fear propaganda which certainly played a part in some patients not accessing healthcare when they needed it.

Behavioural scientists were over represented among advisors to government. When the list of SAGE members was first published in May 2020 there were a total of 148 experts included on committees advising government and a quarter of them were in 'SPI-B' the Scientific Pandemic Insights Group on Behaviours (including the aptly named Professor Fear from King's College London).[691] They aced what they set out to do. In a survey of people from Britain, Italy, USA and China, the British had the highest levels of psychological distress.[692]

NUDGE STRATEGIES

These behavioural scientists used nudging techniques to increase compliance through fear inflation, shaming, peer pressure and guilt. Clinical psychologist, Gary Sidley led 46 other psychologists and therapists to write to the British Psychological Society asking their view on the ethics of using covert psychological strategies.[693] The chair replied that nothing unethical had happened[694] but a member of SPI-B when interviewed by Laura Dodsworth for her book *A State of Fear* said, *"The way we have used fear is dystopian... The use of fear has definitely been ethically questionable. It's been like a weird experiment. Ultimately, it backfired because people became too scared."* [695]

Having unethically scared the public with no plans for how to reverse the damage, no-one from SPI-B showed up to a parliamentary, Science and Technology Committee session when asked, in March 2022,[696] and there was no mention of investigating what happened in the original proposed terms of reference for the UK Covid-19 Public Inquiry when published in March 2022.[697]

There are broadly three strategies used to persuade people to behave in ways the Government wanted: firstly making changes to the immediate environment and limiting choice; secondly informing and persuading and thirdly, manipulating group identity by defining what was considered to be virtuous. Restricting choices, and altering the environment are the crudest methods. Clearly, the laws and regulations had a huge impact but the one way systems, stickers and endless scary posters all created the covid environment that altered what we did and how we interacted.

A more conventional form of persuasion is to inform. However, it is not enough to say there is a new respiratory virus. The nudge team ensured the message was delivered by the right person in a way that seemed relevant to the listener. Certain messages included an incentive or a threat to increase the chance the message would penetrate. Messages that can create an emotional reaction were thought to be more likely to influence behaviour and the 'go to' emotion was fear. The frequent press conferences achieved much of this with a series of authority figures rolled out to deliver messaging about how critical they believed changing people's behaviour would be and how simple: "*Stay home. Protect the NHS. Save lives.*"

Exaggerated fear of death induced by fear propaganda became the driver of policy rather than the actual threat of the disease. The threat from a pandemic is loss of life. However, the focus presented by politicians and the media moved to whichever measure was most frightening. When overall mortality returned to normal levels, the threat moved to become any covid death even if these deaths were expected. When covid deaths were low the threat became case numbers or the rate of spread. When these figures were disappointing long covid or hypothetical future variants became the focus of fear. Being a journalist, politician, scientist or doctor did not make you immune to this fear.

The controversial third level of persuasion proposed by the nudge unit was to manipulate identity. It takes advantage of the fact that we will act in ways that make us feel better about ourselves. At its most basic most people do not like to stand out from the crowd and will copy others in order to not draw attention to themselves. The behavioural unit found that the most powerful way to influence is to create a group identity. If a particular behaviour can be presented as being kind and ethical then anyone who believes themselves to be kind and ethical would need a good reason not to comply. Likewise non-compliant behaviours could be demonised.

TRIBES

The idea that those who were not compliant with lockdown rules were *"killing granny"* took hold early on.[698] This created a critical division. One group was doing all the right things and any dissenters were being dismissed as granny killers. This inverted the no blame strategy seen with HIV in the 1980s. For HIV everyone (regardless of risk) was encouraged to take responsibility for not catching it. Huge care was rightly taken to avoid the risk of demonising a subgroup of the population as disease spreaders. With covid, shame was used and the government told people *"act like you've got it"* regardless of risk.[699]

With covid instead of feeling responsible if you caught it, a finger could be pointed blaming others for giving it to you. The groundswell of feeling around infections being the fault of the unvaccinated or unmasked became so embedded during the Delta wave that people would announce their infections, blame the unvaccinated and then not even spare a thought for who they themselves might have spread it to. The vaccinated could not possibly be granny killers – it was an oxymoron in their thinking.

The government posters were deliberately designed to make people feel not only a threat to their own life but guilt that they might be responsible for others losing their lives. Guilt brings with it shame and the easiest way to alleviate any shame is to find someone else to blame. The government messaging built a tinder box of fear and it only needed a spark of guilt to turn that into a fire of anger and blame. That spark may have come from being separated from loved ones at a critical time, particularly if they were ill or dying. Trish Greenhalgh writes about how she felt when her mother died, *"Anger towards everyone who had ignored masking and social distancing guidance, resulting in the surge of cases from which my mother had become infected. Guilt that I had so rarely found time in my busy schedule to visit her before the pandemic cut us off."* [700] She would not have been the only one who felt this way. The nudge unit campaigns channelled such anger and guilt into blame.

The 'good' people saw themselves as 'pro-science', meaning believers of the government narrative. They followed *'The Science'*. It did not take much effort to include 'pro-vaccine' as part of the identity of the 'pro-science' crowd, not least because 'anti-vaccine' already had such negative connotations in the minds of the public. Keeping a distance, wearing a mask, testing incessantly and taking a vaccine were all portrayed as virtuous behaviours.

There were plenty of people who would comply with distancing and mask wearing even when they knew they made no impact, for fear of upsetting someone. Many people I knew, who understood full well the futility of the things they were being asked to do, nevertheless complied. They did so because of the fear of being accused of not having done everything possible should someone they know become infected with negative consequences.

The pressure people felt to comply was so overwhelming that for the majority there was no feeling of choice. For the compliant, who felt under this pressure, the mere existence of a group who seemed resistant to this pressure exposed a difference. How did it not affect them in the same way? If I am doing all this to show I care (even if I know it might not work), does that mean these people are uncaring?

Mask wearing became part of identifying your tribe and became political in the USA, with the unmasked being assumed to be Republicans. The denigration of people who were not conforming was the perfect way to achieve compliance. People would rather be liked than be right. Psychology experiments have shown this clearly. The classic example is the Asch conformity experiment.[701] People were asked to compare the length of lines on a board. They were put with a group of seven actors pretending to be participants in the study but who deliberately gave incorrect results. Only a quarter of the participants always stated the truth and avoided conforming with the incorrect group answer at some point. Afterwards, they said they had agreed with the consensus for fear of being ridiculed. These are people who knew they were participating in a psychology experiment. How much worse might this be in the real world and with a more ambiguous question?

If people will lie just to conform, what more would they do to be seen as ethical? If what they are being told is ethical causes harm, how much harm would a person be prepared to cause in order to be seen as virtuous? If they closed their eyes and ears to the harms being caused, how long could the harm be sustained?

As well as wanting to be seen as ethical, there is an even stronger drive to not be seen as unethical. The power of the 'unethical' label arises because our brains associate immorality with feelings of disgust. It is therefore a small step for the same group to be considered unclean and a threat to your safety. Causing a minority to be hated and disgusted is a very dangerous strategy. For decades, modern societies have successfully prevented any minority being considered inferior or undeserving let alone immoral. Humans are tribal and ensuring people do not discriminate takes effort. With covid the rules were inverted. Apparently, because compliance was presented as a choice, discrimination was no longer unethical. At last, there was a group that it was acceptable to hate. When a government is actively trying to breed hate among its citizens, it is worth taking a step back and asking, why?

BEING ETHICAL

Who were these people who were prepared to risk being labelled as unethical? What was motivating them? In every instance I can think of, these people could see the harm that restrictions were causing especially to the most disadvantaged both in the short term and with the longer term economic fallout. They could see the harm that comes from a faceless society where fear is perpetuated. They were concerned about the effects on children's development from not seeing how we talk or how facial expressions convey our meaning. The harm that comes from spending £37 billion on a testing programme that made no impact on the trajectory of the disease but created mass paranoia and inflated case numbers. For these people the coming economic Armageddon as a result of the deliberate economic destruction of lockdowns would clearly cause more harm than covid ever could. Ultimately, these people were also identifying as ethical.

The drive to be ethical is hugely powerful. The government changed the definition of ethical and managed to co-opt celebrities and even religious leaders to promote their new definition. Making people stay in cramped, overcrowded accommodation in the heat while shutting the only local park, as happened in Tower Hamlets, was now declared 'ethical' by people who had not only their own garden but access to less crowded public open spaces. Staying at home on full pay while others took what you believed were risks to deliver your every convenience to your door was now 'ethical'. Shutting small shops while letting big business continue was 'ethical'. Separating the dying from their loved ones and restricting their funerals was 'ethical'.

So long as the stories of harms caused by policy were suppressed there was no problem with continuing to identify as ethical. As those stories slowly leaked out there have already been attempts from high profile people trying to distance themselves from past statements and make out that they did not support harmful policies. They did. Almost everyone did.

The government's use of behavioural psychology and nudge strategies to influence people's behaviour during the covid years has been extensive and controversial. They aimed to manipulate people's choices and identities to promote compliance with lockdown rules and vaccination programs. This approach involved the use of fear and guilt to create a sense of group identity and virtue. The government's messaging led to a groundswell of pressure to comply, making it difficult for individuals to feel that they had a choice. The harm caused by the policies was often ignored or suppressed, and those who raised concerns were labelled as unethical or immoral. The long-term impact of these policies and strategies on the population's behaviour and identity is yet to be fully understood.

THE OPPOSITE OF FEAR

Courage is feeling fear, not getting rid of fear, and taking action in the face of fear. **Roy T. Bennett, 2016**[702]

The opposite of fear is courage. Someone shows courage when they have something to defend and the confidence to do so. There were principles that needed defending including medical ethics, the rights of the individual and their freedom, whether of expression or movement or the ability to earn. For some it was more personal and they were defending people they loved who were being harmed by restrictions. The courageous protested. Many NHS workers defended bodily autonomy for themselves and their patients by refusing to comply with mandates, despite fear of losing their livelihoods. The government was forced to fold.[703] Others, despite fear of confrontation, spoke out courageously in public challenging the narrative.

There are, however, two types of fear. Firstly, there is physiological fear which is a healthy response proportionate to the situation. Courage is the opposite of this fear because the fear is real and courage is required. The second type of fear is pathological. The propagandised fear created by the government and the media bore no relation to reality. Telling people who have been terrified by propagandised fear to be courageous would only reinforce the idea that there was a reason to be scared. That kind of courage would also be pathological because it could be present in the absence of a genuine risk.

Pathological fear stripped people of their sense of confidence, their inner peace, their ability to plan ahead rationally. The search for safety that it provoked stripped people of their ability to trust their own instincts and interact naturally. A disconnect occurred between people and those they loved. Any contact was presented as a risk and those with a differing view were presented as untrustworthy. Relationships that were built on love and trust over many years were torn apart as a result.

Courage could not help healing from this pathological fear. Nor can it help people reconnect with those they love or restore their confidence and inner peace.

More fear is not the answer either. Fear of covid has been displaced with fear of the war in Ukraine and the pending economic crisis. For some people this displacement seems to work as a distraction but they are still living in fear. Fear is the route of division and gives power to the people directing it.

How can pathological fear be countered. If a child wakes after a nightmare, in a state of misplaced fear, parents comfort them. Firstly, they listen, acknowledging that nightmares are scary. Then they seek to reassure them. They might hold the child tight, giving them a surge of the bonding and love hormone oxytocin, letting them know they were loved, not alone and could depend on their parents. The opposite of pathological fear is comfort. Comfort comes from a place of love.

Psychiatrist Elizabeth Kubler-Ross emphasises the importance of love in overcoming fear saying, *"There are only two emotions: love and fear. All positive emotions come from love, all negative emotions from fear. From love flows happiness, contentment, peace, and joy. From fear comes anger, hate, anxiety and guilt. It's true that there are only two primary emotions, love and fear. But it's more accurate to say that there is only love or fear, for we cannot feel these two emotions together, at exactly the same time. They're opposites. If we're in fear, we are not in a place of love. When we're in a place of love, we cannot be in a place of fear." 704* Reconnecting with family and friends and providing comfort is a way to replace fear with love.

Where some sought comfort from friends and families others turned to God. Many of the people I have met because of covid were immune to the fear because of their religious faith. Others have converted to a religious faith in response to seeing society so harmed by the artificial covidian cult and what they perceived as the evil against humanity that it perpetrated. Charles de Foucauld a catholic hermit said, *"Faith strips the mask from the world and renders meaningless such words as anxiety, danger and fear, so the believer goes through life calmly and peacefully, with profound joy, like a child, hand in hand with his mother." 705* Every religion offers faith as an antidote to fear. The bible also talks of comfort, *"Yea, though I walk through the valley of the shadow of death, I will fear no evil: for thou art with me; thy rod and thy staff they comfort me." 706*

In Cloud-Covid-Land, religious leaders did not offer faith or love as an antidote to fear, they offered vaccination. The Dalai Lama urged the world to *"be brave and get vaccinated." 707* The Archbishop of Canterbury said *"to love one another, as Jesus said, get vaccinated, get boosted." 708* The Pope said getting vaccinated was, *"an act of love." 709* Feelings of love and worship were instead directed towards safety, the NHS and the high priests and this played an important role, alongside fear, in the emotional capture of people's thinking.

More than ever before, families, friends and communities needed to reconnect to provide comfort to those most harmed by pathological fear. Many of those worst affected continue to lock themselves away from the human connections that could help them heal. Loneliness is strongly linked to depression and certain diseases.[710] The UN considers solitary confinement of more than 15 days to be torture.[711] What lonely people need is comfort.

Even with the love of friends and family and even with religious faith, some struggle to find inner peace because of a new focus on their mortality. There is a choice when we are faced with mortality. Either be fearful and restrict our lives in an attempt to avoid fate or accept its inevitability and ensure life is lived to its fullest. The closer one is to death the more that principle should apply. Fear of death and love of life are also opposites.

There were attempts to distort the meaning of love equating it to compliance with policy. By complying, people could say they did everything they could to prevent the spread and thus they cannot be accused of being responsible for harm. Living like this is living in fear of being accused of spreading. Compliance is not a display of love. Even if a first encounter with a variant can be delayed, eventual exposure is virtually inevitable. Compliance did not stop the spread. The vaccinated, the masked, the public health officials and the politicians themselves have all contracted and spread disease. Real love would mean blaming the virus for your illness, not your nearest and dearest. To love someone means ensuring they would not be fearful of your response if they unwittingly passed a virus to you.

As well as those who continue to be trapped by fear of covid, those who have lost their livelihoods or been injured or killed by vaccines, however many of them there are (which will be discussed further in *Spiked: A Shot in the Covid Dark*), deserve our comfort and love. What kind of a society would we be if such people were left voiceless and shunned as has happened for at least two years? Regardless of your religious faith or lack of faith, acts of kindness to others decrease stress and promote feelings of optimism, connection and satisfaction. We need to return to a society where people are not judged for their performance of the covid rituals, or their opinion on how covid should be dealt with. The caricatures that the powerful have painted to create and maintain division should be seen for the oversimplified hate messaging that they are.

The opposite of fear is indeed courage but the antidote is faith or comfort from a place of love.

BELIEF TWELVE: ZERO COVID IS ACHIEVABLE

We have the opportunity to do something no other country has achieved – elimination of the virus. **Jacinda Ardern, April 2020**[712]

The most extreme pro-lockdowners pushed the concept that an earlier harder lockdown would have prevented covid deaths. They claimed that instigating early, draconian restrictions was the best way to avoid… restrictions. Their Orwellian description was that, *"Zero-Covid is liberty and safety rolled into one."* [713] There were influential proponents of this strategy including ex-Health Secretary Jeremy Hunt, First Minister of Scotland, Nicola Sturgeon, trade unions and Independent SAGE. The chair of the latter, Sir David King said, *"I don't believe the schools should be opened until we have approached zero covid."* [714]

The premise was based on the hypothesis that the whole of South East Asia had avoided covid, not because of differences in immunity or seasonal triggers but because every one of those countries had proved better at interventions than any of the countries that had covid waves.

A fundamental belief of zero-covidists was that we had not locked down hard enough or long enough to reach zero covid. Once we had endured a *"real lockdown"* to get to zero, tightly sealed borders and the continual threat of a sudden lockdown would eliminate the virus. Airborne or surface spread cannot be controlled in the way they imagine. One doctor described covid as like dropping red ink into a bath. There is a short period where you might be able to scoop it out but it rapidly turns all the water pink. Shutting borders, restricting movement and using contact tracing to try and prevent it at that point is useless.

The ability of the virus to find its way into every country despite the strictest of restrictions at borders demonstrates the futility of trying to control a virus this way. Zero-covidists claimed they did not want to ban travel but argued for *"travel bubbles with other Covid-free jurisdictions."* [715]

The belief that covid could be eradicated rests on a total denial of animal hosts including dogs, cats, mink and bats and its airborne nature. David Livermore, Professor of Medical Microbiology, asked how we still have 50,000 cases a year of gonorrhoea despite all efforts to eradicate it. *"You can remember far better who you slept with, I hope, than who you breathed next to on the tube."* He also points out that we have only managed, after decades of mass vaccination, *"to eradicate one viral disease, Smallpox, caused by a particularly stupid, unchanging virus."*[716] Professor Donald Henderson who directed the international campaign against smallpox, co-wrote a paper which put the success down to the fact that every case could be accurately diagnosed; interventions to prevent transmission were 100 percent effective and there was no animal host.[717] The only other disease to have been eradicated, rinderpest, affected only even-toed ungulates, like buffalo. Even the bacterium that caused the Black Death still causes illness to this day.

For periods between waves where we have approached zero covid, the inaccuracy of covid testing meant that, with mass testing and non-infectious positives being described as cases, zero covid could never be reached even when there was minimal chance of spread.

Zero-covidists insulted huge numbers of people by suggesting that their policies *"would be mentally good for the population,"* [718] when the Royal College of Psychiatrists was warning of a *"tsunami"* of mental illness caused by lockdowns.[719]

They were also happy to create perverse incentives for failing or fraudulent businesses, proposing firehosing of government money at businesses to justify repeated and destructive closures and showing a total disregard for how that money could have been better spent.

Numerous restrictions which, even in the best light, were likely to have only minimal impact, were argued for on the basis of the Swiss Cheese Model. The idea being that each layer of cheese provides a barrier but the flaws in each barrier are represented by a hole. Only when all the holes align is there a risk of infection. The irony of describing a mask as Swiss cheese seemed to be lost on them. The analogy was taken from the aviation industry where each layer of cheese represented an incredibly effective intervention with a tiny hypothetical failure rate. For example, it is used to justify the presence of a second pilot to mitigate any sudden incapacitation of the primary pilot. The layers for covid were more like a series of superstitions with more holes than cheese. Ironically, the same people who pitched this model that justified the use of leaky interventions, argued that the same interventions could lead us to zero covid.

Among the zero covid proponents was a group who adopted the name of the public health hero, John Snow, who deduced the source of a cholera epidemic in Soho, London, which he halted by removing the handle from a water pump.[720] They claimed that the answer to covid is to provide clean air. It is hard to comprehend how they fail to see the difference between supplying a separate clean water supply for public consumption and the impossibility of separating clean air from dirty air.

Those that thought they would be John Snow-like heroes by calling for the air to be cleansed were missing the point that the susceptible were very likely to be exposed eventually even if some indoor environments were cleansed. Improving the air in hospitals, where people are temporarily more vulnerable, is something to be supported. Calls for the same for care homes are more questionable. Care homes are needed to ensure the last years of people's lives are as dignified and comfortable as possible. Prolonging lives at any cost means scarce resources[721] cannot be spent to improve the quality of life for residents by, for example, increasing average daily interactions with staff from two to eight minutes which reduced aggression and agitation in dementia.[722]

The logical conclusion to zero-covidism played out horribly in China in 2022. Harsh lockdowns were carried out. There was forced testing with even very young children removed from parents and placed in cells alone to quarantine, with a hatch for food. Apartment blocks were sealed with minimal food deliveries leading to protestors shouting *"we want to eat."* [723] Pets were collected in sacks, from apartments where someone had tested positive, to be killed. China lost more than 3 percent of their GDP every month as a result.[724] At one point 71 million people were estimated to be in lockdown or about to be.[725] Even China gave up on the policy after considerable harm, at the end of 2022.

Calls for zero-covid were a fantasy born out of hysteria and have thankfully been abandoned.

TOP THREE MYTHS
1. Covid could have been prevented from spreading globally with more effort
2. Harsh restrictions could prevent more restrictions
3. Covid could be eradicated like smallpox

21

PURITANISM

> *[T]he sin of every individual man, does but repeat and renew the cause of sickness unto him.* **Mens Sane in Corpore Sane, Cotton Mather (Puritan Clergyman), 1698**[726]

The parallels with covidism and puritanism aren't subtle. Early puritans believed that disease was a result of sin. Hairshirts were worn to cause a repentant sinner discomfort as an instrument of penance. Liars, gluttons and gossips were punished by being made to wear masks, such as a scold's bridle, as a sign of submission and reducing their ability to communicate. Puritans banned choral music and instruments from their worship, mixed sex dancing and theatre. Boxing, blood sports and gambling were banned as was football on Sundays. Alehouses were closely regulated with licensing requirements and restricted opening times and drunkenness was publicly punished. Christmas was banned in 1647 as the festivities were thought to be impious but people secretly gathered to mark the holy day and sing carols.

An interesting aspect of those virtue-signalling about their participation in restrictions was that the harder the restrictions were the more they believed they elevated their status by demonstrating compliance. Someone who pointed out that they had spent ten hours straight in a face mask in the heat, was not interested in what impact that had had on those around them. The point was that it had been unpleasant for them and therefore, their sacrifice for the greater good, made them more worthy. It was more important to be righteous than to be right.

The government published over 7,000 words of guidance for worship which banned an indoor congregation from *"communal singing, shouting and chanting"* and suggested, where musical instruments were used, they should be *"positioned in a way that avoids face-to-face performance."*[727] Dancing was banned with people arrested for dancing in Scotland as late as 31st December 2021. Theatres were shut. Sports were banned in person during lockdowns in the UK – even outdoors – even where there was no close contact e.g. tennis. Pubs had restrictions on opening times and the number allowed to mix as a group. The most ludicrous rule was that drinks could only be purchased with a 'substantial meal', causing much harm to pubs in working class areas while allowing gastro-pub establishments in middle class areas a chance at recovery. After several MPs publicly stated that they considered a scotch egg to be a 'substantial meal', the demand for them rocketed but publicans were still fined for not ensuring that those eating them were seated while doing so. Christmas gatherings were banned in 2020 but many people ignored this rule and had family reunions. Rather than surging as a consequence, cases peaked within days and started to fall.

The restrictions were sometimes cruel and nonsensical. Unfortunately, the desire to control others seems to be a prevalent human trait. Covid regulations gave people with an officious nature a chance to exercise it. In the second lockdown in Wales, shops were stopped from selling non-essential items. The supermarkets cordoned off items such as books and birthday cards. Parents were stopped from buying shoes for their growing children and old ladies from buying gloves to keep their hands warm.

In England, Police officers searched shopping bags to ensure only 'essential items' had been bought[728] and admonished children for building a snowman in the park. Lone hill walkers were harassed by the police. Vicky Hutchinson was fined £10,000 for holding an outside balloon release after her father-in-law's death.[729] Two students at Leeds University were also fined £10,000 for organising a snowball fight.[730] Three other students were given a criminal record after refusing to pay a £40,000 fine for hosting a party in October 2020.[731] An elderly man was fined for talking to people at his allotment.[732] A 72 year old man was jailed for six months after selling mince pies at Christmas in a region that had stricter rules to neighbouring regions at the time.[733] A homeless man was brought to court after being arrested outside Liverpool Street Station. The arresting officer said, *"I was arresting him for breaching coronavirus conditions because he had no address."* [734]

A pub landlord was fined £4,000 for holding a party for employees of his business in December 2020. The people at these parties were working together during the day but it was illegal to enjoy the same company after work. The law was officious. That same day Boris Johnson was also attending one of six parties he attended for which he was fined £50.[735] Of course the lawmakers should not break their laws. More importantly they were not scared unlike those they had scared.

Trish Greenhalgh said, *"I think the key behavioural change we need is SILENCE while unmasked for eating / drinking, since it's vocalising that releases viruses."* [736] How much of these additional restrictions were really aimed at decreasing the spread of disease and how much were about piety? Was this right or righteousness?

In defence of the original puritans, they might have been horrified by the way people have been treated since covid. They believed that no-one should be deprived of liberty without due process. The Body of Liberties was a puritan legal code listing liberties rather than restrictions and including the rights to freedom of speech, jury trial and the prevention of cruel and unusual punishment.

22

FINDING A MORAL COMPASS

Those are my principles, and if you don't like them… well I have others.
Groucho Marx, c1970s[737]

Two percent of people in the UK were opposed to lockdowns from the outset. These people were largely those who believed in important principles and human rights and were horrified to see them overturned with so little protest. I regret not being of them. It was a select group of people who could see the mistakes playing out from the very beginning. Most human rights lawyers said nothing. However, former supreme court judge, Lord Sumption,[738] barrister Francis Hoar[739] and law lecturer Robert Craig[740] (who happens to be my brother), saw how the 1984 Public Health Act, which allowed for quarantining of a 'group' of people who were sick, was being abused to legislate that the healthy should be quarantined and that the entire country could be considered to be a group.[741] It was somewhat ironic that the premise of the legislation was that restrictions must be imposed to protect the vast majority who were not infected and yet the legislation was dependent on treating everyone as if they were sick. Francis Hoar fought a case, funded by businessman Simon Dolan, on the legality of lockdown in the High Court. Judges are not immune to fear and, despite the strength of the legal arguments the case was lost. Lord Sumption summed up what happened saying that the courts *"shamefully failed to intervene"* adding, *"The deafening silence of the human rights lobby in the face of this really is quite remarkable."* [742]

Over time others joined the sceptics falling into two categories. There were the data nerds who could see that the government narrative did not add up and there were the rebels who thought the laws were unjustified regardless. Every skin colour, every age, every religious group, every political persuasion, every background and skill set and every region of the country were represented. Protest marches looked, as one person described, *"like a supermarket in town had just had a fire alarm go off."* There was no easy way to predict what side people would take.

Many had strong religious convictions and perhaps their existing beliefs immunised them from the new government beliefs. The adage *"if you don't stand for something you'll fall for anything"* seems apt. Nick Hudson, founder of PANDA, talks of a God-shaped hole, *"by cancelling the God, by wiping him out because the bearded man in the sky is too far-fetched for many, you cancel the value system, creating a hole. What I like to refer to as the God-shaped hole. And into it comes Fauci with a spreadsheet. A utilitarian system."* [743] In Latin, spiritus (from which we get expired) means both breath and spirit. Consequently the phrase, the breath leaving the body, is the same as for the spirit leaving the body. In Cloud Covid Land, the spirit had left society.

I did not have the strength of principle or religious belief that the two percent had. I would say that I was ethical without being principled. In retrospect that was a dangerous place to be. Ethics can be argued in different ways in different situations such that important principles can be overridden. One medical ethicist once said to me, *"just tell me what you want it to say and I'll write the argument that supports that."*

I did not oppose the initial lockdown. I accepted the premise that they were needed to stop the NHS being overwhelmed as an ethical one. I did not even stop to think about people's rights being overridden or the irreversible harm being done to ethical principles that had held our society together for so long. I did not stop to question the way harm from lockdown was pitted against supposed benefits as a societal balance even though the benefits were for a different sector of society to the ones being harmed. For all those omissions I am sorry.

A fundamental breach of principle came with the overriding of bodily autonomy and informed consent through coercion and vaccine mandates. I cannot do justice to that discussion here and have included my thoughts on that in my second book, *Spiked: A Shot in the Covid Dark.*

RE-DISCOVERING PRINCIPLES

Our society was built on a set of values. These values created a strong social fabric over hundreds of years, and were refined with time. They are based around respect for the individual and their agency but they include the principle that adults have a duty to protect children and not the other way around. One principle of democracy is that more powerful people should not assume they know what is best for other people. As the actor Clifton Duncan, who lost his career to his stand against mandates, said, *"Their arrogance has deluded them into thinking their credentials or their position qualify them to dictate to me what is for my own good, as if they know and care for me better than I know and care for myself."* [744]

I had thought that, if restrictions did prevent death, then implementing harmful lockdowns could be justified so long as the overall years of life lost by intervention were fewer than those saved. Looking back I think this was utterly wrong. Government should never deliberately legislate in a way they know will cause substantial harm. I am ashamed of having thought such an equation could be justified. All the more so when it was based only on hypothetical lives saved and excluded unmeasurable harm.

Writer, Tom Moran made an important point to me, *"If it would be wrong to breach the rights of a minority of the population by imposing restrictions on basic liberties, like sitting on a park bench, why is it alright when they are imposed on everybody? Oppression is oppression is oppression. Whether it is being targeted at a minority, the majority or indeed everyone, no moral society should tolerate it from their government, no matter the alleged justification."* Had we all, at a time of fear, held strong to basic principles could it have prevented government overreach?

Covid saw an attack on freedom of speech, freedom of movement, the right to protest, the right to earn, the right to a family life. It is easy to defend rights when the powerful support doing so. The strength of human rights comes not from the paper they are written on but in the willingness of people to defend them when the powerful do not. Human Rights lawyers remain happy to attack violations committed by certain governments abroad but have been silent on the attacks on rights perpetrated by governments in the West. Were the rest of us wrong to think we could rely on human rights lawyers to defend rights? If we had all taken responsibility for upholding these principles then perhaps aboriginal Australians would not have been arrested after breaking out of a quarantine camp, even though they tested negative?[745] Perhaps, the families would not have been stopped from being at the bedside when their relatives were ill or dying? Perhaps Canada would not have frozen the bank accounts of those protesting lockdowns and vaccine mandates, even those who had made trivial donations to the cause? What is the point of having human rights if they are not defended when they are most needed, when the threat comes from those in power?

Fundamental principles and rights are meant to be immovable, not shifting according to general opinion of the balance of risks and benefits. I get that now. It has been an education.

23

A LETTER TO MY CHILDREN

Dear Children,

Your mum has been very busy over the last couple of years. Sometimes you have rightly accused me of being obsessed with covid and I wasn't able to fully explain to you why I thought it was so important. You couldn't understand why I was working so hard for no pay. I am writing this to you now so you might understand one day. This book only introduces some of the issues, but I think they are enough to let you understand how I started on this journey. My next book will show quite how serious things became.

Lockdown caused enormous harm. Some of it you were painfully aware of but much of it was hidden from you. Children were deprioritised thanks to covid policy but blame for that does not lie only with those in power. My generation failed to either speak out or to refuse to comply with the harm. Currently that is my generation's legacy to yours.

Never before has a country confined people to their homes. Never before has fear been so successfully and deliberately manufactured. Never before has so much been spent without thought to the outcome of that spending. Never before have leading voices been universally censored, people fired for not taking medical treatment and bank accounts frozen without legal process.

These serious mistakes are how my generation will be remembered for evermore. It is shameful. Our parents' generation boast of putting people on the moon and we put people in their homes. Each generation has an opportunity to improve society for their children. In terms of doing better than your parents, your generation has a low bar to clear. My generation is currently leaving you a world with shrinking horizons and debt, where traditional values have been trashed and thrown aside in the name of an emergency.

People prioritised protecting their reputations for fear of being cancelled. I have first-hand experience of what they feared. This led to compliance and silence from others. I do not know how long it will take for this to be overcome. One day I hope that people will care more about protecting their values than their supposed reputations. As Edmund Burke said (if only apocryphally), *"The only thing necessary for evil to triumph in the world is that good men do nothing."* [746]

If our systems had failed us a little the solution would have been to patch them up. They have failed us totally and patching up and ignoring the underlying issues will not be adequate. What an opportunity that presents for gifting you something better than we inherited. A return to and strengthening of the traditional values that have served our society so well and for so long.

The most fundamental part of that process must be a recognition that our values as a society are needed most at times of emergency. Our society needs high ethical standards based on foundational moral principles. Without these a desire to appear ethical can allow core principles of western society to be swept aside. One lesson from this period, not to be forgotten, is that the deceptive notion of it being ethical to do anything 'for the greater good' is dangerous because it makes people lose sight of the inherent sacrifices of minorities. Moral frameworks that have served humanity for eons should not be bent or broken even in the service of a perceived higher moral purpose. Individuals deserve respect and agency and the right to decide what is best for themselves. Adults have a duty to protect children. The right to earn, to buy, to move and to have meaningful human relationships are not privileges. The powerful must be restrained from thinking they can legislate to remove these rights even in an 'emergency'.

POLITICS

My generation inherited liberal democratic values and somehow, in a very short time, we relinquished them, giving governments control over our bodies, money, speech and even relationships with each other. Reversing this shift and ensuring it cannot be repeated will not be simple but we have a responsibility to try and do it.

Politicians from every party have failed to call out the unethical manufacturing of an environment of fear. None have provided any reassurance that lockdown will not be used again (even for covid). Similarly there has been no reassurance that the worst aspects of digital currencies and the coming WHO treaty will be dealt with to protect human rights.

We need to protect ourselves from future overreach of power, whether by national government, the heads of institutions or international bodies. Japan has a post-war constitution to stop authoritarian overreach and it is illegal there to deny people their livelihood and liberty en masse. Those in charge of institutions including care homes and hospitals were also guilty of overreach of their power. It may be that the only way to stop this is to legislate our way back. Laws to force care homes and hospitals to allow visitor access and end forced mask wearing are needed along with legal protections preventing insurance companies dictating safety minutiae. New legislation is needed to restrict the powerful and empower and ensure the independence of those meant to protect people such as the Children's Commissioner. Furthermore, schools could be designated as part of our Critical National Infrastructure so they cannot be shut down.

I want you to recognise the dangers in the abuse of language and the philosophy of collectivism. Covid policy was a good illustration of both. Freedom and democracy are not to be taken for granted. For many, the last few years have brought into focus the importance of the right to protest. That right is not there for when a benign government is in power. The reason we need those rights is because of the risk of a malign government in the future. Any government who tries to remove them is by definition not benign and you should protest against them.

Given the utter failure of human rights legislation to protect people when it was most needed, with no apparent negative consequences for those responsible, there needs to be a rethink about where power lies and how your rights are defended. How do we make people stick to ethical rules during a period of hysteria? How do we stop another episode of fear driven decision making? These are huge challenges to solve and we might need to enlist your generation's help to complete that task.

MEDICINE

The loss of trust in public health and doctors provides an opportunity for people to take more responsibility for their own health with doctors as advisors. I want to see a cultivated culture of whistleblowing in medicine including systems for anonymous reporting of concerns. Surgeons already have monthly meetings where post-surgical injuries and deaths are discussed with colleagues in an attempt to mitigate future harm. Trainee doctors currently have to provide evidence of participation in audits, research and teaching. Participation in the critical investigation of harm could be added to this list to help cultivate a healthier culture of error correction.

I want a future for you where if you are sick, a doctor will greet you with an outstretched hand and a smile. Where your doctor can decide on care that best suits your needs, rather than what fits the structure of a centrally and anonymously prescribed algorithm. I want it to be laughable that catching a virus could be considered a punishment for sinful behaviour or be blamed on other people, particularly children. I want you to live in a world where people with a chronic cough are not treated as pariahs. I want the NHS to have increased intensive care capacity and stop using a business model where operations are cancelled even in quiet winters.

The UK can be proud of its relative openness in sharing statistics and we should encourage a culture of further transparency. Importantly this should include the publication of data on deaths referred to the coroner. Currently those deaths most likely to be preventable are not reported on for months or even years. The coroner's system as a whole needs updating to help members of the public refer deaths and appeal a decision.

Scientific debate needs to be revitalised so that people can contribute honestly without fear of being cancelled. All voices must be heard if we want to avoid the dangers of high priests with claims of omniscience. Scientific publishing needs to be freed from the biased filtering that currently occurs. Thought must also be given to ways of allocating research funding to reward those whose work challenges the orthodoxy and encourage breakthroughs.

COMMUNITY

I want you to grow to appreciate the importance of community and real world social networks. Your experiences of the last two years will, I hope, mean that you consume media and propaganda, if at all, with considerable scepticism. You saw challenges in our relationships with family and friends but also saw us uphold our values of tolerating other beliefs and having open debate. In a world of division I hope you manage the same.

You have seen us meet people from all over the world with totally different approaches to life and views of how the world should work. Learning from them or debating them gave me a much deeper understanding of my own beliefs and principles. I hope you manage to see opposing views as enriching in this way, not a threat.

I want you to know the power of speaking up when something is wrong. It is fundamental to error correction in all aspects of life and is what makes democracies strong. A culture of free speech can start with you. Confronting the boss, speaking up when something looks unethical, speaking out when free speech is under threat, questioning things that don't seem right and defending colleagues and friends who do so are all ways you can contribute. That culture was lacking before covid and meant that speaking out and correcting errors was slower than it otherwise might have been. It also means you or anyone can help reinvigorate a culture of openness.

An admission that no-one is omniscient is urgently needed. If you do not listen to others you will be more likely to get things wrong. It takes wisdom to know that no-one has all the answers. Everyone is flawed and, without a willingness to listen to all voices our errors – and those of others – will go uncorrected.

I hope you find your voice, take responsibility for learning and forming independent opinions and use that power for good. Speak up loudly, alert people to inhumane treatment especially when the victims have not been heard and act humanely regardless of what is asked of you.

I love you all,

Mum

xxx

24

LEAVING CLOUD-COVID-LAND

Cloud-Covid-Land was a cruel place. One illness was made a priority over every other aspect of life overshadowing births, marriages and deaths. Families were separated, the elderly in care homes were stripped of everything that makes life worthwhile and children's wellbeing was sacrificed for the Greater Good. It is no good thinking of these harms as sunk costs. As well as ensuring there is no repeat, those who were harmed need their voices heard and efforts must be made to repair the damage caused to them.

In a society where religion had withered and where scientific endeavour had led to a massive expansion in knowledge and innovation, scientists were trusted to be omniscient. Those in trusted positions of authority failed to challenge anything that was said and instead simply amplified the oversimplified messaging of the covid high priests and false prophets. Far too much unquestioning faith was put into stories based on assumptions that were dressed up as science by modellers.

Into the void that religion left behind, Nick Hudson's *"God-shaped-hole,"* flooded fear. In a world where spiritual health had no worth, and where baby boomers were reaching an existential crisis, physical health and fear of death became a prime concern. Covid deaths alone became the measure of a successful health service despite a lack of evidence that interventions would prevent them. Meanwhile policies that would predictably cause deaths were pushed through under the banner of keeping people safe.

In Cloud-Covid-Land people's perception of risk was distorted and their fear of death exaggerated. People developed obsessive compulsive behaviour traits for fear of what was in the air or on a surface – air and surfaces that had always been chock full of microbes that our bodies are designed to defend us from.

Scientific 'evidence' was produced to support policy by creating fantasy graphs based on unjustified assumptions. Science is based on real world measurements and modelling based on assumptions is the antithesis of that. Other 'science' dependent on modelling, where dissent is not allowed and where there is political interference should be treated with as much respect as *'The Science'* modelling presented for covid.

Apparently, such modelling was produced to order, with Graham Medley of SAGE saying, *"We generally model what we are asked to model. There is a dialogue in which policy teams discuss with the modellers what they need to inform their policy."* [747] Evidence that contradicted the mainstream narrative was rejected even for hosting on pre-print websites as well as for peer review. Science requires openness, debate even polite conversations among friends and all of those were missing. The structures that allow for career progression in science did not help. Nor did the fear of being cancelled for being wrong. No-one is never wrong. The Evidence Manipulation Triad of extrapolating, excusing and excluding was used repeatedly to maintain a one sided and shaky narrative. New unproven hypotheses were given prominence over ideas that had been established after decades of research. The needs of institutions and politics were antithetical to science and have strangled it.

FALLING INTO CLOUD-COVID-LAND

If we are to leave Cloud-Covid-Land we need to understand how we arrived there. False beliefs were born out of an environment of deliberately cultivated fear. *Expired* has only scratched the surface of what was believed in Cloud-Covid-Land. There was much more to the Cloud-Covid-Land belief system that will be covered in *Spiked: A Shot in the Covid Dark* including the origin of the virus, covid treatments, the vaccines and long covid. *Expired* was largely the story of 2020. The story has not run its course because most have only heard a very distorted version of events.

Expired covered three broad topics. Primarily, *Expired* was the story of aerosols in exhaled breath which no amount of draconian legislation can control. Secondly, it was the story of a fear of death which led to policy-induced spikes of death which are hard to disentangle from the deaths due to a virus. There is no doubt such deaths from policy existed even if arguments can be had about quantifying them. Finally, the whole situation was heavily dependent on beliefs and models that continued to be used long after evidence demonstrated their foundational assumptions had floundered. The belief that covid would have brought with it a tsunami of infections rather than seasonal waves continues to be pervasive despite real world evidence showing it was contradicted long ago.

To fully leave Cloud-Covid-Land we must first fully enter it. To really understand Cloud-Covid-Land we have to dive deeper into a darker place. In *Spiked: A Shot in the Covid Dark* the high priests go a step further in 2021. After maximising fear of the unknown, a saviour narrative was born in the form of the vaccines. Anyone who expressed doubts about the claims made of the saviour vaccine was a blasphemer and a few were singled out for witch-hunts. Questions about how we got so lost and where we are headed are also addressed. Other questions cannot be answered for certain yet. Why was the UK Parliament bypassed and kept virtual even when schools were back to normal? Why were policies repeatedly adopted in total synchrony across the world? Why were dissenting voices systematically censored across the globe? Why was fear maximised, even after the initial threat had passed?

Further revelations will doubtless come, but no one can start to unveil what has happened without risking being wrong. Whichever explanations prove to be the more correct over time, the current widely held narrative does not stand up to scrutiny and its cracks are ever widening.

The beliefs covered in *Expired* stemmed from the core tenet that spread could only occur through close contact. From that came the beliefs that masking would work and zero covid was possible but also that asymptomatic spread was a significant problem. The only way to hold on to the belief in close contact being the only route of transmission was to believe that each infected person infected on average more strangers they interacted with than people they actually knew or lived with. Even when sources could not be traced and when the hypothetical asymptomatic 'cases' did not develop antibodies the belief was maintained. Complete faith in an ideologically perfect medical test fuelled the illusion of asymptomatic 'cases'. This belief persisted despite evidence that the results were not reproducible during periods of high prevalence when almost all positive cases were sick, and accumulating evidence indicating that viral production was much higher in sick individuals.

With symptomatic people already isolating, a belief in asymptomatic spread provided part of the supposed rationale for lockdowns. The belief that everyone was susceptible to each wave and fantasy modelling completed the rationale. Critical feedback by way of risk benefit analysis, evidence of actual harm from the first lockdown and evidence that the peak would have occurred without lockdown was excluded from people's thinking. The juggernaut carried on into a second lockdown with all its foreseeable harm.

Ultimately, the entire Cloud-Covid-Land belief system was predicated on the bizarre and mistaken belief that tiny aerosols would plummet to the ground. On a cold day, you can see your breath because these aerosols condense and become briefly large enough to be visible to the human eye before evaporating and disappearing again. Everyone has seen their own breath hang in the air at head height and yet the belief that it could not travel beyond a couple of metres led to not only social distancing, one way systems and perspex screens but to several other beliefs. The whole belief system was based on a human error in transcribing a number and the fear of being branded as a miasma theory proponent from a debate that ended over a hundred years ago.

The evidence against close contact spread being the only means of transmission kept accumulating. The first Omicron wave took only three or four weeks to reach every corner of the world and peak, far faster than the modelled 14 weeks it would take for close contact spread. Even then the consequences of long distance aerosol transmission were not taken on board. The virus was ubiquitous with every genetic subtype present in the community being detectable in Cambridge care homes and reaching ships at sea and an Antarctic research base. The final nail in the coffin for the idea that close contact droplet transmission was the driver of spread was the failure of masking to impact on the trajectory of viral spread. There remains total denial of this reality.

The claims that interventions worked were heavily dependent on pointing to large geographical areas which had not been affected significantly in spring 2020 by covid, like South East Asia and Eastern Europe. These places had not had a seasonal trigger, for whatever reason, be that immunity or environmental. They have had many since and this 'evidence of effectiveness' of interventions expired. Still there was denial.

The failure of interventions has been blamed for too long on people not carrying them out properly – they needed to be earlier and longer. They needed more people to do them better. In reality, the reason interventions failed is because it is not possible to stop a ubiquitous virus spread at a distance through the air.

WHERE WERE THE BRAKES?

Cloud-Covid-Land was born of more than just fear, ignorance and incompetence. Whether or not there was any deliberate intent there was a total failure of checks and balances resulting in multiple destructive positive feedback loops. Positive feedback occurs when a speaker picks up its own sound in the microphone that it is amplifying until that sound rapidly becomes very loud and someone pulls the plug. Cloud-Covid-Land was created by a series of similarly destructive positive feedback loops.

Policy makers and the media surrounded themselves in an echo chamber such that they repeated each other's messaging, convinced each other they were right and dismissed questioning voices without even engaging. So-called fact checkers, media capture and social media censorship scuppered any chance that these voices could help correct for errors.

Governments copied one another with more and more draconian restrictions. Each restriction made the next one easier. The more that had been sacrificed the easier it was to ask for more sacrifice, all for the Greater Good. There was a driving belief that not doing enough would make all the earlier sacrifices pointless. Political opposition was worse than absent. Every draconian policy was met with accusations that more should have been done creating an incentive for ever more draconian policy. A Rubicon was crossed in terms of the social contract when legislation was introduced which affected the most intimate, most mundane and most human aspects of our day to day lives.

Processes to implement policies were put in place with no consideration for how they could be paused, corrected or stopped. The result was civil service juggernauts that ran out of control. This included those created to purchase protective equipment, mass testing, and vaccination.

China set the pace with daily reports of cases and deaths and every other country fell in line producing daily data. Putting hospitals under pressure to produce a daily figure sounds like a small intervention, however, it had serious consequences. In an epidemic the only meaningful definition of a case is that they have disease and so the presence of symptoms at some point must be a fundamental requirement for such a definition. However, the pressure to produce a daily figure meant there was no time for a sense check. The doctors caring for the patients had no input whatsoever into the daily case numbers. A 'case' was defined by a positive test result using a single, sometimes flawed, test. The result was a maximising of the case numbers and this then fed back into driving policies, including mass testing which created another vicious circle.

Mass testing led to higher case numbers which led to a demand for more testing. This was particularly harmful in schools where numerous children had their education disrupted because of positive test results among their classmates.

Fear was born in a vacuum of knowledge and from the foundation of Chinese videos of sudden deaths in public places. It grew through a continued concern as we all sought more information. Was it also driven exponentially by a feedback loop of policy makers and media being frightened by their own propaganda into pushing yet more fear-inducing material? The co-founder of the Nudge Unit, Simon Ruda, thought so, *"Initially encouraged to boost public compliance, that fear seems to have subsequently driven policy decisions in a worrying feedback loop."* [748] Fear was amplified by a hysterical media exaggerating the risk particularly to the young. The more fear was created the more money was thrown at campaigns to induce more fear. Covid went from being a disease that would be *"mild in the majority"* according to Chris Whitty to being treated like the Black Death. The exaggerated fear led to a vicious circle of ever more restrictive policies, which were justified by appeals to the Greater Good and a sense of moral duty. This dynamic created a pseudo-religious atmosphere where dissent was silenced and critical thinking was discouraged. The result was a loss of nuance and an erosion of democratic values.

Individuals have been stripped of power and people who were meant to represent them have failed to. The resulting failure in checks and balances meant that mistakes were bound to be made. Normally checks and balances expose mistakes which can then be corrected. Harmful errors were not corrected because those responsible deliberately destroyed the fundamental error correction systems that are necessary in a democratic society. Free speech, a free press, the right to protest, parliamentary debate, and cabinet collective responsibility are not just nice to have. They are all fundamental to a functioning democratic society and need defending and strengthening. Of course every leader is responsible for their actions but those who deliberately break these mechanisms of error correction were choosing a path where harm would be more likely to result. Therefore they cannot defend themselves by saying they *"did not know."* They must therefore take full responsibility for the consequences of their actions.

There were numerous ways in which we could have avoided Cloud-Covid-Land. All of them failed. When people in power believe that those around them are infallible and while systems that correct for error are put on hold, then terrible harm can result. Harm resulted from abuse and inversion of the precautionary principle and the belief that something *must* be done. Oversimplifications, no acknowledgement of nuance and the managed presentation of a simple narrative was even more destructive.

The path we took may have been paved with good intentions. It is therefore all the more important that steps are taken, which cannot be bypassed, to prevent the same mistakes in future. There is much to be done to repair the damage caused and prevent a repeat of the same mistakes, especially now a precedent has been set.

To leave Cloud-Covid-Land we need to reintroduce error correction mechanisms to stop dangerous positive feedback loops by ensuring there is political opposition and providing an alternative to the biases of mainstream media. More than anything else people need to speak to others. Many people are holding on to a lot of harmful beliefs that can only be unpicked through gentle questioning and persistently sharing evidence-based interpretations.

In spring 2020, I was scared and believed all that I was told, even things I ought to have been more sceptical about. It took me six months to feel I had a handle on the overall story and writing this book has taught me even more. The mistakes of Cloud-Covid-Land could have been avoided if we had been more sceptical, nuanced, and open to dissenting views. In the coming years, we must take steps to reintroduce error correction mechanisms to stop dangerous positive feedback loops by ensuring there is political opposition and providing an alternative to the biases of mainstream media. People need to speak to family, friends, and colleagues to challenge harmful beliefs and promote evidence-based interpretations. This process may require a long period of time, but it is essential for repairing the damage caused and preventing a repeat of the same mistakes. We all have a responsibility to educate ourselves and speak out against injustice. I do not intend to stop.

INVESTIGATIONS

Many people seem to be expecting change to come from the Covid Public Inquiry. Perhaps it will, eventually. However, the statements that people most want to hear recorded were already set out in black and white in Dame Deirdre Hine's 2010 report into the response to the Swine Flu outbreak.[749] They can already be quoted to hammer home the lessons that need to be learnt.

"Use of the phrase 'reasonable worst case' should be reconsidered in future. It suggests that the outcome is relatively likely, whereas this is usually quite the opposite."

"Ministers should determine early in a pandemic how they will ensure that the response is proportionate to the perceived level of risk."

"[They should] ensure that clear guidance is set out to enable the rapid adjustment of... policy as more is learned about the nature of the virus."

"Ministers and senior officials should receive training on the strengths and limitations of scientific advice as part of their induction"

"SAGE [should release] papers to a wider group of scientists than that engaged in SAGE, who would be bound by confidentiality but who would have greater freedom to speak to the media. It was put to me that such a group would be able to comment authoritatively on the overall government strategy and give the media greater assurance about the approach being taken, as well as being able to challenge this if necessary and reduce the chance of group-think clouding views."

It criticised the use of modelling saying, *"modellers are not 'court astrologers'"* and said models were *"dependent on a number of important assumptions, particularly about the number of fatalities prevented in a severe outbreak. Based on what we now know about the relatively mild nature of this outbreak, the actual benefits were lower. This raises two questions. Were the costs of the 2009 response value for money? And should the government resource the same approach again?"*

The only recommendation that might be changed this time was, *"the behavioural scientists… were not used as effectively as they might have been."*

What would be gained by writing yet another report highlighting all these same issues again? More inquiries, reports, codes of practice and recommendations are not what is needed. There must be clear legislation with consequences to prevent a further repeat.

WHAT ABOUT COVID?

Predicting the future is a way to look foolish and covid has demonstrated that repeatedly. I am going to carefully hedge my bets.

If you have not had covid yet then it is possible that your immune system is remarkably well educated from a lifetime of service to you, but there remains a risk that you will succumb to a future variant. Reinfections have occurred but remain rare for those who had symptoms and developed antibodies, the vast majority of whom have resilient immunity. The risk of serious illness or death becomes incredibly low in those with antibodies from previous infection.

Remember the worst case scenario risk of dying if you caught a pre-Omicron variant is in the table in *Belief Three*. Given the low chance of catching it in any specific wave your overall risk of dying during any one wave would be more like that of the next youngest category. It can then be halved for a variant as mild as Omicron. Enough time has been wasted by people trying not to die and consequently failing to live. Everyone needs to put that behind us and live a full life again.

When a new influenza strain has emerged it has taken around a decade for it to work its way slowly through the whole population, causing a cyclical rise in hospital admissions. (Sometimes a season is skipped by a particular strain before it returns the following year). It could be that we see a similar pattern with *SARS-CoV-2* and it remains the dominant winter virus, surging most in December, for a few more winters to come. In the past a quiet winter season was often followed by one with more deaths because of the build-up of vulnerable people in the community. Future years may prove to be more challenging than the 2021-22 Omicron winter season.

Covid and influenza appear to have adopted a reciprocal relationship.[750] Globally covid waves have been interrupted by influenza waves making them M-shaped and vice versa. While influenza was dominant at the end of 2022, it has rapidly disappeared since, as it did in 2020, and that might portend a future 2023 covid wave.

On the other hand, influenza strains are replaced after around 90 percent of the population have developed antibodies and we have reached that threshold for the younger age groups against *SARS-CoV-2* already. Influenza has also returned in certain countries albeit in a competitive relationship with *SARS-CoV-2*. The 1889–1891 Russian Flu was thought to be a coronavirus pandemic and, after three seasons of unusual deaths around the world, death counts returned to normal. Perhaps that will be the case here. Perhaps it will be between these two extremes.

There is no reason to think that there would come a point where no *SARS-CoV-2* would be detectable if testing remained extensive. The influenza virus that caused the 1918 pandemic and the bacteria that caused Black Death have not disappeared. Therefore, those who benefit from covid policies may seek to continue testing to prolong the situation long after any problems with excess mortality have passed.

SHIFTING MAINSTREAM NARRATIVE

People are starting to change their stance on lockdowns. Even mainstream media have praised the *"remarkable bravery"* of the anti-lockdown protesters in China, having previously described western protestors as *"extremely selfish."* [751] This would make it much harder to condemn western anti-lockdown protests in future. Many voices have demanded punishment of those responsible for the harm caused by lockdown policy. Where political power continues to mislead the public there is no possibility of that. Yet there is likely to come a time, it might be months or years away, when there may be a race for politicians to express regret and commit to never carrying out lockdowns again. Those who lose that race may be the ones who risk being held accountable.

The Cloud-Covid-Land illusion has been sustained for so long that it is sometimes hard to believe it can ever be defeated. I wrote this book in an attempt to reach beyond the bubble of people who are already questioning. If you have friends and family who you believe could benefit from it, please consider sharing this book with them. Some people will realise they have been misled. They will read for hours to understand why and how and to rewire their belief system. They will likely feel anger at the breach of trust from those responsible but will also regain a sense of personal responsibility and agency. Others will be too lazy. The size of those two groups will have an important bearing on our future.

TIMELINE OF EVENTS

2019

31st December 2019 – Chinese report to WHO that they have a *"pneumonia of unknown cause"*

2020

11th January 2020 – First death reported in China

21st January 2020 – Protocol for PCR test published

30th January 2020 – WHO declares a global emergency

31st January 2020 – First two people test positive in UK

23rd February 2020 – Lombardy go into lockdown

11th March 2020 – WHO declares a pandemic

23rd March 2020 – First lockdown is announced

26th March 2020 – First clap for carers

27th March 2020 – Italian deaths peak

28th March 2020 – WHO call aerosol spread misinformation

5th April 2020 - Boris Johnson admitted to hospital

8th April 2020 – Total deaths peak in UK. Wuhan lockdown ends.

12th April 2020 –Three weeks to flatten the curve ends

14th April 2020 – Prime Minister Boris Johnson discharged from hospital

1st June 2020 - Phased reopening of schools

15th June 2020 - Non-essential shops reopen

Summer 2020 – Various lockdowns for Leicestershire, Blackburn, Luton and Bradford

14th August 2020 - Indoor theatres, bowling alleys and soft play reopen

14th September 2020 – Rule of Six – gatherings of more than six illegal

22nd September 2020 – 10pm curfew and working from home

14th October 2020 – Three tier system of restrictions begins

31st October 2020 – Second lockdown announced

5th November 2020 – Second lockdown begins

25th November 2020 – Autumn variant period peak all cause daily deaths

2nd December 2020 – Second lockdown ends

19th December 2020 – London and South East enter 'tier 4' restrictions

2021

6th January 2021 – Third national lockdown begins

19th January 2021 – Alpha variant period peak all cause daily deaths

8th March 2021 – Schools reopen

29th March 2021 - Rule of six reintroduced but two households of any size can meet

12th April 2021 – Shops, gyms and outdoor venues reopen

17th May 2021 – Indoor venues reopen and rule of six or two households for indoor gatherings

14th June 2021 – No more restrictions for weddings and funerals

19th July 2021 – Freedom Day - no restrictions on gatherings and nightclubs reopen

31st October 2021 – Delta period peak all cause daily deaths

10th December 2021 – Masks compulsory in public indoor venues

15th December 2021 – Covid pass mandatory in nightclubs and large events

19th December 2021 – Netherlands enter lockdown and other countries increase restrictions

23rd December 2021 – WHO admit that long-range aerosol spread is possible

2022

1st January 2022 – Omicron period winter peak all cause daily deaths

31st January 2022 – Government drops NHS vaccine mandate

25th March 2022 – Coronavirus Act expires

6th April 2022 – Omicron period spring peak all cause daily deaths

12th May 2022 – Covid pass *"dropped for domestic use"*

19th July 2022 – Omicron period summer peak all cause daily deaths

31st August 2022 – NHS ends asymptomatic testing of staff

31st October 2022 – Omicron period autumn peak covid deaths

2023

21st January 2023 – Omicron period winter peak covid deaths

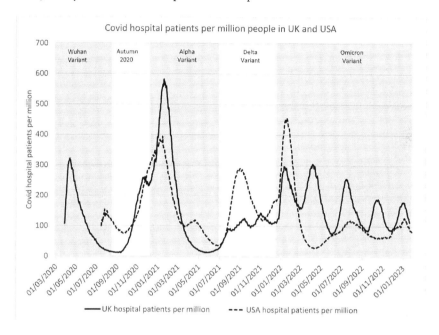

Figure 4: *Daily number of covid patients in hospital per million people in the population.*[752]

NOTES

1 Johnson, B. (2020, March 23). Prime Minister's statement on Coronavirus (COVID-19): 23 March 2020. Gov.uk. https://www.gov.uk/government/speeches/pm-address-to-the-nation-on-coronavirus-23-march-2020

2 NowThis News. (2020, February 26). Chinese 'SWAT' team shows off aggressive coronavirus protocol. NowThis News. YouTube. https://www.youtube.com/watch?v=6o6ZzBULLAs

3 Frias, L. (2020, January 30). Wuhan is running low on food, hospitals are overflowing, and foreigners are being evacuated as panic sets in after a week under coronavirus lockdown. https://www.businessinsider.com/no-food-crowded-hospitals-wuhan-first-week-in-coronavirus-quarantine-2020-1

4 Anandaciva, S. (2021, May 5). Was building the NHS Nightingale hospitals worth the money? The King's Fund. https://www.kingsfund.org.uk/blog/2021/04/nhs-nightingale-hospitals-worth-money

5 Bainbridge, R. (2020, March 31). The UK's healthcare industry ramps up its Covid19 response. The International Travel & Health Insurance Journal (ITIJ). https://www.itij.com/latest/news/uks-healthcare-industry-ramps-its-covid-19-response

6 Schulz, K. (2011). Being Wrong: Adventures in the Margin of Error (p.4). Portobello Books.

7 How To Avoid Catching And Spreading Coronavirus (COVID-19). (n.d.). nhs.uk. https://www.nhs.uk/conditions/coronavirus-covid-19/how-to-avoid-catching-and-spreading-coronavirus-covid-19/

8 Chamary, JV. (2021, May 4). WHO finally admits coronavirus is airborne. It's too late. Forbes. https://www.forbes.com/sites/jvchamary/2021/05/04/who-coronavirus-airborne/

9 UK Health Security Agency. (2021, March 4). Ventilation to reduce the spread of respiratory infections, including COVID-19. Gov.uk. https://www.gov.uk/guidance/ventilation-to-reduce-the-spread-of-respiratory-infections-including-covid-19

10 Plaut, M. (2020). "The USA advice during the 'Spanish Flu' of 1918 – Original document". (martinplaut.com). https://martinplaut.com/2020/03/19/the-usa-advice-during-the-spanish-flu-of-1918-original-document

11 Cabinet Office. National Risk Register of Civil Emergencies 2017 Edition. (2017). Cabinet Office, London. Available at: https://assets.publishing.service.gov.uk/government/uploads/system/uploads/attachment_data/file/644968/UK_National_Risk_Register_2017.pdf

12 Department of Health and Social Care and The Rt Hon Matt Hancock MP. (2020, March 1). Health Secretary sets out government 'battle plan' for COVID-19. Gov.uk. https://www.gov.uk/government/news/health-secretary-sets-out-government-battle-plan-for-covid-19

13 DH Pandemic Influenza Preparedness Team. (2011). UK influenza pandemic preparedness strategy 2011. Gov.uk. https://assets.publishing.service.gov.uk/government/uploads/system/uploads/attachment_data/file/213717/dh_131040.pdf

14 Adams, D. L. (2008). Putting Pandemics In Perspective: England and the Flu, 1889-1919. Doctoral thesis. University of Kansas.

15 Allen, T. R. et al. (1973). An outbreak of common colds at an Antarctic base after seventeen weeks of complete isolation. The Journal of Hygiene [Epidemiology and Infection], 71(4), 657–667. https://doi.org/10.1017/s0022172400022920

16 Coronavirus Pandemic: Antarctic Outpost Hit By Covid-19 Outbreak. (2022, January 5). BBC News. https://www.bbc.co.uk/news/world-europe-59848160

17 Wickramasinghe et al., N. C. (2020). Seasonality of respiratory viruses including SARS-CoV-2. 4(2). https://www.hilarispublisher.com/open-access/seasonality-of-respiratory-viruses-including-sarscov2-51923.html

18 Gregory, A. (2020, July 16). Nearly 60 fishermen infected aboard Argentinian trawler at sea for five weeks. The Independent. https://www.independent.co.uk/news/world/americas/coronavirus-argentina-fishermen-trawler-ushuaia-covid19-echizen-maru-a9621716.html

19 Reche, I., D'Orta, G., Mladenov, N. et al. (2018) Deposition rates of viruses and bacteria above the atmospheric boundary layer. The ISME Journal: Multidisciplinary Journal of Microbial Ecology, 12, 1154–1162. https://doi.org/10.1038/s41396-017-0042-4

20 Sharoni, S. et al. (2015). Infection of phytoplankton by aerosolized marine viruses. Proceedings of the National Academy of Sciences, 112(21), 6643–6647. https://doi.org/10.1073/pnas.1423667112

21 Molteni, M. (2021, May 13). The 60-year-old scientific screwup that helped Covid kill. WIRED. https://www.wired.com/story/the-teeny-tiny-scientific-screwup-that-helped-covid-kill

22 Yang, W., Elankumaran, S. and Marr, L. C. (2011). Concentrations and size distributions of airborne influenza A viruses measured indoors at a health centre, a day-care centre and on aeroplanes. Journal of the Royal Society, Interface, 8(61), 1176– 1184. https://doi.org/10.1098/rsif.2010.0686

23 Greenhalgh T., Ozbilgin M. and Contandriopoulos D. (2021). Orthodoxy, illusion, and playing the scientific game: A Bourdieusian analysis of infection control science in the COVID-19 pandemic. Wellcome Open Res 2021, 6, 126. https://doi.org/10.12688/wellcomeopenres.16855.3

24 Randall, K. et al. (2021) How did we get here: What are droplets and aerosols and how far do they go? a historical perspective on the transmission of respiratory infectious diseases (April 15). SSRN. Available at: https://ssrn.com/abstract=3829873

25 Molteni, M. (2021, May 13). The 60-year-old scientific screwup that helped Covid kill. WIRED. https://www.wired.com/story/the-teeny-tiny-scientific-screwup-that-helped-covid-kill

26 Wells, W.F. (1934). On air-borne Infection: Study II. Droplets and droplet nuclei. American Journal of Epidemiology, 20(3), 611–18, November. https://doi.org/10.1093/oxfordjournals.aje.a118097

27 Ibid

28 Wells, W. F. and Wells, M. W. (1936) Air-borne infection. Journal of the American Medical Association (JAMA), 107(21): 1698–1703, 1936. https://jamanetwork.com/journals/jama/article-abstract/273913

29 Fennelly, K. P. (2020). Particle sizes of infectious aerosols: Implications for infection control. The Lancet Respiratory Medicine, 8(9), 914–924. https://doi.org/10.1016/S2213-2600(20)30323-4

30 Langmuir A.D. (1980, December) Changing concepts of airborne infection of acute contagious diseases: A reconsideration of classic epidemiologic theories. Annals of the New York Academy of Sciences, 353(1), 35-44. https://doi.org/10.1111/j.1749-6632.1980.tb18903.x

31 Riley, R. L. (1961). Infectiousness of air from a tuberculosis ward ultraviolet irradiation of infected air: Comparative infectiousness of different patients. 85(4), 511–525. https://www.atsjournals.org/doi/pdf/10.1164/arrd.1962.85.4.511

32 Randall, K. et al. (2021) How did we get here: What are droplets and aerosols and how far do they go? a historical perspective on the transmission of respiratory infectious diseases (April 15). SSRN. Available at: https://ssrn.com/abstract=3829873

33 Chapin, C. (1912) [1910]. The sources and modes of infection, pp. 188, 297. John Wiley & Sons. https://archive.org/details/sourcesmodesofin00ch/page/n5/mode/2up

34 Ibid

35 Ibid, p. 314

36 Ibid, p. 297

37 Ibid, p. 313

38 Ibid, p. 296

39 Chapin, C. (1914, February 7). The air as a vehicle of infection. Journal of the American Medical Association, LXII(6), 423-430. https://doi.org/10.1001/jama.1914.02560310001001

40 Chapin, C. (1912) [1910]. The sources and modes of infection, pp. 189 John Wiley & Sons. https://archive.org/details/sourcesmodesofin00ch/page/n5/mode/2up

41 Chapin, C. (1914, February 7). The air as a vehicle of infection. Journal of the American Medical Association, LXII(6), 423-430. https://doi.org/10.1001/jama.1914.02560310001001

42 Ibid, p. 313

43 Ibid, p. 282

44 Chen, W. et al. (2020, April 10). Short-range airborne route dominates exposure of respiratory infection during close contact. Building and Environment 176: 106859. https://doi.org/10.1016/j.buildenv.2020.106859

45 Ibid.

46 Jimenez, J. L. et al. (2022, August 21). What were the historical reasons for the resistance to recognizing airborne transmission during Indoor Air, 32(8), e13070. https://doi.org/10.1111/ina.13070

47 Ibid

48 Ibid

49 Kutter, J.S et al. (2021, March 12). SARS-CoV and SARS-CoV-2 are transmitted through the air between ferrets over more than one meter distance. Nature Communications 12, 1653. https://doi.org/10.1038/s41467-021-21918-6

50 Environmental Modelling Group, SAGE. (2021, July 30). Role of screens and barriers in mitigating covid-19 transmission. gov.uk. https://assets.publishing.service.gov.uk/government/uploads/system/uploads/attachment_data/file/1007489/S1321_EMG_Role_of_Screens_and_Barriers_in_Mitigating_COVID-19_transmission.pdf

51 Morawska, L. and Milton, D. K. (2020). It Is time to address airborne transmission of Coronavirus disease 2019 (COVID-19). Clinical Infectious Diseases, 71(9), 2311– 2313. https://doi.org/10.1093/cid/ciaa939

52 Lewis, D. (2022). Why the WHO took two years to say COVID is airborne. Nature, 606: 26–31. https://www.nature.com/articles/d41586-022-00925-7

53 Ash, S. and Kuenssberg, L. (2021, May 20). Covid: Test and trace failure helped Indian variant spread, report says. BBC News. https://www.bbc.co.uk/news/uk-politics-57186059

54 Public Health Agency of Canada. (2021, July 23). Canada covid-19 weekly epidemiology report 11 July to 17 July 2021 (week 28). Canada.ca. https://web.archive.org/web/20210729084701/https://www.canada.ca/content/dam/phac-aspc/documents/services/diseases/2019-novel-coronavirus-infection/surv-covid19-weekly-epi-update-20210723-en.pdf

55 Qian, H. et, al. (2021). Indoor transmission of SARS-CoV-2. Indoor Air, 31(3), 639–645. https://doi.org/10.1111/ina.1276639 Chapin, C. (1914, February 7). The air as a vehicle of infection. Journal of the American Medical Association, LXII(6), 423-430. https://doi.org/10.1001/jama.1914.02560310001001

56 William, L. H. (2021). Genomic epidemiology of COVID-19 in care homes in the east of England. ELife, 10, e64618. https://doi.org/10.7554/eLife.64618

57 Foster, A. (2021, June 8). Covid: Victoria finds source of highly infectious Delta strain. NZ Herald. https://www.nzherald.co.nz/world/covid-19-coronavirus-victoria-finds-source-of-highly-infectious-delta-variant/5727WBZ2NMB324ZC5SYMAO4CP4/

58 Dunstan, J. (2021, June 8). A hotel quarantine leak is likely behind the Delta strain outbreak in Melbourne. ABC News. https://www.abc.net.au/news/2021-06-08/melbourne-covid-outbreak-delta-strain-link-hotel-quarantine/100183468

59 Davey, M. (2021, May 31). Victoria reports 11 new cases across state as outbreak hits two Melbourne aged care homes. The Guardian. https://www.theguardian.com/australia-news/2021/may/31/victoria-covid-update-11-new-cases-across-state-as-outbreak-hits-two-melbourne-aged-care-homes

60 Ibfelt, T. et al. (2015). Presence of pathogenic bacteria and viruses in the daycare environment. PubMed: Journal of Environmental Health, 78(3), 109–115. https://pubmed.ncbi.nlm.nih.gov/26591334/

61 Onakpoya, I. J. (2022). Viral cultures for assessing fomite transmission of SARSCoV-2: A systematic review and meta-analysis. Journal of Hospital Infection, 130, 63– 94. https://doi.org/10.1016/j.jhin.2022.09.007

62 Onakpoya, I. J. et, al. (2021). SARS-CoV-2 and the role of fomite transmission: A systematic review. F1000Research 2021, 10(233), 233 https://doi.org/10.12688/f1000research.51590.3

63 Food Standards Agency. (2022, November 29). Survival Of SARS-CoV-2 On Food Surfaces: Discussion. Food Standards Agency. https://web.archive.org/web/20221220133216/https://www.food.gov.uk/research/survival-of-sars-cov-2-on-food-surfaces-discussion

64 Fennelly, K. P. (2020). Particle sizes of infectious aerosols: Implications for infection control. The Lancet Respiratory Medicine, 8(9), 914–924. https://doi.org/10.1016/S2213-2600(20)30323-4

65 Regan, H. (2020, February 12). How can the coronavirus spread through bathroom pipes? Experts are investigating in Hong Kong. CNN. https://edition.cnn.com/2020/02/12/asia/hong-kong-coronavirus-pipes-intl-hnk/index.html

66 Senatore, V. et, al. (2021). Indoor versus outdoor transmission of SARS-COV-2: Environmental factors in virus spread and underestimated sources of risk. EuroMediterranean Journal for Environmental Integration, 6(1), 30. https://doi.org/10.1007/s41207-021-00243-w

67 Fears, A.C. et al. (2020). Persistence of severe acute respiratory syndrome coronavirus 2 in aerosol suspensions. Emerging Infectious Diseases, 26(9), 2168 - 2171. https://doi.org/10.3201/eid2609.201806

68 Garrod, L. P. (1955). Obituary Notices Of Fellows Of The Royal Society: Volumes 1-9 (1932-1954). Royal Society.

69 Alonso, C. (2014). Evidence of infectivity of airborne porcine epidemic diarrhea virus and detection of airborne viral RNA at long distances from infected herds. Veterinary Research, 45(1), 73. https://doi.org/10.1186/s13567-014-0073-z

70 Department of Agriculture, Environment and Rural Affairs. (n.d.). Foot And Mouth Disease. DAERA. Retrieved January 23, 2023, from https://www.daera-ni.gov.uk/articles/foot-and-mouth-disease

71 Fennelly, K. P. (2020). Particle sizes of infectious aerosols: Implications for infection control. The Lancet Respiratory Medicine, 8(9), 914–924. https://doi.org/10.1016/S2213-2600(20)30323-4

72 Stadnytskyi, V. et al. (2020). The airborne lifetime of small speech droplets and their potential importance in SARS-CoV-2 transmission. Proceedings of the National Academy of Sciences of the United States of America, 117, 11875 - 11877. https://doi.org/10.1073/pnas.2006874117

73 Zwart, M.P. et al. (2009). An experimental test of the independent action hypothesis in virus–insect pathosystems. Proceedings of the Royal Society B: Biological Sciences, 276, 2233 - 2242.

74 Yezli, S. and Otter, J.A. (2011). Minimum infective dose of the major human respiratory and enteric viruses transmitted through food and the environment. Food and Environmental Virology, 3, 1–30.

75 Department of Health and Social Care. (2022, December 1). Technical Report on the COVID-19 Pandemic In The UK. Gov.UK. https://www.gov.uk/government/publications/technical-report-on-the-covid-19-pandemic-in-the-uk

76 Nikitin, N.A., Petrova, E., Trifonova, E., and Karpova, O.V. (2014). Influenza virus aerosols in the air and their infectiousness. ADVANCES IN VIROLOGY, VOL 2014. https://doi.org/10.1155/2014/859090

77 Lednicky, J. A. (2020). Viable SARS-CoV-2 in the air of a hospital room with COVID-19 patients. International Journal of Infectious Diseases, 100, 476–482. https://doi.org/10.1016/j.ijid.2020.09.025

78 Ignatius, T. L. et al. (2004, April 22). Evidence of airborne transmission of the Severe Acute Respiratory Syndrome Virus. New England Journal of Medicine 350, 1731–1739. https://doi.org/10.1056/NEJMoa032867

79 Suñer, C. et al. (2022). Association between two mass-gathering outdoor events and incidence of SARS-CoV-2 infections during the fifth wave of COVID-19 in northeast Spain: A population-based control-matched analysis. The Lancet Regional Health - Europe, 15,100337. https://doi.org/10.1016/j.lanepe.2022.100337

80 Chirizzi, D. et al. (2020). SARS-CoV-2 concentrations and virus-laden aerosol size distributions in outdoor air in north and south of Italy. Environment International, 156, 10625. https://doi.org/10.1016/j.envint.2020.106255

81 Linillos-Pradillo, B. et al. (2021). Determination of SARS-CoV-2 RNA in different particulate matter size fractions of outdoor air samples in Madrid during the lockdown. Environmental Research, 195, 110863. https://doi.org/10.1016/j.envres.2021.110863

82 Setti, L. (2020). SARS-Cov-2RNA found on particulate matter of Bergamo in Northern Italy: First evidence. Environmental Research, 188, 109754. https://doi.org/10.1016/j.envres.2020.109754

83 Kayalar, Ö. (2021, October 1). Existence of SARS-CoV-2 RNA on ambient particulate matter samples: A nationwide study in Turkey. Science of the Total Environment, 1;789:147976 https://doi.org/10.1016/j.scitotenv.2021.147976

84 Veronesi G, De Matteis S, Calori G, et al. (2022). Long-term exposure to air pollution and COVID-19 incidence: a prospective study of residents in the city of Varese, Northern Italy Occupational and Environmental Medicine 2022, 79(3): 192-199. https://doi.org/10.1136/oemed-2021-107833

85 Bourdrel, T. et al. (2021). The impact of outdoor air pollution on COVID-19: A review of evidence from in vitro, animal, and human studies. European Respiratory Review, 30(159), 200242. https://doi.org/10.1183/16000617.0242-2020

86 McAloon, C.G. et al. (2020). Incubation period of COVID-19: A rapid systematic review and meta-analysis of observational research. BMJ Open 2020; 10:039652. https://doi.org/10.1136/bmjopen-2020-039652

87 Gov.uk. (2020, n.d.). Addendum to fourteenth SAGE meeting on Covid-19, 10th March 2020 held in 10 Victoria St, London, SW1H 0NN. https://assets.publishing.service.gov.uk/government/uploads/system/uploads/attachment_data/file/888782/S0382_Fourteenth_SAGE_meeting_on_Wuhan_Coronavirus__Covid-19__.pdf

88 Klepac, P., Kissler, S.M., and Gog, J.R. (2018). Contagion! The BBC Four Pandemic - The model behind the documentary. Epidemics, 24, 49-59. https://doi.org/10.1016/j.epidem.2018.03.003

89 Li, Y. (2020). Indoor air: A short history of holistic and reductionistic approaches. Indoor Air, 30(1), 3–6. https://doi.org/10.1111/ina.12544

90 Davenport, F. M. (1977). Reflections on the epidemiology of myxovirus infections. Medical Microbiology and Immunology, 164, 69–76 https://doi.org/10.1007/BF02121303

91 Thorp, H., Vinson, V. and Ash, C. (2020, June 23). Modeling herd immunity. Science. https://web.archive.org/web/20200629082015/https:/blogs.sciencemag.org/editors-blog/2020/06/23/modeling-herd-immunity/

92 Pfizer. (2020, April 15). Science will win - Ask science. YouTube. https://www.youtube.com/watch?v=Xl0tEfLve1U

93 Ferguson, N. F. et al. for Imperial College COVID-19 Response Team. (2020, March 16). Report 9: Impact of non-pharmaceutical interventions (NPIs) to reduce COVID19 mortality and healthcare demand (pp. 1–20). Spiral: Imperial College London Repository. https://doi.org/10.25561/77482

94 Steenbergen, S., et al. (2015, October 12) Long-term treated intensive care patients outcomes: The one-year mortality rate, quality of life, health care use and long-term complications as reported by general practitioners. BMC Anesthesiology, 15, 142. https://doi.org/10.1186/s12871-015-0121-x

95 Ferguson, N. F. et al. for Imperial College COVID-19 Response Team. (2020, March 16). Report 9: Impact of non-pharmaceutical interventions (NPIs) to reduce COVID19 mortality and healthcare demand (pp. 1–20). Spiral: Imperial College London Repository. https://doi.org/10.25561/77482

96 Howdon, D., Oke, J. and Heneghan, C. (2020, June 29). COVID-19: Declining admissions to intensive care units.The Centre For Evidence-Based Medicine. https://www.cebm.net/covid-19/covid-19-declining-admissions-to-intensive-care-units

97 Ferguson, N. F. et al. for Imperial College COVID-19 Response Team. (2020). Report 9: Impact of non-pharmaceutical interventions (NPIs) to reduce COVID19 mortality and healthcare demand (pp. 1–20). Spiral: Imperial College London Repository. https://doi.org/10.25561/77482

98 Ibid

99 Weinreich, D.M. et al. (2021, December 2). REGEN-COV antibody combination and outcomes in outpatients with Covid-19. New England Journal of Medicine, 385: e81. https://www.nejm.org/doi/full/10.1056/NEJMoa2108163

100 Kirkwood, D. (2022, July 10). "It's difficult to see how anyone could be more wrong" – new code review of Neil Ferguson's "amateurish" model. The Daily Sceptic. https://dailysceptic.org/2022/07/10/its-difficult-to-see-how-anyone-could-be-more-wrong-new-code-review-of-neil-fergusons-amateurish-model

101 Ibid

102 Walker, P. G. T., and Imperial College COVID-19 Response Team. (2020). Report 12 - The global impact of COVID-19 and strategies for mitigation and suppression (pp. 1–18). Imperial College: MRC Centre for Global Infectious Disease Analysis. https://doi.org/10.25561/77735

103 Cummings, D. (2021, May 23). @Dominic2306. Twitter. https://twitter.com/Dominic2306/status/1396493979064094720

104 Seely, B. (2022, January 18). Covid-19: Forecasting and modelling. Part of the debate – in Westminster Hall at 4:30 pm on 18th January 2022. TheyWorkForYou. https://www.theyworkforyou.com/whall/?id=2022-01-18a.67.1

105 Petrova, V. and Russell, C. (2017, October 30). The evolution of seasonal influenza viruses. Nature Reviews Microbiology 16, 47–60. https://doi.org/10.1038/nrmicro.2017.118

106 Klotz, L. C., and Sylvester, E. J. (2012, August 7). The Unacceptable Risks of a Manmade Pandemic. Bulletin Of The Atomic Scientists. https://thebulletin.org/2012/08/the-unacceptable-risks-of-a-man-made-pandemic/

107 Whitty, C. (2020). Oral evidence: Preparations for the Coronavirus, HC 36. House of Commons. https://committees.parliament.uk/oralevidence/113/pdf

108 Ionnadis, J. P. A. (2020). Infection fatality rate of COVID-19 inferred from seroprevalence data. Bulletin of the World Health Organization, 99, 19–33F. http://dx.doi.org/10.2471/BLT.20.265892

109 Ionnadis, J. P. A. (2021). Reconciling estimates of global spread and infection fatality rates of COVID-19: An overview of systematic evaluations. European Journal of Clinical Investigation, 51(5: e13554). https://doi.org/10.1111/eci.13554

110 Pezzullo, A. et al. (2022, 13 October). Age-stratified infection fatality rate of COVID-19 in the non-elderly population. Environmental Research, 216(3): 114655. https://doi.org/10.1101/2022.10.11.22280963.

111 Keeling, M. J. et al., Road Map Scenarios and Sensitivity: Steps 3 and 4 (2021). gov.uk.

https://assets.publishing.service.gov.uk/government/uploads/system/uploads/attachment_data/file/984533/S1229_Warwick_Road_Map_Scenarios_and_Sensitivity_Steps_3_and_4.pdf

112 Eikenberry, S. E. et al. (2020). To mask or not to mask: Modeling the potential for face mask use by the general public to curtail the COVID-19 pandemic. Infectious Disease Modelling, 5, 293–308. https://doi.org/10.1016/j.idm.2020.04.001

113 WHO Director-General's opening remarks at the media briefing 3 March 2020. On COVID-19. (2020, March 3). World Health Organization. https://www.who.int/director-general/speeches/detail/who-director-general-s-opening-remarks-at-the-media-briefing-on-covid-19---3-march-2020

114 Dycus, K. (2021). Peter Panum and the "geography of disease." Hektoen International. https://hekint.org/2021/08/18/peter-panum-and-the-geography-of-disease

115 Hope-Simpson, R. E. (2013) [1992]. The transmission of epidemic influenza. Springer.

116 Walker, P. G. T. and Imperial College COVID-19 Response Team. (2020). Report 12 - The global impact of COVID-19 and strategies for mitigation and suppression (pp. 1–18). Imperial College: MRC Centre for Global Infectious Disease Analysis. https://doi.org/10.25561/77735

117 Ferguson, N. F. et al. for Imperial College COVID-19 Response Team. (2020). Report 9: Impact of non-pharmaceutical interventions (NPIs) to reduce COVID-19 mortality and healthcare demand (pp. 1–20). Spiral: Imperial College London Repository. https://doi.org/10.25561/77482

118 House of Commons. (2020, June 10). Science and Technology Committee oral evidence: UK science, research and technology capability and influence in global disease outbreaks, HC 136. https://committees.parliament.uk/oralevidence/539/pdf

119 Levitt, M., Scaiewicz, A. and Zonta, F. (2020, June 30). Predicting the trajectory of any COVID19 epidemic from the best straight line. MedXriv. https://doi.org/10.1101/2020.06.26.20140814

120 Herby, J., Jonung, L. and Hanke, S. H. (2022). A literature review and metaanalysis of the effects of lockdowns on COVID-19 mortality. Studies in Applied Economics 200.

121 Muller, M. (2022, March 30). Never had covid? You may hold key to beating the virus. Bloomberg UK. https://www.bloomberg.com/news/articles/2022-03-30/never-had-covid-you-may-hold-key-to-beating-the-virus?

122 Hope-Simpson, R. E. (2013) [1992]. The transmission of epidemic influenza. Springer.

123 Petrova, V. and Russell, C. (2018). The evolution of seasonal influenza viruses. Nature Reviews Microbiology 16, 47–60 https://doi.org/10.1038/nrmicro.2017.118

124 Public Health England. (2021, June 3). SARS-CoV-2 variants of concern and variants under investigation in England Technical briefing 14. https://assets.publishing.service.gov.uk/government/uploads/system/uploads/attachment_data/file/991343/Variants_of_Concern_VOC_Technical_Briefing_14.pdf

125 Jones, W. (2021, June 13). Claims the Indian variant is "hyper-transmissible" are nonsense – And here's the graph that proves it. The Daily Skeptic. https://dailysceptic.org/2021/06/13/claims-the-indian-variant-is-hyper-transmissible-are-nonsense-heres-the-graph-that-proves-it

126 Hall, J. A. et al. (2021, June). HOSTED–England's household transmission evaluation dataset: Preliminary findings from a novel passive surveillance system of COVID-19, International Journal of Epidemiology, 50(3), 743– 752. https://doi.org/10.1093/ije/dyab057

127 Public Health England. (2021, 6 August). SARS-CoV-2 variants of concern and variants under investigation in England Technical briefing 20. https://assets.publishing.service.gov.uk/government/uploads/system/uploads/attachment_data/file/1009243/Technical_Briefing_20.pdf

128 Public Health England. (2022). SARS-CoV-2 variants of concern and variants under investigation in England Technical briefing 38. https://assets.publishing.service.gov.uk/government/uploads/system/uploads/attachment_data/file/1060337/Technical-Briefing-38-11March2022.pdf

129 de Laval, F. et al.(2022). Investigation of a COVID-19 outbreak on the Charles de Gaulle aircraft carrier, March to April 2020: A retrospective cohort study. Eurosurveillance 27(21). https://doi.org/10.2807/1560-7917.ES.2022.27.21.2100612

130 Coronaheadsup.com. (2021, December 9). Omicron: 73% attack rate at Oslo restaurant superspreader event. Coronaheadsup.com. https://www.coronaheadsup.com/science/variants/omicron/omicron-73-attack-rate-at-oslo-restaurant-superspreader-event

131 Milton, D.K. et al. (2010). Influenza virus aerosols in human exhaled breath: Particle size, culturability, and effect of surgical masks. PLOS PATHOGENS, 9. https://doi.org/10.1371/journal.ppat.1003205

132 Drummond, H. (2022). The Face Mask Cult. CantusHead Books citing Milton, D.K. et al.(2010). Influenza virus aerosols in human exhaled breath: Particle size, culturability, and effect of surgical masks. PLOS Pathogens, 9(3): e1003205. https://doi.org/10.1371/journal.ppat.1003205

133 Swinkels, K. (2020). SARS-CoV-2 Superspreading events around the world [Google Sheet]. Retrieved from www.superspreadingdatabase.com. https://covid19settings.blogspot.com/p/about.html

134 Sarles, C. (2020, December 14). Airway Heights Corrections COVID outbreak quickly approaching 1,000 infections. www.kxly.com. https://web.archive.org/web/20201215032317/https://www.kxly.com/airway-heights-corrections-covid-outbreak-quickly-approaching-1000-infections/

135 Garibay, C. (2020, December 14). SLO County reports more than 20 COVID outbreaks. These facilities have been hit the hardest. The Tribune. https://web.archive.org/web/20201228200837/https://www.sanluisobispo.com/news/coronavirus/article247780835.html

136 Read, R. (2020, March 29). A choir decided to go ahead with rehearsal. Now dozens of members have COVID-19 and two are dead. Los Angeles Times. https://www.latimes.com/world-nation/story/2020-03-29/coronavirus-choir-outbreak

137 C.J. Axon, et al. (2023) The Skagit County choir COVID-19 outbreak – have we got it wrong?, Public Health, 214: 85–90. https://doi.org/10.1016/j.puhe.2022.11.007

138 Cheemarla, N. R., et al. (2021). Dynamic innate immune response determines susceptibility to SARS-CoV-2 infection and early replication kinetics. Journal of Experimental Medicine 218(8): e20210583. https://doi.org/10.1084/jem.20210583

139 Killingley, B. et al. Safety, tolerability and viral kinetics during SARS-CoV-2 human challenge in young adults. Nature Medicine 28, 1031–1041 (2022). https://doi.org/10.1038/s41591-022-01780-9

140 Yezli S., and Otter J. A. Minimum infective dose of the major human respiratory and enteric viruses transmitted through food and the environment. Food and Environmental Virology. 3(1):1-30. https://doi.org/10.1007/s12560-011-9056-7

141 Wageningen University and Research Centre. (2009, March 14). One virus particle is enough to cause infectious disease. https://www.sciencedaily.com/releases/2009/03/090313150254.htm

142 Chadha, J. (2020, April 11). New York City's most crowded neighborhoods are often hardest hit by coronavirus. Politico. https://www.politico.com/states/new-york/albany/story/2020/04/11/new-york-citys-most-crowded-neighborhoods-are-often-hardest-hit-by-coronavirus-1274875

143 Ward, H. et al. (2021). Prevalence of antibody positivity to SARS-CoV-2 following the first peak of infection in England: Serial cross-sectional studies of 365,000 adults. The Lancet Regional Health – Europe, 4: 100098. https://doi.org/10.1016/j.lanepe.2021.100098

144 Ibid

145 World Health Organization. (2020, December 1). Coronavirus disease (COVID19: Herd immunity, lockdowns and COVID-19. World Health Organization. https://www.who.int/news-room/questions-and-answers/item/herd-immunity-lockdowns-and-covid-19

146 Gold, J. E. et al. (2020). Analysis of measles-mumps-rubella (MMR) titers of recovered Covid-19 patients. mBio, 11(6), e02628-20. https://doi.org/10.1128/mBio.02628-20

147 Public Health England. (2021a). Weekly national influenza and COVID-19 surveillance report: Week 10 report (up to week 9 data). gov.uk. https://assets.publishing.service.gov.uk/government/uploads/system/uploads/attachment_data/file/968513/Weekly_Flu_and_COVID-19_report_w10.pdf

148 Le Bert, N. et al. (2020). SARS-CoV-2-specific T cell immunity in cases of COVID-19 and SARS, and uninfected controls. Nature 584, 457–462. https://doi.org/10.1038/s41586-020-2550-z

149 Kundu, R. et al. (2022). Cross-reactive memory T cells associate with protection against SARS-CoV-2 infection in COVID-19 contacts. Nature Communications 13, 80. https://doi.org/10.1038/s41467-021-27674-x

150 Craig, C. (2022, July 18). In reality, hospital acquired infections peak *before* admissions peak. This happened in autumn 2020: @clarecraigpath, Twitter. https://twitter.com/ClareCraigPath/status/1549057964698488833 Source data available at: https://www.england.nhs.uk/statistics/statistical-work-areas/covid-19-hospital-activity/

151 Suárez-García, I. (2021). In-hospital mortality among immunosuppressed patients with COVID-19: Analysis from a national cohort in Spain. PLoS ONE 16(8): e0255524. https://doi.org/10.1371/journal.pone.0255524

152 Tassone, D. et al. (2021). Immunosuppression as a risk factor for COVID-19: A meta-analysis. Internal Medicine Journal, 51(2): 199–205. https://doi.org/10.1111/imj.15142

153 Connelly, J. A. et al. (2021). Impact of COVID-19 on pediatric immunocompromised patients. Pediatric Clinics of North America, 68(5), 1029–1054. https://doi.org/10.1016/j.pcl.2021.05.007

154 Hope-Simpson, R. E. (2013) [1992]. The transmission of epidemic influenza, p. 82. Springer.

155 Ibid, p. 194.

156 Evans, A. S. (1989). Viral infections of humans: epidemiology and control (3rd edn.), p. 8. Plenum Medical Book Company.

157 World Health Organization (n.d.) Global Influenza Surveillance and Response System (GISRS) https://www.who.int/tools/flunet

158 Lydall, R. (2020, September 15). First UK coronavirus death may have occurred in January, new ONS data reveals. Evening Standard. Retrieved January 24, 2023, from https://www.standard.co.uk/news/health/first-uk-coronavirus-death-end-january-a4547606.html

159 Gallagher, M. (2021, June 21). Letter on world military games to Lloyd J. Austin III and General Mark A. Milley in Washington Post (June 22). https://www.washingtonpost.com/context/june-21-letter-on-world-militarygames/ef61f9b3-a814-432b-afd7-ba57970c1186/

160 Ward, H. (2021, March 1). REACT-2 Round 5: Increasing prevalence of SARSCoV-2 antibodies demonstrate impact of the second wave and of vaccine roll-out in England (pre-print). https://doi.org/10.1101/2021.02.26.21252512

161 Public Health England. (2020). Weekly coronavirus disease 2019 (COVID-19) surveillance report Summary of COVID-19 surveillance systems. Year: 2020. Week: 18. gov.uk. https://assets.publishing.service.gov.uk/government/uploads/system/uploads/a ttachment_data/file/882420/COVID19_Epidemiological_Summary_w18_FI NAL.pdf

162 Dopico, X. et al. (2015). Widespread seasonal gene expression reveals annual differences in human immunity and physiology. Nature Communications. 6, 7000. https://doi.org/10.1038/ncomms8000

163 Fujito, A. B. (2021, March 6). Rapid response: Susceptibilities to COVID-19 severity and complications are driven largely by vitamin D deficiency. British Medical Journal, 202, 372: n544. https://doi.org/10.1136/bmj.n544

164 Di Domenico, L. et al. (2021, December 6). Adherence and sustainability of interventions informing optimal control against the COVID-19 pandemic. Communications Medicine 1, 57. https://doi.org/10.1038/s43856-021-00057-5

165 Killerby M. E. et al. (2018). Human coronavirus circulation in the United States 2014–2017. Journal of Clinical Virology, 101: 52-56. https://doi.org/10.1016/j.jcv.2018.01.019

166 World Health Organization (n.d.) Global Influenza Surveillance and Response System (GISRS). https://www.who.int/tools/flunet

167 Hope-Simpson, R. E. (2013) [1992]. The transmission of epidemic influenza, p. 82. Springer.

168 Fourie, J. and Jayes, J. (2021, January 4). Health inequality and the 1918 influenza in South Africa. University of Warwick. https://warwick.ac.uk/fac/soc/economics/research/centres/cage/manage/public ations/wp532.2021.pdf

169 Subject Access. (2022, July 12). Matt Hancock challenged by Heiko Khoo. YouTube. https://www.youtube.com/watch?v=S97_YfgJvJw

170 Neher, R.A. et al.(2020, March 16). Potential impact of seasonal forcing on a SARS-CoV-2 pandemic. Swiss Medical Weekly, 150, w20224. https://doi.org/10.4414/smw.2020.20224

171 Nikitin, N. et al. (2014). Influenza virus aerosols in the air and their infectiousness. Advances in Virology, 2014: 859090. https://doi.org/10.1155/2014/859090

172 BBC News. (2020, March 12). It's not possible to stop everybody getting it and it's also not desirable because you want some immunity in the population Sir Patrick Vallance … Twitter. https://twitter.com/BBCNews/status/1238151926379880449

173 Abbey, E. (2015) [1989]. A voice crying in the wilderness. RosettaBooks.

174 Cited in, Cecil, N., Sophia, M. and Sleigh, S. (2020, March 4). Coronavirus epidemic 'highly likely' as it spreads 'person to person.' Evening Standard. https://www.standard.co.uk/news/health/uk-coronavirus-epidemic-highly-likelyspreading-a4378196.html

175 SPI-B (prepared for) discussion at SAGE #18 meeting 23 March 2020. (2020, March, 22). "Options for increasing adherence to social distancing measures". Gov.uk:

https://assets.publishing.service.gov.uk/government/uploads/system/uploads/a ttachment_data/file/887467/25-options-for-increasing-adherence-to-social-distancing-measures-22032020.pdf

176 Ruda, S. (2022, January 13). Will nudge theory survive the pandemic? UnHerd. https://unherd.com/2022/01/how-the-government-abused-nudge-theory/

177 Nesta. (2021, December 13). Nesta acquires behavioural insights team to help tackle uk's biggest social challenges. Nesta. Retrieved January 24, 2023, from https://www.nesta.org.uk/press-release/nesta-acquires-behavioural-insights-team Quinn, B. (2018, November 10). The "nudge unit": the experts that became a prime UK export. The Guardian. https://www.theguardian.com/politics/2018/nov/10/nudge-unit-pushed-way-private-sector-behavioural-insights-team

178 London South Bank University. (2021, July 6). The pandemic's mental toll: New survey finds one in five suffer from covid-19 anxiety syndrome. https://www.lsbu.ac.uk/about-us/news/the-pandemics-mental-toll-new-survey-finds-one-in-five-suffer-from-covid-19-anxiety-syndrome

179 Department of Health and Social Care. (2020, March 16). Controlling the spread of COVID-19: Health Secretary's statement to Parliament. Gov.uk. https://www.gov.uk/government/speeches/controlling-the-spread-of-covid-19-health-secretarys-statement-to-parliament

180 Foad, C.M. et al. (2021, July 7). The limitations of polling data in understanding public support for COVID-19 lockdown policies. Royal Society Open Science, 8(7): 210678. https://doi.org/10.1098/rsos.210678

181 USC Dornsife. (n.d.). Understanding Coronavirus In America. Understanding America Study. https://covid19pulse.usc.edu

182 Birrell, P. et al. (2021, November 25)., COVID-19: Nowcast and forecast. MRC Biostatistics Unit, University of Cambridge. https://web.archive.org/web/20220101224305/https://www.mrc-bsu.cam.ac.uk/now-casting/report-on-nowcasting-and-forecasting-9th-december-2021/

183 Sky News Australia. (2021, July 1). "Fear factor" over COVID Is 'out of hand". YouTube. https://www.youtube.com/watch?v=DOJOj3Pmz_s

184 Woolhouse, M. (2022). The year the world went mad. Sandstone Press.

185 Coronavirus: 'Virus does not discriminate' - Gove. (2020, March 27). Sky News. https://news.sky.com/video/coronavirus-virus-does-not-discriminate-gove-11964771

186 Kadlec, D. (2013, August 14). A good death: How boomers will change the world a final time. Time. https://business.time.com/2013/08/14/a-good-death-how-boomers-will-change-the-world-a-final-time/

187 Birrell, P. et al. (2021, December 12). COVID_19: nowcast and forecast. MRC Biostatistics Unit, University of Cambridge. https://web.archive.org/web/20220101224305/https://www.mrc-bsu.cam.ac.uk/now-casting/report-on-nowcasting-and-forecasting-9th-december-2021/

188 UK Parliament Coronavirus. (2021, July 12) Death. Full question for Department of Health and Social Care. https://questions-statements.parliament.uk/written-questions/detail/2021-07-12/31381

189 Cited in Knapton, S. (2020, March 25). Two thirds of coronavirus victims may have died this year anyway, government adviser says. The Telegraph. https://www.telegraph.co.uk/news/2020/03/25/two-thirds-patients-die-coronavirus-would-have-died-year-anyway

190 NHS England. (n.d.). COVID-19 Deaths. England.nhs.uk. https://www.england.nhs.uk/statistics/statistical-work-areas/covid-19-deaths

191 Ghisolfi, S. et, al. (2020). Predicted COVID-19 fatality rates based on age, sex, comorbidities and health system capacity. British Medical Journal Global Health 2020, 5:e003094. http://dx.doi.org/10.1136/bmjgh-2020-003094

192 Oliver, D. (2021). Deaths from hospital acquired covid are everyone's problem. British Medical Journal, 373:n1492. https://doi.org/10.1136/bmj.n1492

193 Fleming D. M., Pannell R. S. and Cross K.W. (2005). Mortality in children from influenza and respiratory syncytial virus. Journal of Epidemiology & Community Health 59: 586-590.

194 Paediatric Intensive Care Audit Network report on COVID-19 confirmed cases in PICU. (2021, July 8). Universities of Leeds and Leicester. https://web.archive.org/web/20211108115957/https://www.picanet.org.uk/wp-content/uploads/sites/25/2021/07/PICANet_COVID_report_2021-06-21_final.pdf

195 Deaths in the UK from 1990 To 2020 - Office For National Statistics. (2021, February 5). Office for National Statistics. https://www.ons.gov.uk/aboutus/transparencyandgovernance/freedomofinformationfoi/deathsintheukfrom1990to2020

196 E was 2020. A = 1991, B = 2000, C = 1999, D = 2008, F = 2019, G = 1990, H = 2015, I = 2003.

197 McGowan, J. (2021, January 25). Are COVID death numbers comparing apples and oranges? Mathematical Software. http://wordpress.jmcgowan.com/wp/are-covid-death-numbers-comparing-apples-and-oranges

198 Deaths registered weekly in England and Wales, provisional (2022, December 6). Office for National Statistics. https://www.ons.gov.uk/peoplepopulationandcommunity/birthsdeathsandmarriages/deaths/bulletins/deathsregisteredweeklyinenglandandwalesprovisional/weekending25november2022

199 Coronavirus: A year like no other. (2021, March 15). Office for National Statistics. https://www.ons.gov.uk/peoplepopulationandcommunity/healthandsocialcare/conditionsanddiseases/articles/coronavirusayearlikenoother/2021-03-15

200 Senger, M. (2021). Snake oil: How Xi Jinping shut down the world. Plenary Press.

201 Irvine, B. (2022). The truth about the Wuhan lockdown. And how the world became Wuhan. Oldspeak Publishing.

202 Fact check: NYC is not planning to use trenches in parks as burial grounds. (2020, April 11). USA Today News. https://eu.usatoday.com/story/news/factcheck/2020/04/11/fact-check-nyc-wont-uselocal-parks-temporary-burial-grounds/5133460002

203 Slotnik, D. E. (2021, March 25). Up to a tenth of New York City's coronavirus dead may be buried in a potter's field. The New York Times. https://www.nytimes.com/2021/03/25/nyregion/hart-island-mass-graves-coronavirus.html

204 Hart Island. (n.d.). Data Team, New York City Council. https://council.nyc.gov/data/hart-island/

205 Dalton, K. F. Highest single-year death toll in NYC was recorded pre-pandemic. (2022, January 6). Silive.Com. https://www.silive.com/news/2022/01/highest-single-year-death-toll-in-nyc-was-recorded-pre-pandemic.html

206 New York City Department of Health and Mental Hygiene. (2020, December). Summary of vital statistics 2018 the city of New York. https://www1.nyc.gov/assets/doh/downloads/pdf/vs/2018sum.pdf

207 Pope, A. (1711). An Essay on Criticism, pt. 1: 9.–10. See https://www.gutenberg.org/cache/epub/7409/pg7409-images.html

208 Hassan, A., Barber, S.J. The effects of repetition frequency on the illusory truth effect. Cognitive. Research: Principles and Implications 6, 38 (2021). https://doi.org/10.1186/s41235-021-00301-5

209 See https://shmaltzandmenudo.wordpress.com/2017/04/07/famous-sayings-56-repeat-a-lie-often-enough/comment-page-1 and Blagden, I. (1869) The Crown of a Life. Hurst and Blackett.

210 Staton, T. (2014, November 6). New numbers back old meme: Pharma does spend more on marketing than R&D. Fierce Pharma. https://www.fiercepharma.com/sales-and-marketing/new-numbers-back-old-meme-pharma-does-spend-more-on-marketing-than-r-d

211 "CDC Director Rochelle Walensky: Too little caution and too much optimism… " (2022, March 5). Philip Smith, YouTube. https://www.youtube.com/watch?v=8DPS4nBFXBo

212 Schulz, K. (2011). Being wrong: Adventures in the margin of error. Portobello Books.

213 Browne, T. (1643).from the Religio Medici. https://www.bartleby.com/library/prose/919.html

214 Cory, G.A. (1999). MacLean's triune brain concept: In praise and appraisal. In: The reciprocal modular brain in economics and politics, pp. 13–27. Springer. https://doi.org/10.1007/978-1-4615-4747-1_3

215 London South Bank University. (2021, July 6). The pandemic's mental toll: New survey finds one in five suffer from covid-19 anxiety syndrome. https://www.lsbu.ac.uk/about-us/news/the-pandemics-mental-toll-new-survey-finds-one-in-five-suffer-from-covid-19-anxiety-syndrome

216 The Conservative Woman. (2020, November 14). TCW poll reveals ignorance and exaggerated fear about covid. Conservativewoman.co.uk. https://www.conservativewoman.co.uk/tcw-poll-reveals-ignorance-and-exaggerated-fear-about-covid

217 Single year of age and average age of death of people whose death was due to or involved coronavirus (COVID-19). Office of National Statistics Dataset. https://www.ons.gov.uk/peoplepopulationandcommunity/birthsdeathsandmarriages/deaths/datasets/singleyearofageandaverageageofdeathofpeoplewhosedeathwasduetoorinvolvedcovid19

218 Kulldorff, M., and Bhattacharya, J. (2020, November 2). Lockdown isn't working. The Spectator. https://www.spectator.co.uk/article/lockdown-isn-t-working

219 KEKST CNC. (2020, July 27). Covid-19 opinion tracker, Edn. 4. 1 https://www.kekstcnc.com/media/2793/kekstcnc_research_covid-19_opinion_tracker_wave-4.pdf

220 Nikčević, A. V. (2021, January 15). Journal of Affective Disorders 279, 578–584. https://doi.org/10.1016/j.jad.2020.10.053

221 Breton-Provencher, V. et al. (2022). Spatiotemporal dynamics of noradrenaline during learned behaviour. Nature 606, 732–738. https://doi.org/10.1038/s41586-022-04782-2

222 BBC 5 Live. (2020, 29 December), transcribed by author.

223 Vollmer, M.A.C. et al. (2021). The impact of the COVID-19 pandemic on patterns of attendance at emergency departments in two large London hospitals: an observational study. BMC Health Services Research 21, 1008. https://doi.org/10.1186/s12913-021-07008-9

224 Heneghan, Carl. (2020, June 4). Dying of neglect: The other covid care home scandal. The Spectator. https://www.spectator.co.uk/article/dying-of-neglect-the-other-covid-care-home-scandal

225 Ramezani, M. et al. (2020). The role of anxiety and cortisol in outcomes of patients with Covid-19. Basic and Clinical Neuroscience, 11(2), 179–184. https://doi.org/10.32598/bcn.11.covid19.1168.2

226 Care Quality Commission, General Medical Council, & Healthcare Improvement Scotland. (2020, May 6). Joint statement on death certification during the COVID19 pandemic. General Medical Council. https://www.gmc-uk.org/news/news-archive/joint-statement-on-death-certification-during-the-covid-19-pandemic

227 World Health Organization. (2020, June 7). Medical certification, ICD mortality coding, and reporting mortality associated with COVID-19: Technical Note. https://apps.who.int/iris/bitstream/handle/10665/332297/WHO-2019-nCoV-Mortality_Reporting-2020.1-eng.pdf

228 Fritz, J. (2021, April 14). A Covid death? The bureaucracy decides. The American Mind. https://americanmind.org/salvo/a-covid-death-the-bureaucracy-decides

229 CDC COVID-19 Vaccine Breakthrough Case Investigations Team. (2021). COVID-19 Vaccine breakthrough infections reported to CDC – United States, January 1–April 30, 2021. Morbidity and Mortality Weekly Report, 70(21): 792–793.

230 Coronavirus Act 2020. Gov.uk. https://www.legislation.gov.uk/ukpga/2020/7/section/1/enacted

231 Department of Health and Social Care. (2020, August 12). New UK-wide methodology agreed to record COVID-19 deaths. https://web.archive.org/web/20200819023354/https://www.gov.uk/government/news/new-uk-wide-methodology-agreed-to-record-covid-19-deaths

232 UK Health Security Agency. (2022, February 24). COVID-19 vaccine surveillance report Week 8. Gov.uk. https://assets.publishing.service.gov.uk/government/uploads/system/uploads/attachment_data/file/1057599/Vaccine_surveillance_report_-_week-8.pdf

233 Coronavirus.data.gov.uk. England Summary. (n.d.).
https://coronavirus.data.gov.uk

234 Struyf, T. et al. (2020). Signs and symptoms to determine if a patient presenting in primary care or hospital outpatient settings has COVID-19 disease. Cochrane Database of Systematic Reviews.
https://doi.org/10.1002/14651858.CD013665

235 Karanikolos, M. and McKee, M. (2020). How comparable is covid-19 mortality across countries? Eurohealth 26(2), 45–50.
https://apps.who.int/iris/bitstream/handle/10665/336295/Eurohealth-26-2-45-50-eng.pdf

236 US Today News (2020, April 24). The claim: Hospitals get paid more if patients are listed as COVID-19, and on ventilators. USA Today News.
https://eu.usatoday.com/story/news/factcheck/2020/04/24/fact-check-medicarehospitals-paid-more-covid-19-patients-coronavirus/3000638001

237 The Associated Press. (2021, December 26). FEMA wants to give families up to $9,000 for COVID funerals, but many don't apply. NPR.org.
https://www.npr.org/2021/12/26/1068103241/fema-wants-to-give-families-up-to-9-000-for-covid-funerals-but-many-dont-apply

238 Craig, C., Engler, J. and Smalley, J. (2021, January 7). What does endemic covid look like? The Daily Sceptic
https://dailysceptic.org/archive/what-does-endemic-covid-look-like

239 Craig, C. (2021, December 5). This year excess acute respiratory infection deaths exactly match those where covid was the cause. it's almost as if we were overdiagnosing it last year. Twitter.
https://twitter.com/ClareCraigPath/status/1467415563328753668

240 Navascués, A. et al. (2018). Detection of respiratory viruses in deceased persons, Spain, 2017. Emerging Infectious Diseases, 24(7), 1331–1334.
https://doi.org/10.3201/eid2407.180162

241 SARS-CoV-2 Positive Deaths Surveillance Group. (2021, October 5). Characteristics of SARS-CoV-2 patients dying in Italy report based on available data on October 5th, 2021. Istituto Superiore di Sanità.

242 Office for National Statistics (2022, January 17). COVID-19 deaths and autopsies Feb 2020 to Dec 2021. Gov.uk.
https://www.ons.gov.uk/aboutus/transparencyandgovernance/freedomofinformationfoi/covid19deathsandautopsiesfeb2020todec2021

243 Office for National Statistics. (2022, October 21). Pre-existing conditions of people who died due to COVID-19, England and Wales. Gov.uk.
https://www.ons.gov.uk/peoplepopulationandcommunity/birthsdeathsandmarriages/deaths/datasets/preexistingconditionsofpeoplewhodiedduetocovid19englandandwales

244 Turnbull, A., Osborn, M. and Nicholas, N. (2015). Hospital autopsy: Endangered or extinct? Journal of Clinical Pathology, 68, 601–604.
https://jcp.bmj.com/content/jclinpath/68/8/601.full.pdf

245 Goldman, L. (2018). Autopsy 2018: Still necessary, even if occasionally not sufficient. Circulation, 37, 2686–2688. https://doi.org/10.1161/circulationaha.118.033236

246 Osborn, M. et al. (2020). Briefing on COVID-19: Autopsy practice relating to possible cases of COVID-19 (2019-nCov, novel coronavirus from China 2019/2020). Royal College of Pathologists. https://www.rcpath.org/uploads/assets/d5e28baf-5789-4b0f-acecfe370eee6223/fe8fa85a-f004-4a0c-81ee4b2b9cd12cbf/Briefing-on-COVID-19-autopsy-Feb-2020.pdf

247 Coronavirus Act 2020. Schedule 13. (2020). Gov.uk. https://www.legislation.gov.uk/ukpga/2020/7/schedule/13/enacted

248 Ibid

249 NHS England & NHS Improvement. (2020, March 31). Coronavirus Act – excess death provisions: Information and guidance for medical practitioners. https://web.archive.org/web/20200512135005/https:/improvement.nhs.uk/documents/6590/COVID-19-act-excess-death-provisions-info-and-guidance-31-march.pdf

250 NHS England. (2020, March 31). Coronavirus Act – excess death provisions: Information and guidance for medical practitioners. NHS England. https://www.england.nhs.uk/coronavirus/documents/coronavirus-act-excess-death-provisions-information-and-guidance-for-medical-practitioners

251 Covid19 Assembly (2021, March 23). https://archive.is/OX1nD saved from https://www.covid19assembly.org

252 Krans, B. (2021, June 4). Alameda County's new COVID death toll is 25% lower than thought. The Oaklandside. https://oaklandside.org/2021/06/04/alameda-countys-new-covid-death-toll-is-25-lower-than-thought/

253 Reese, M. (2021, July 8). Santa Clara county revises total COVID deaths by over 20%. San José Spotlight. https://sanjosespotlight.com/santa-clara-county-revises-total-covid-deaths-by-over-20/

254 Weisman, R. (2021, April 15). The Mass. COVID-19 death toll is about to drop as state adopts new way to report long-term care deaths. The Boston Globe. https://web.archive.org/web/20210415190141/https://www.bostonglobe.com/2021/04/15/metro/covid-19-death-toll-is-about-drop-state-adopts-new-way-report-long-term-care-deaths/

255 Clancy, D. and Anthony, B. (2021, April 27). State disclosure in official count of longterm care deaths. Pioneer Institute. https://pioneerinstitute.org/blog/massachusetts-should-disclose-more-information-about-its-recent-reduction-in-the-official-count-of-long-term-care-deaths

256 Kenyon, T. (2021, July 28). COVID-19 Deaths: Underreported or overestimated? PANDA. https://www.pandata.org/covid-19-deaths-underreported-or-overestimated

257 Reuters. (2022, March 18). CDC reports fewer COVID-19 pediatric deaths after data correction. Reuters. https://www.reuters.com/business/healthcare-pharmaceuticals/cdc-reports-fewer-covid-19-pediatric-deaths-after-data-correction-2022-03-18/

258 Kerr, A. (2022, March 18). Reported pediatric COVID-19 deaths plummet 24% after CDC fixes "coding logic error'". Washington Examiner. https://www.washingtonexaminer.com/news/reported-pediatric-covid-19-deaths-plummet-24-after-cdc-fixes-coding-logic-error

259 Wang, H. et al. (2022). Estimating excess mortality due to the COVID-19 pandemic: A systematic analysis of COVID-19-related mortality, 2020–21. Lancet 399, 1513 - 1536.

260 Blake, W. (1975) [1790] The marriage of heaven and hell, p. xxiv. (ed. Keynes, G.). Oxford University Press.

261 Cohen, T. and Southwood, B. (2019, August 5). Is the rate of scientific progress slowing down? GMU Working Paper in Economics No. 21-13, SSRN. http://dx.doi.org/10.2139/ssrn.3822691

262 Kozlov, M. (2023, January 4). 'Disruptive' science has declined – And no one knows why. Nature, News. https://www.nature.com/articles/d41586-022-04577-5

263 Toole, A. A., Grimpe, C. and Czarnitzki, D. (2014, August 31). Does industry sponsorship restrict the disclosure of academic research? Blog.OUP.com (Oxford University Press).

264 Relman A. S. and Angell, M. America's other drug problem: How the drug industry distorts medicine and politics. New Republic. 2002; 227(25):27-41.

265 Several sources however note that this was likely to have been an eloquent paraphrase the actual quote being: "A new scientific truth does not triumph by convincing its opponents and making them see the light, but rather because its opponents eventually die and a new generation grows up that is familiar with it …". Planck, Max K. (1950). Scientific autobiography and other papers. New York: Philosophical Library.

266 Azoulay, P., Fons-Rosen, C. and Zivin, J. S. G. (2019). Does science advance one funeral at a time? The American Economic Review, 109(8), 2889–2920. https://doi.org/10.1257/aer.20161574

267 Festinger, L., Riecken, H. W. and Schachter, S. (2013). When Prophecy Fails, Ch. 1. Start Publishing.

268 Sagan, Carl (1995) The Demon-Haunted World: Science as a Candle in the Dark. Random House

269 Wallace, A. R. 2011. [1908] My life: A record of events and opinion, p. 362. Cambridge University Press.

270 Boyd, C. (2021, March 16). Another covid variant 'from the Philippines' spotted In England. Mail Online. https://www.dailymail.co.uk/news/article-9368755/ANOTHER-Covid-variant-Philippines-spotted-two-people-England.html

271 Allen, H. et al. (2022, February 17). Comparative transmission of SARS-CoV-2 Omicron (B.1.1.529) and Delta (B.1.617.2) variants and the impact of vaccination: National cohort study, England. MEDRXIV. https://doi.org/10.1101/2022.02.15.22271001

272 Jones, W. (2021, June 13). Claims the Indian variant is "hyper-transmissible" are nonsense – And here's the graph that proves it. The Daily Sceptic. https://dailysceptic.org/2021/06/13/claims-the-indian-variant-is-hyper-transmissible-are-nonsense-heres-the-graph-that-proves-it/

273 Department of Health and Social Care. (2022, December 1). Technical report on the COVID-19 pandemic in the UK, Chapter 1: Understanding the pathogen. Gov.uk. https://www.gov.uk/government/publications/technical-report-on-the-covid-19-pandemic-in-the-uk/chapter-1-understanding-the-pathogen

274 Eccles, R. (2021). Why is temperature sensitivity important for the success of common respiratory viruses? Reviews in Medical Virology, 31, e02153. https://doi.org/10.1002/rmv.2153

275 Speaking on CNBC International TV. (2022, February 19). Munich Security Conference: Finding A way out of the pandemic. YouTube. https://www.youtube.com/watch?v=U70Q9WqbMFM

276 Benza, B. (2021, December 3). Botswana diplomats with omicron 'European'. The Canberra Times. https://www.canberratimes.com.au/story/7537679/botswanadiplomats-with-omicron-european

277 Harding, A. (2022, January 20). Was South Africa ignored over mild omicron evidence? BBC News. https://www.bbc.co.uk/news/world-africa-60039138

278 White, S. (2021, December 29). WHO warns of a 'tsunami of cases' from Omicron and Delta variants. Financial Times. https://www.ft.com/content/74e328e1-4868-46cd-b4f7-b78af65bc030

279 Helfand, M. (2022). Risk for reinfection after SARS-CoV-2: A living, rapid review for American College of Physicians practice points on the role of the antibody response in conferring immunity following SARS-CoV-2 Infection. Annals of Internal Medicine, 175(4), 547–555. https://doi.org/10.7326/M21-4245

280 Immunisation and Vaccine Preventable Diseases Division, UK Health Security Agency. (2021). Weekly national Influenza and COVID-19 surveillance report Week 51 report (up to week 50 data) 23 December 2021. https://assets.publishing.service.gov.uk/government/uploads/system/uploads/attachment_data/file/1043583/Weekly_Flu_and_COVID-19_report_w51.pdf

281 Immunisation and Vaccine Preventable Diseases Division, UK Health Security Agency. (2022, April 21). Weekly national Influenza and COVID-19 surveillance report Week 16 report (up to week 15 data). https://assets.publishing.service.gov.uk/government/uploads/system/uploads/attachment_data/file/1070353/Weekly_Flu_and_COVID-19_report_w16.pdf

282 Abu-Raddad, L.J., Chemaitelly, H., and Bertollini, R. (2021, December 23). Severity of SARS-CoV-2 reinfections as compared with primary infections. The New England Journal of Medicine, 385: 2487-2489. https://www.nejm.org/doi/full/10.1056/NEJMc2108120

283 Altarawneh, H.N. et al. (2022, March 31). Protection against the Omicron variant from previous SARS-CoV-2 infection. The New England Journal of Medicine, 385: 2487-2489. https://www.nejm.org/doi/full/10.1056/NEJMc2108120

284 Doucleff, M. (2022, February 7). the future of the pandemic is looking clearer as we learn more about infection. NPR.org. https://www.npr.org/sections/goatsandsoda/2022/02/07/1057245449/the-future-of-the-pandemic-is-looking-clearer-as-we-learn-more-about-infection

285 Le Bert, N. et al. (2020, 15 July) SARS-CoV-2-specific T cell immunity in cases of COVID-19 and SARS, and uninfected controls. Nature 584, 457–462. https://doi.org/10.1038/s41586-020-2550-z

286 Orwell, G. (2021) 1984, p. 233. Oxford University Press.

287 Konishi, T. (1993). The semantics of grammatical gender: A cross-cultural study. Journal of Psycholinguistic Research, 22(5), 519–534. https://doi.org/10.1007/BF01068252

288 Sera M. D. et al. (2002). When language affects cognition and when it does not: An analysis of grammatical gender and classification. Journal of Experimental Psychology: General, 31(3): 377-397. https://doi.org/10.1037/0096-3445.131.3.377

289 Greco D., Stern E.K., Marks, G. (2011, August). Review of ECDC's response to the influenza pandemic 2009–2010. European Centre for Disease Prevention and Control. https://www.ecdc.europa.eu/sites/default/files/media/en/publications/Publications/241111COR_Pandemic_response.pdf

290 HM Government (2020). National Risk Register. Gov.uk. https://assets.publishing.service.gov.uk/government/uploads/system/uploads/attachment_data/file/952959/6.6920_CO_CCS_s_National_Risk_Register_2020_11-1-21-FINAL.pdf

291 Cabinet Office (2017). National Risk Register of Civil Emergencies 2017 Edition. Available at: https://assets.publishing.service.gov.uk/government/uploads/system/uploads/attachment_data/file/644968/UK_National_Risk_Register_2017.pdf

292 @neil_ferguson. (2020, March 26). 4/4 - Without those controls, our assessment remains that the UK would see the scale of deaths reported in our study (namely, up to approximately 500 thousand). Twitter. https://twitter.com/neil_ferguson/status/1243294819952230402

293 House of Commons Health and Social Care, and Science and Technology Committees. (2021, 12 October). Coronavirus: Lessons learned to date Sixth Report of the Health and Social Care Committee and Third Report of the Science and Technology Committee of Session 2021–22. House of Commons. https://committees.parliament.uk/publications/7496/documents/78687/default

294 Fox, F. (2021, August 8). Britain's covid experts are under attack, but they are just doing their jobs. The Guardian. https://www.theguardian.com/world/2021/aug/08/britains-covid-experts-neil-ferguson-sage-are-under-attack-but-they-are-just-doing-their-jobs

295 Addendum to eleventh SAGE meeting on Covid-19, 27th February 2020 held in 10 Victoria St, London, SW1H 0NN. (2020). Gov.uk. https://assets.publishing.service.gov.uk/government/uploads/system/uploads/attachment_data/file/888778/S0379_Eleventh_SAGE_meeting_on_Wuhan_Coronavirus__Covid-19__.pdf

296 Health Advisory and Recovery Team. (2021, December 9). What are the symptoms of covid? https://www.hartgroup.org/full-article-what-are-the-symptoms-of-covid

297 Ibid

298 Ealy, H. et al. (2020, 12 October). COVID-19 data collection, comorbidity & federal law: A historical retrospective. Science, Public Health Policy, and The Law 2: 4- 22. https://www.researchgate.net/publication/344753727_COVID-19_Data_Collection_Comorbidity_Federal_Law_A_Historical_Retrospective

299 Government of Canada. (2020, January 6 updated February 6). Interim national case definition: Novel Coronavirus (2019-nCoV). Government of Canada. https://web.archive.org/web/20200211164225/https://www.canada.ca/en/public-health/services/diseases/2019-novel-coronavirus-infection/health-professionals/national-case-definition.html

300 World Health Organization. (2003, February 2). Pandemic preparedness. Who.int. http://web.archive.org/web/20030202145905/http://www.who.int/csr/disease/influenza/pandemic/en/

301 Doshi, P. (2008). The elusive definition of pandemic influenza. Bulletin of the World Health Organization, 89(7):532-8. https://archive.hshsl.umaryland.edu/bitstream/handle/10713/6669/Doshi_ElusiveDefinitionPandemicInfluenza_2011.pdf

302 Cabinet Office (2017). National Risk Register of Civil Emergencies 2017 edition. Available at: https://assets.publishing.service.gov.uk/government/uploads/system/uploads/attachment_data/file/644968/UK_National_Risk_Register_2017.pdf

303 Ibid

304 Public Health England. (2014, August). Pandemic Influenza Response Plan 2014. Gov.uk. https://assets.publishing.service.gov.uk/government/uploads/system/uploads/attachment_data/file/344695/PI_Response_Plan_13_Aug.pdf

305 Tiplady, C. (2018, December 19). Blog – Uncertainty in diagnosis: What to do when you don't know what to do. Royal College of Pathologists. https://www.rcpath.org/discover-pathology/news/blog-uncertainty-in-diagnosis-what-to-do-when-you-don-t-know-what-to-do.html

306 BBC News. (2020, October 22). Covid: Teacher trapped In Italy quarantine happy to be home. BBC.co.uk. https://www.bbc.co.uk/news/uk-wales-54628221

307 BBC News. (2020a, September 17). Coronavirus: Pembrokeshire man 'stuck' In Italian facility. BBC.co.uk. https://www.bbc.co.uk/news/uk-wales-54188163

308 Barnacle, J. R. et al.(2021). Changes in the hospital admission profile of COVID19 positive patients at a central London trust. The Journal of infection,82(1), 159–198. https://doi.org/10.1016/j.jinf.2020.07.022

309 Vermeulen, J. et al. (2009, 1 November). External oligonucleotide standards enable cross laboratory comparison and exchange of real-time quantitative PCR data, Nucleic Acids Research, 37(21), page e138, https://doi.org/10.1093/nar/gkp721

310 Arévalo-Rodriguez, I. et al. (2020, December 10). False-negative results of initial RT-PCR assays for COVID-19: A systematic review. PLoS ONE 15(12): e0242958 https://doi.org/10.1371/journal.pone.0242958

311 Josephs, J. (2020, October 7). Government to pay £2m to settle coronavirus testing case. BBC News. https://www.bbc.co.uk/news/business-54455666

312 Ibid

313 Zhang, L. et al. (2020, December 13). SARS-CoV-2 RNA reverse-transcribed and integrated into the human genome. BIORXIV. https://doi.org/10.1101/2020.12.12.422516

314 Borger, P. et al. (2021, January 12). Addendum - Corman Drosten review report by an international consortium of scientists in life sciences (ICSLS). https://doi.org/10.5281/zenodo.4433503

315 World Health Organization. (2012). The use of PCR in the surveillance and diagnosis of influenza. Report of the 5th meeting of the WHO working group on polymerase chain reaction protocols for detecting subtype influenza a viruses Geneva, Switzerland 26–27 June 2012. WHO. https://web.archive.org/web/20210917021057/https:/www.who.int/influenza/gisrs_laboratory/report_2012pcrwg5thmeeting.pdf

316 Bullard, J. et al. (2020, November 15). Predicting infectious SARS-CoV-2 from diagnostic samples. Clinical Infectious Diseases, 71(10), 2663–2666. https://doi.org/10.1093/cid/ciaa638

317 Jaafar, R. et al. (2021, June 1) Correlation between 3790 quantitative polymerase chain reaction–positives samples and positive cell cultures, including 1941 severe acute respiratory syndrome Coronavirus 2 isolates, Clinical Infectious Diseases, 72(11):e921. https://doi.org/10.1093/cid/ciaa1491

318 Jefferson, T. et al. (2022, February 2). PCR testing in the UK during the SARSCoV-2 pandemic – evidence from FOI requests. Collateral Global. https://collateralglobal.org/wp-content/uploads/2022/02/CG-Report-7-PCR-Testing-in-the-UK-During-the-SARS-CoV-2-Pandemic-%E2%80%93-Evidence-From-FOI-Requests.pdf

319 Instructions For Use LightMix® Modular Wuhan CoV RdRP-gene 530 (n.d.). TibMOLBIOL. https://www.roche-as.es/lm_pdf/MDx_53-0777_96_Wuhan-R-gene_V200204_09155376001%20%282%29.pdf

320 Matheeussen, V. et al. International external quality assessment for SARS-CoV-2 molecular detection and survey on clinical laboratory preparedness during the COVID-19 pandemic, April/May 2020. Euro surveillance: Bulletin Europeen sur les maladies transmissibles 25(27), 2001223. https://doi.org/10.2807/1560-7917.ES.2020.25.27.2001223

321 Office for National Statistics. (2020, December 18). Coronavirus (COVID-19) Infection survey: Cycle threshold and household transmission analysis - Office For National Statistics. Gov.uk. https://www.ons.gov.uk/peoplepopulationandcommunity/healthandsocialcare/conditionsanddiseases/adhocs/12683coronaviruscovid19infectionsurveycyclethresholdandhouseholdtransmissionanalysis

322 Juanola-Falgarona, M. et al. (2022) Ct values as a diagnostic tool for monitoring SARS-CoV-2 viral load using the QIAstat-Dx® Respiratory SARS-CoV-2 panel. International Journal of Infectious Diseases,122, 930-935, https://doi.org/10.1016/j.ijid.2022.07.022

323 This Week in Virology [podcast] (2020, July 16). Available at: https://justthenews.com/politics-policy/coronavirus/newly-surfaced-video-july-fauci-tests-dead-virus

324 Public Health England. (2020, September 28). Guidance COVID 19. Management of staff and exposed patients or residents in health and social care settings. Gov.uk. https://web.archive.org/web/20210309002848/https://www.gov.uk/government/publications/covid-19-management-of-exposed-healthcare-workers-and-patients-in-hospital-settings/covid-19-management-of-exposed-healthcare-workers-and-patients-in-hospital-settings

325 World Health Organization. (2020, December 14). WHO Information notice for IVD users nucleic acid testing (NAT) technologies that use real-time polymerase chain reaction (RT-PCR) for detection of SARS-CoV-2. https://web.archive.org/web/20201215013928/https://www.who.int/news/item/14-12-2020-who-information-notice-for-ivd-users

326 Pollock, A. (2020, September 22). Yes extraordinary that UK. National Screening Committee not involved. Twitter. https://twitter.com/AllysonPollock/status/1308438843356569600

327 The Royal College of Pathologists. (2020). College statement: Standalone virus testing facilities. Rcpath.org. https://www.rcpath.org/static/e8e2e803-8f27-46c3-b7ba3bc0b9d46e00/G220-RCPath-statement-on-standalone-virus-testing-facilities.pdf

328 Navascués A. et al. Detection of respiratory viruses in deceased persons, Spain, 2017. (2018, July). Emerging Infectious Diseases, 24(7):1331-1334. https://doi.org/10.3201/eid2407.180162

329 Koetz, A. et al. (2006). Detection of human coronavirus NL63, human metapneumovirus and respiratory syncytial virus in children with respiratory tract infections in south-west Sweden. Clinical Microbiology and Infection, 12(11), 1089– 1096. https://doi.org/10.1111/j.1469-0691.2006.01506.x

330 Patrick, D. M. et al. (2006). An outbreak of human coronavirus OC43 infection and serological cross-reactivity with SARS Coronavirus. The Canadian Journal of Infectious Diseases & Medical Microbiology, 17(6), 330–336. https://doi.org/10.1155/2006/152612

331 Channel 4. (2020, November 16). Dispatches uncovers serious failings at one of UK's largest COVID-testing labs. Channel4.Com. https://www.channel4.com/press/news/dispatches-uncovers-serious-failings-one-uks-largest-covid-testing-labs

332 BBC. (2021, April 4). BBC One – Panorama, Undercover: Inside the Covid testing lab. BBC.co.uk. https://www.bbc.co.uk/programmes/m000tqjj

333 BBC Panorama. (2021, March 29). BBC Panorama sent an undercover reporter inside one of the biggest UK Covid testing labs. Twitter. https://twitter.com/BBCPanorama/status/1376639962649665540

334 Bethell, Lord. (2021, April 7). Letter to Fiona Bruce, MP. https://dailysceptic.org/wp-content/uploads/2021/04/DHSC-response-Letter-Apr2021.pdf

335 Te Whatu Ora – Health New Zealand. (2022, November 24,updated). Case definition and clinical testing guidelines for COVID-19. Te Whatu Ora. https://www.tewhatuora.govt.nz/for-the-health-sector/covid-19-information-for-health-professionals/case-definition-and-clinical-testing-guidelines-for-covid-19

336 Deeks, J. (2020, December 21). Our estimate of overall sensitivity Is 3.2% (95% CI 0.6% To 15.6%). Twitter. https://twitter.com/deeksj/status/1340975402391662592

337 Gov.UK. (2020, November 11). Oxford University and PHE confirm highsensitivity of lateral flow tests. Gov.UK. https://www.gov.uk/government/news/oxford-university-and-phe-confirm-high-sensitivity-of-lateral-flow-tests

338 @Beepr5 (2020, November 12). Replying to @StopCoronavir12 @alanmcn1 It Is Not A Separate Debate! It's In Context! If You're Making Fake Claims That No False Positives Exist … Twitter . https://twitter.com/Beepr5/status/1326960304950026246

339 Paltiel, A. D. and Wolensky, R. (2020, September 11). Screening to prevent SARS-CoV-2 Outbreaks: Saliva-based antigen testing is better than the PCR Swab. Health Affairs, Forefront. https://www.healthaffairs.org/do/10.1377/forefront.20200909.430047/full

340 Wymant, C. et al. (2021, May 12). The epidemiological impact of the NHS COVID-19 app. Nature 594, 408–412 https://doi.org/10.1038/s41586-021-03606-z

341 Adams, J. (1850) The Works of John Adams, Second President of the United States, p. 113. Little, Brown

342 Intensive Care, National Audit and Research Centre. (n.d.). COVID-19 Report. Icnarg.org. Retrieved January 11, 2023, from

https://www.icnarc.org/Our-Audit/Audits/Cmp/Reports

343 Oke, J., Howdon, D., and Heneghan, C. (2020, September 9). Declining COVID19 case fatality rates across all ages: Analysis of German data. The Centre For Evidence-Based Medicine. https://www.cebm.net/covid-19/declining-covid-19-case-fatality-rates-across-all-ages-analysis-of-german-data/

344 Heneghan, C. and Oke, J. (2020, August 12). Public Health England has changed its definition of deaths: Here's what it means. The Centre For Evidence-Based Medicine. https://www.cebm.net/covid-19/public-health-england-death-data-revised

345 Sayers, F. (2020, July 17). Prof Carl Heneghan: Can we trust the Covid-19 death numbers? UnHerd.Com. https://unherd.com/thepost/prof-carl-heneghan-can-we-trust-the-covid-19-death-numbers

346 Clare, C.(2020, September 11). Coronaviruses are seasonal… Twitter. https://twitter.com/ClareCraigPath/status/1304299112804888576

347 Gold, J. E. et al. (2020). Analysis of measles-mumps-rubella (MMR) titers of recovered Covid-19 patients. mBio, 11(6), e02628-20. https://doi.org/10.1128/mBio.02628-20

348 Drummond, H. (2020, October 29). The correlation between weekly excess deaths and covid deaths by region, then and now. Hectordrummand.com. https://hectordrummond.com/2020/10/29/the-correlation-between-weekly-excess-deaths-and-covid-deaths-by-region-then-and-now

349 Public Health England. (2020). Weekly national Influenza and COVID19 surveillance report week 4 report (up to week 3 data) 28 January 2021. PHE. https://assets.publishing.service.gov.uk/government/uploads/system/uploads/attachment_data/file/956709/Weekly_Flu_and_COVID-19_report_w4_FINAL.PDF

350 Kolata, G. (2007, January 22). Faith In quick test leads to epidemic that wasn't. The New York Times. https://www.nytimes.com/2007/01/22/health/22whoop.html

351 NHS Digital. (2020, September 17). Hospital admitted patient care activity 2019- 20: Summary report-ACC-Days. digital.nhs.uk. https://digital.nhs.uk/data-and-information/publications/statistical/hospital-admitted-patient-care-activity/2019-20/summary-reports---acc---days

352 BBC Newsnight. (2020, April 2). "We used to admit patients to ICU who were at risk of deterioration, just to observe them …" @BBC Newsnight, Twitter. https://twitter.com/BBCNewsnight/status/1245839392478040065

353 Shovlin, C. L. and Vizcaychipi, M.P. (2020) Implications for COVID-19 triage from the ICNARC report of 2204 COVID-19 cases managed in UK adult intensive care units. Emergency Medicine Journal, 37, 332-333. http://dx.doi.org/10.1136/emermed-2020-209791

354 Instituto de Salud Carlos III. (2020). Estudio ene-covid: Cuarta ronda estudio nacional de sero-epidemiología de la infección por sars-COV-2 en España. Portalcne.Isciii.Es. https://portalcne.isciii.es/enecovid19/informes/informe_cuarta_ronda.pdf

355 Knapton, S. (2022, January 7). Almost half of all Covid hospital patients in some areas are 'incidental' cases. The Telegraph. https://www.telegraph.co.uk/news/2022/01/07/incidental-covid-cases-make-half-nhs-hospital-admissions

356 Engler, J. (2021, January 4). More bonkers data - why are case rises in lockstep across the whole country. This is positivity data normalised to 21 dec with thanks to @surbitonsteve.Not natural. Twitter. https://twitter.com/jengleruk/status/1346068727574499328

357 Ward, H. et al. (2021,March 1). REACT-2 Round 5: Increasing prevalence of SARS-CoV-2 antibodies demonstrate impact of the second wave and of vaccine rollout in England. Medrxiv. https://doi.org/10.1101/2021.02.26.21252512

358 Jefferson T. et al. (2022, February 2) PCR testing in the UK during the SARSCoV-2 pandemic – Evidence from FOI requests. Collateral Global. https://collateralglobal.org/wp-content/uploads/2022/02/CG-Report-7-PCR-Testing-in-the-UK-During-the-SARS-CoV-2-Pandemic-%E2%80%93-Evidence-From-FOI-Requests.pdf

359 Health Advisory and Recovery Team. (2021, October 11). Increasing numbers of PCR Positives are asymptomatic https://www.hartgroup.org/increasing-numbers-of-pcr-positives-are-asymptomatic

360 Tribunal da Relação de Lisboa. (2020, November 11). Habeas corpus interesse em agir sars-cov-2 testes rt-pcr privação da liberdade detenção ilegal. http://www.dgsi.pt.

http://www.dgsi.pt/jtrl.nsf/33182fc732316039802565fa00497eec/79d6ba338dcbe5e28025861f003e7b30

361 Attributed by Riis, J. (1900). The American Monthly Review of Reviews, 22(2), p. 184.

362 Craig, C. (2020, September 7). Waiting for zero, in 'Logic in the time of COVID' (blog) . https://logicinthetimeofcovid.com/blog

363 The parties' joint statement on discovery disputes. State of Missouri ex rel. Eric S. Schmitt, Attorney General, State of Louisiana ex rel. Jeffrey M. Landry, Attorney General, et al., Plaintiffs, v. Joseph R. Biden, Jr., In his official capacity as President of the United States, et al. Case No. 3:22-cv-01213. (2022, August 31). United States District Court for the Western District Of Louisiana Monroe Division. New Civil Liberties Alliance. https://nclalegal.org/wp-content/uploads/2022/09/Joint-Statement-on-Discovery-Disputes-Combined.pdf

364 European Commission. (2021, July 29). Reports on June actions - Fighting COVID-19 Disinformation Monitoring Programme.

https://digital-strategy.ec.europa.eu/en/library/reports-june-actions-fighting-covid-19-disinformation-monitoring-programme

365 European Commission . (2022, November 14, updated). Questions and answers: Digital Services Act. https://ec.europa.eu/commission/presscorner/detail/en/QANDA_20_2348

366 Transparency Center, Twitter. (2022, July 28). COVID-19 misinformation. Twitter. https://transparency.twitter.com/en/reports/covid19.html

367 Coronavirus: Staying safe and informed on Twitter (blog). (2020, July 14). Protecting the public conversation: Clarifying how we assess misleading information. Twitter. https://blog.twitter.com/en_us/topics/company/2020/covid-19

368 Craig, C. (2022, August 24). Two days of Twitter exile. Substack. https://drclarecraig.substack.com/p/two-days-of-twitter-exile

369 Jimenez, J.-L. (2022, October 6). Shockingly, @Twitter has done what @WHO Has REFUSED to do for 2.5 Years. Tell us that the "FACT: COVID-19 Is NOT AIRBORNE …". @jljcolorado, Twitter. https://twitter.com/jljcolorado/status/1578063884400971777

370 Bitnun, A et al. (2020, May 9) Evidence shows COVID-19 is almost exclusively spread by droplets. https://www.thestar.com/opinion/letters_to_the_editors/2020/05/09/evidence-shows-covid-19-is-almost-exclusively-spread-by-droplets.html

371 Health Advisory and Recovery Team. (2021, October 6). No effect of school closures on COVID-19 in children. https://www.hartgroup.org/no-effect-covid-school-closures

372 Fuller, J. (2022, December 13). "DeSantis hosts 'COVID accountability roundtable' …". FISM TV. https://fism.tv/desantis-hosts-covid-accountability-roundtable-petitions-for-grand-jury-to-investigate-covid-vaccine-wrongdoing The full conference can be found on Local10news: https://www.local10.com/news/local/2022/03/07/gov-ron-desantis-surgeon-general-discuss-what-they-call-failure-of-lockdowns-mandates

373 Science and Technology Committee Health and Social Care Committee. (2021, June 10). Oral evidence: Coronavirus: Lessons learnt, HC 95. House of Commons. https://committees.parliament.uk/oralevidence/2318/pdf

374 Chapin, C. (1912) [1910]. The sources and modes of infection, p. 129. John Wiley & Sons. https://archive.org/details/sourcesmodesofin00ch/page/n5/mode/2up

375 Ibid

376 Ibid, p. 99

377 Tian, S. et al. (2020). Characteristics of COVID-19 infection in Beijing. The Journal of Infection,80(4), 401–406. https://doi.org/10.1016/j.jinf.2020.02.018

378 Chen, C. (2020, April 2). What we need to understand about asymptomatic carriers if we're going to beat Coronavirus. ProPublica. https://www.propublica.org/article/what-we-need-to-understand-about-asymptomatic-carriers-if-were-going-to-beat-coronavirus

379 Edelstein, P. and Ramakrishnan, L.(2020, July 7). Report on face masks for the general public - An update. DELVE Addendum MAS-TD1. Rs-delve.github.io. https://rs-delve.github.io/addenda/2020/07/07/masks-update.html

380 Lewis, D. (2022, April 6). Why the WHO took two years to say COVID is airborne. Nature, News Feature. https://www.nature.com/articles/d41586-022-00925-7

381 Davey, M. (2021, May 31). Victoria reports 11 new cases across state as outbreak hits two Melbourne aged care homes. The Guardian. https://www.theguardian.com/australia-news/2021/may/31/victoria-covid-update-11-new-cases-across-state-as-outbreak-hits-two-melbourne-aged-care-homes

382 Unknown. (1993). Kary Mullis speaks to misuse of PCR (1993) [video]. Internet Archive. https://archive.org/details/kary-mullis-speaks-to-misuse-of-pcr-1993

383 National Institute For Health And Clinical Excellence. (2008, February). Bacterial meningitis and meningococcal septicaemia in children. NICE. https://www.nice.org.uk/guidance/cg102/documents/meningococcal-disease-and-meningitis-in-children-and-young-people-final-scope2

384 O'Grady H.M., et al. (2022, July 7). Asymptomatic severe acute respiratory syndrome coronavirus-2 (SARS-CoV-2) infection in adults is uncommon using rigorous symptom characterization and follow-up in an acute-care adult hospital outbreak. Infection Control & Hospital Epidemiology, 1–3. https://doi.org/10.1017/ice.2022.168

385 Killingley, B, et al. (2022, March 31) Safety, tolerability and viral kinetics during SARS-CoV-2 human challenge in young adults. Nature Medicine 28, 1031–1041. https://doi.org/10.1038/s41591-022-01780-9

386 Ibid

387 Extended Data Fig. 5: Human SARS-CoV-2 challenge infection causes a range of mild-moderate symptoms without remdesivir treatment from Killingley, B. et al., (FN above). https://www.nature.com/articles/s41591-022-01780-9/figures/10

388 Killingley, B. et al. (2022, March 31). Safety, tolerability and viral kinetics during SARS-CoV-2 human challenge in young adults. Nature Portoflio (Supplementary Information), 28, 1031–1041. https://static-content.springer.com/esm/art%3A10.1038%2Fs41591-022-01780-9/MediaObjects/41591_2022_1780_MOESM1_ESM.pdf

389 Zhou, X. (2020, March 28) Letter to the Editor: Follow-up of asymptomatic patients with SARS-CoV-2 infection. Clinical Microbiology and Infection, 26, 957–959. https://doi.org/10.1016/j.cmi.2020.03.024

390 Petersen I, and Phillips A. (2020) Three quarters of people with SARS-CoV-2 infection are asymptomatic: Analysis of English household survey data. Clinical Epidemiology, 12, 1039-1043 https://doi.org/10.2147/CLEP.S276825

391 The Telegraph. (2020, March 5). Watch again: Chris Whitty answers coronavirus questions from MPs. YouTube. https://www.youtube.com/watch?v=IfJcwDaZrsA

392 10 Downing Street. (2020, March 25). United Kingdom Prime Minister Boris Johnson hosts a coronavirus press conference 25/03/2020. YouTube. https://www.youtube.com/watch?v=3Kjc7R1yGMc

393 Supplement to: Letizia A. G. et al. (2020) SARS-CoV-2 transmission among Marine recruits during quarantine. New England Journal of Medicine, 383(25): 2407- 16. https://doi.org/10.1056/NEJMoa2029717

https://www.nejm.org/doi/suppl/10.1056/NEJMoa2029717/suppl_file/nejmoa2029717_appendix.pdf

394 U.S. Department of Health and Human Services. (2020, January 28). Update on the new coronavirus outbreak first identified in Wuhan, China. YouTube. https://www.youtube.com/watch?v=w6koHkBCoNQ&t=2642s

395 Viwer, T. (2020, June 9). WHO: Asymptomatic carriers of coronavirus are not contagious. U.S. News. https://web.archive.org/web/20210425143005/https:/usnewslatest.com/who-asymptomatic-carriers-of-coronavirus-are-not-contagious

396 Craig, C. and Engler, J. (2021, August 21). Covid: The woeful case for asymptomatic transmission. The Conservative Woman. https://www.conservativewoman.co.uk/covid-the-woeful-case-for-asymptomatic-transmission-2

397 Wong, J. et al. (2020). Asymptomatic transmission of SARS-CoV-2 and implications for mass gatherings. Influenza and Other Respiratory Viruses, 14(5), 596– 598. https://doi.org/10.1111/irv.12767

398 Cao, S. et al.(2020, November 20). Post-lockdown SARS-CoV-2 nucleic acid screening in nearly ten million residents of Wuhan, China. Nature Communications, 11, 5917. https://doi.org/10.1038/s41467-020-19802-w

399 Letizia A. G. et al. (2020, December 17) SARS-CoV-2 transmission among Marine recruits during quarantine. New England Journal of Medicine, 383(25): 2407- 16. https://www.nejm.org/doi/pdf/10.1056/NEJMoa2029717

400 Nakajo, K., and Nishiura, H. (2021, February 19). Transmissibility of asymptomatic COVID-19: Data from Japanese clusters. International Journal of Infectious Diseases,105, 236–238. https://doi.org/10.1016/j.ijid.2021.02.065

401 Sim, W. (2020, February 13). Coronavirus: Japan confirms first death, but unclear if virus is direct cause of death. The Straits Times. https://www.straitstimes.com/asia/east-asia/coronavirus-japan-confirms-first-death-but-unclear-if-virus-is-direct-cause-of-death

402 Hershaw, R. B. (2021, March 26). Low SARS-CoV-2 transmission in elementary schools – Salt Lake County, Utah, December 3, 2020–January 31, 2021. Morbidity and Mortality Weekly Report 2021;70:442–448. Centers For Disease Control and Prevention. https://stacks.cdc.gov/view/cdc/104805

403 Cordery, R. et al. (2022, November 1). Transmission of SARS-CoV-2 by children to contacts in schools and households: A prospective cohort and environmental sampling study in London. The Lancet Microbe, 3(11), e814 - e823. https://doi.org/10.1016/S2666-5247(22)00124-0

404 Rothe, C. et al. (2020, March 5). Transmission of 2019-nCoV infection from an asymptomatic contact in Germany. The New England Journal of Medicine, 382, 970 - 971. https://www.nejm.org/doi/full/10.1056/nejmc2001468

405 Kupperschmidt, K. (2020, February 3). Study claiming new coronavirus can be transmitted by people without symptoms was flawed. Science. https://www.science.org/content/article/paper-non-symptomatic-patient-transmitting-coronavirus-wrong

406 de Laval, F. et al. (2021, May 19). Lessons learned from the investigation of a COVID-19 cluster in Creil, France: Effectiveness of targeting symptomatic cases and conducting contact tracing around them. BMC Infectious Diseases21, 457 https://doi.org/10.1186/s12879-021-06166-9

407 Wei, W. E. et al. (2020, April 10). Presymptomatic transmission of SARS-CoV-2 – Singapore, January 23–March 16, 2020. Morbidity and Mortality Weekly Report 69. https://www.cdc.gov/mmwr/volumes/69/wr/pdfs/mm6914e1-h.pdf

408 Kant, I. [1784] An answer to the question: What is enlightenment?. In Humphrey T. (trans) Perceptual peace and other essays (1983), Hackett Publishing Co.

409 Ghosh, Pallab. (2011, May 6) BBC News. https://www.bbc.co.uk/news/science-environment-13299666

410 Simanowitz, S. (2020, July 16). 6/. At the SAGE meeting on 27 Feb, the reasonable worst case for deaths if action wasn't taken, was estimated at 500,000. Twitter. https://twitter.com/StefSimanowitz/status/1283818009325379590

411 Breuninger, K. (2021, June 9). Fauci blasts 'preposterous' covid conspiracies, accuses his critics of 'attacks on science'. CNBC.com. https://www.cnbc.com/2021/06/09/fauci-blasts-preposterous-covid-conspiracies-accuses-critics-of-attacks-on-science.html

412 CBS News. (2021, November 28).Transcript: Dr. Anthony Fauci on "Face The Nation," November 28, 2021. CBSnews.com. https://www.cbsnews.com/news/transcript-dr-anthony-fauci-on-face-the-nation-november-28-2021

413 Georgiou, A. (2020, October 7). Fauci says he told Trump to 'shut the country down.' Newsweek. https://www.newsweek.com/fauci-trump-united-states-coronavirus-lockdown-1536999

414 Brown, S. (2022, July 26). Fauci: 'I didn't recommend locking anything down' during COVID. Townhall.com. https://townhall.com/tipsheet/spencerbrown/2022/07/26/fauci-claims-he-didnt-recommend-locking-anything-down-here-are-9-times-he-did-n2610782

415 Stern, D. (2021, February 15). Fauci awarded $1 Million Israeli prize for 'speaking truth to power' amid pandemic. NPR.org. https://www.npr.org/sections/coronaviruslive-updates/2021/02/15/968059128/fauci-awarded-1-million-israeli-prize-forspeaking-truth-to-power-amid-pandemic

416 Dr Simon Goddek (2020, December 8) As you can see in the tweet below, the peer review process of the Corman-Drosten paper represents an extreme outlier..." Twitter. https://twitter.com/goddeketal/status/1336112820484706307

417 Wouter Aukema (2020, December 7) "@eurosurveillance published an Editorial note, see..." Twitter https://twitter.com/waukema/status/1336032577916989449

418 Kutter, S. (2014, May 16). Virologe Drosten im gespräch 2014 "Der Körper wird ständig von Viren angegriffen". Wirtschafts Woche . https://www.wiwo.de/technologie/forschung/virologe-drosten-im-gespraech-2014- der-koerper-wirdstaendig-von-viren-angegriffen/9903228-all.html

419 D'Alessio, V. (2020, April 20). 'I don't see any other way out': Diagnostic testing and smartphone contact tracing to beat pandemic. Horizon Magazine. https://ec.europa.eu/research-and-innovation/en/horizon-magazine/i-dont-see-any-other-way-out-diagnostic-testing-and-smartphone-contact-tracing-beat-pandemic

420 Hayes, A. (2022, March 2). COVID-19 expert claims he was told to 'correct his views' after criticising 'implausible graph' shown during official briefing. Sky News. https://news.sky.com/story/covid-19-expert-claims-he-was-told-to-correct-his-views-after-criticising-implausible-graph-shown-during-official-briefing-12555800

421 Office for Statistics Regulation. (2020, November 5). OSR statement regarding transparency of data related to COVID-19. OFSR. https://osr.statisticsauthority.gov.uk/news/osr-statement-regarding-transparency-of-data-related-to-covid-19

422 House of Commons. (2020, November 3). Science and Technology Committee Oral evidence: UK science, research and technology capability and influence in global disease outbreaks, HC 136. https://committees.parliament.uk/oralevidence/1122/pdf

423 Ibid

424 Rowlatt, J. (2021, February 15). Bill Gates: Solving Covid easy compared with climate. BBC News. https://www.bbc.co.uk/news/science-environment-56042029

425 Still I rise motivation. (2022, June 24). Why I dropped out of college | Bill Gates - Best advice and regret. YouTube. https://www.youtube.com/watch?v=w2-HXLkqOq4

426 OFCOM. (2020, March 23). Note to Broadcasters. Ofcom.org.uk. https://www.ofcom.org.uk/__data/assets/pdf_file/0025/193075/Note-to-broadcasters-Coronavirus.pdf

427 Iu, M. (2021, March 2). Govt spent more than £184m on Covid comms in 2020. Campaign. https://www.campaignlive.co.uk/article/govt-spent-184m-covid-comms2020/1708695

428 Gov.uk. (2021, April 30). COVID 19 - Media Buying Services. Contracts Finder, Gov.Uk. https://www.contractsfinder.service.gov.uk/notice/6043d1fd-1f8c-4232-a32a-a658e19abcb1

429 Mayhew, F. (2020, April 22). Poll: Most say journalists not doing good job of holding govt to account during daily Covid-19 briefings. Press Gazette. https://pressgazette.co.uk/news/poll-journalists-have-not-donea-good-job-at-covid-19-briefings-majority-of-respondents-say/

430 Farrell, K. (2021, September 4). Robert Peston put on the spot during heated clash with JHB. Express.co.uk. https://www.express.co.uk/news/politics/1486225/Robertpeston-julia-hartley-brewer-itv-latest-covid-press-conferences-boris-johnson-vn

431 Liles, J. (2022, July 26). Did New Zealand PM Jacinda Ardern once say, 'unless you hear it from us, it is not the truth'? Snopes.com. https://www.snopes.com/fact-check/jacinda-ardern-truth

432 Guardian staff and agency. (2022, May 19). US Homeland security pauses new disinformation board amid criticism. The Guardian.

https://www.theguardian.com/us-news/2022/may/19/homeland-security-disinformation-board-nina-jankowicz-resigns

433 Gov.uk. (2020, March 29). Government cracks down on spread of false Coronavirus information online. Gov.uk. https://www.gov.uk/government/news/government-cracks-down-on-spread-of-false-coronavirus-information-online

434 Media Centre, BBC. (2020, December 10). Trusted News Initiative (TNI) to combat spread of harmful vaccine disinformation and announces major research project. BBC.com. https://www.bbc.com/mediacentre/2020/trusted-news-initiative-vaccine-disinformation

435 Pfizer.com. (n.d.). Board member. James C. Smith. Pfizer.Com. Retrieved January 13, 2023, from https://www.pfizer.com/people/leadership/board_of_directors/james_smith

436 Attkisson, S. (2021, April 28). (READ) The true conflicts of interest behind self appointed "fact checkers". Sharylattkisson.com. https://sharylattkisson.com/2021/04/read-the-true-conflicts-of-interest-behind-self-appointed-fact-checkers/

437 Adams, B. (2021, May 22). PolitiFact retracts Wuhan Lab theory 'fact-check.' Washington Examiner. https://www.washingtonexaminer.com/opinion/politifact-retracts-wuhan-lab-theory-fact-check

438 Leonhardt, D. (2022, March 9). Do covid precautions work? The New York Times. https://www.nytimes.com/2022/03/09/briefing/covid-precautions-red-blue-states.html

439 Camus, A. (1955) Speech at a banquet for President Eduardo Santos. https://wist.info/camus-albert/27884

440 Raj, R. and Brown, K. K. (2016). Mortality related to surgical lung biopsy in patients with interstitial lung disease. The devil is in the denominator. American Journal of Respiratory and Critical Care Medicine, 193 10, 1082-4. https://doi.org/10.1164/rccm.201512-2488ED

441 Ouriel, K. et al. (1990). Factors determining survival after ruptured aortic aneurysm: The hospital, the surgeon, and the patient. Journal of Vascular Surgery,11(4), 493–496. https://doi.org/10.1016/0741-5214(90)90292-I

442 Vsauce. (2017, December 6). The trolley problem in real life. YouTube. https://www.youtube.com/watch?v=1sl5KJ69qiA

443 Kaur, K. (2022, March 8). "We inverted precautionary principle …'— Oxford Prof @SunetraGupta. @dockaurG, Twitter. https://twitter.com/dockaurG/status/1501029167479115784

444 Ghebreyesus, T. A. (2020, March 3). WHO Director-General's opening remarks at the media briefing on COVID-19 - 3 March 2020.

https://www.who.int/director-general/speeches/detail/who-director-general-s-opening-remarks-at-the-media-briefing-on-covid-19---3-march-2020

445 Addendum to fourteenth SAGE meeting on Covid-19, 10th March 2020 held in 10 Victoria St, London, SW1H 0NN. (2020). Gov.uk. https://assets.publishing.service.gov.uk/government/uploads/system/uploads/attachment_data/file/888782/S0382_Fourteenth_SAGE_meeting_on_Wuhan_Coronavirus__Covid-19__.pdf

446 Danon, L. et al. (2020, February 14). A spatial model of CoVID-19 transmission in England and Wales: Early spread and peak timing. Medrxiv.Org. https://doi.org/10.1101/2020.02.12.20022566

447 Danon, L. et al. (2021, July 19). A spatial model of COVID-19 transmission in England and Wales: Early spread, peak timing and the impact of seasonality. Philosophical Transactions of the Royal Society B, 376: 20200272. https://royalsocietypublishing.org/doi/full/10.1098/rstb.2020.0272

448 Ferguson, N. M. (2020, March 16). Report 9 - Impact of non-pharmaceutical Interventions (NPIs) to reduce COVID-19 mortality and healthcare demand. Imperial College London. https://www.imperial.ac.uk/mrc-global-infectious-disease-analysis/covid-19/report-9-impact-of-npis-on-covid-19

449 Ibid

450 Danon, L. et al. (2021, July 19). A spatial model of COVID-19 transmission in England and Wales: Early spread, peak timing and the impact of seasonality. Philosophical Transactions of the Royal Society B, 376: 20200272. https://royalsocietypublishing.org/doi/full/10.1098/rstb.2020.0272

451 Scientific Pandemic Influenza Group on Modelling (SPI-M-O) (2020, March 2). SPI-M-O: Consensus statement on 2019 novel coronavirus (COVID-19). Gov.uk. https://assets.publishing.service.gov.uk/government/uploads/system/uploads/attachment_data/file/873713/01-spi-m-o-consensus-statement-on-2019-novel-coronavirus-_covid-19_.pdf

452 Danon, L. et al. (2021, July 19). A spatial model of COVID-19 transmission in England and Wales: Early spread, peak timing and the impact of seasonality. Philosophical Transactions of the Royal Society B, 376: 20200272. https://royalsocietypublishing.org/doi/full/10.1098/rstb.2020.0272

453 Whitty, C. (2020, March 5). Oral evidence: Preparations for the Coronavirus, HC 36. House of Commons. https://committees.parliament.uk/oralevidence/113/pdf

454 Ibid

455 Scientific Pandemic Influenza Group on Modelling (SPI-M-O) (2020, March 2). SPI-M-O: Consensus Statement on 2019 Novel Coronavirus (COVID-19). https://www.gov.uk/government/publications/spi-m-o-consensus-statement-on-2019-novel-coronavirus-covid-19-2-march-2020

456 House of Commons. (2020, June 10). Science and Technology Committee oral evidence: UK science, research and technology capability and influence in global disease outbreaks, HC 136. https://committees.parliament.uk/oralevidence/539/pdf

457 @Brumby. (2021, January 13). There is no evidence that more restrictive nonpharmaceutical interventions (lockdowns) contributed substantially to bending the curve of new cases in England, France, Germany, Iran, Italy, the Netherlands, Spain or the United States in early 2020. Twitter. https://twitter.com/the_brumby/status/1349478826800136196

458 Flaxman, S., et al. (2020, June 8). Estimating the effects of non-pharmaceutical interventions on COVID-19 in Europe. Nature 584, 257–261. https://doi.org/10.1038/s41586-020-2405-7

459 Sy, K.T.L. et al. (2021, April 21) Population density and basic reproductive number of COVID-19 across United States counties. PLOS ONE 16(4): e0249271. https://doi.org/10.1371/journal.pone.0249271

460 Carozzi, F., Provenzano, S. and Roth, S. (2020, August). Urban density and Covid-19 (CEP Discussion Paper No 1711). Centre for Economic Performance Publications Unit, London School of Economics. https://cep.lse.ac.uk/pubs/download/dp1711.pdf

461 Nickelsburg, M. (2020, April 6). Google location data for Seattle shows decline in work, transit and retail trips – but not park visits. GeekWire. https://www.geekwire.com/2020/google-location-data-seattle-shows-decline-work-transit-retail-trips-not-park-visits

462 Healy, K. (2020, April 23). Apple's COVID mobility data. Kieranhealy.org. https://kieranhealy.org/blog/archives/2020/04/23/apples-covid-mobility-data

463 Brunner, J. and Beekman, D. (2020, January 1). Inslee orders Washingtonians to stay at home to slow spread of coronavirus. The Seattle Times. https://www.seattletimes.com/seattle-news/inslee-to-hold-televised-address-monday-evening-to-announce-enhanced-strategies-on-covid-19

464 Grover, N. (2021, July 18). UK covid cases could hit 200,000 a day, says scientist behind lockdown strategy. The Guardian. https://www.theguardian.com/world/2021/jul/18/uk-covid-cases-could-hit-200000-a-day-says-neil-ferguson-scientist-behind-lockdown-strategy-england

465 ZOE Health Study. (2021, July 21). COVID estimates updated as more people are being vaccinated. ZOE Health Study. https://health-study.joinzoe.com/post/covid-estimates-updated-vaccine

466 Jones, W. (2021, August 8). PHE data confirms that new infections peaked and dropped in the unvaccinated before they did in the vaccinated. The Daily Sceptic. https://dailysceptic.org/2021/08/08/phe-data-confirms-that-new-infections-peaked-and-dropped-in-the-unvaccinated-before-they-did-in-the-vaccinated

467 Zoe Health Study. (2021, July 21) COVID estimates updated as more people are being vaccinated. https://health-study.joinzoe.com/post/covid-estimates-updated-vaccine

468 Office of National Statistics. (2021, August 27) Coronavirus (COVID-19) Infection Survey, UK Coronavirus (COVID-19) Infection Survey, UK https://www.ons.gov.uk/peoplepopulationandcommunity/healthandsocialcare/conditionsanddiseases/bulletins/coronaviruscovid19infectionsurveypilot/27august2021

469 Jonung, L. (2020, June 18). Sweden's constitution decides its exceptional Covid-19 policy. Centre for Economic Policy Research.

https://cepr.org/voxeu/columns/swedens-constitution-decides-its-exceptional-covid-19-policy

470 Walker, P. G. T. et al. (2020, March 26). Report 12 – The global impact of COVID19 and strategies for mitigation and suppression. MRC Centre for Global Infectious Disease Analysis, Imperial College London. https://www.imperial.ac.uk/mrc-global-infectious-disease-analysis/covid-19/report-12-global-impact-covid-19

471 Flaxman, S. et al. (2020, June 8). Estimating the effects of non-pharmaceutical interventions on COVID-19 in Europe. Nature 584, 257–261. Nature. https://www.nature.com/articles/s41586-020-2405-7

472 Anonymous. (2021, February 22). Readers' exchanges with Professor Neil Ferguson. The Daily Sceptic. https://dailysceptic.org/2021/02/22/lockdown-sceptics-exchange-with-professor-neil-ferguson

473 Ahlander, J. (2020, June 25). Loved and loathed, Sweden's anti-lockdown architect is unrepentant. Reuters. https://www.reuters.com/article/us-health-coronavirus-sweden-tegnell-idUSKBN23W22K

474 Sayers, F. (2020, July 23). Why we aren't wearing masks in Sweden. UnHerd. https://unherd.com/2020/07/swedens-anders-tegnell-judge-me-in-a-year

475 Walker, P. T. G. (2020, June 12). The impact of COVID-19 and strategies for mitigation and suppression in low- and middle-income countries. Science. https://www.science.org/doi/10.1126/science.abc0035

476 MRC Centre for Global Infectious Disease Analysis, Imperial College London. (2020, March 18). Impact of non-pharmaceutical interventions (NPIs) to reduce COVID19 mortality and healthcare demand. Gov.uk. https://assets.publishing.service.gov.uk/government/uploads/system/uploads/attachment_data/file/891862/S0057_SAGE16_Imperial_Impact_of_NPIs_to_reduce_Mortality_and_Healthcare_Demand.pdf

477 Flaxman, S. et al. (2020, June 8). Extended data table 1 total forecasted deaths since the beginning of the epidemic up to 4 may 2020 in our model and in a counterfactual model that assumes no interventions had taken place. Nature, 584, 257–261 https://www.nature.com/articles/s41586-020-2405-7/tables/2

478 Whitty, C. (2020, March 5). Oral evidence: Preparations for the Coronavirus, HC 36. House of Commons. https://committees.parliament.uk/oralevidence/113/pdf

479 Ibid

480 Science and Technology Committee, House of Commons. (2020). Oral evidence: UK science, research and technology capability and influence in global disease outbreaks, HC 136. https://committees.parliament.uk/oralevidence/539/pdf

481 Davies, G. (2021, December 28). European countries tighten COVID-19 restrictions as Omicron spreads. ABC News. https://abcnews.go.com/International/european-countries-tighten-covid-19-restrictions-amicron-spreads/story?id=81969759

482 Herby, J., Jonung, L. and Hanke, S. H. (2022). A literature review and metaanalysis of the effects of lockdowns on COVID-19 mortality. Studies in Applied Economics 200 https://sites.krieger.jhu.edu/iae/files/2022/06/A-Systematic-Review-and-Meta-Analysis-of-the-Effects-of-Lockdowns-of-COVID-19-Mortality-II.pdf

483 Public Health England. (2020, July 14). Excess mortality In England, week ending 03 July 2020. Public Health England. https://fingertips.phe.org.uk/static-reports/mortality-surveillance/excess-mortality-in-england-week-ending-03-Jul-2020.html

484 Public Health England (2021). Statutory notifications of infectious diseases week 2020/53 week ending 03/01/2021. Gov.uk. https://assets.publishing.service.gov.uk/government/uploads/system/uploads/a ttachment_data/file/949867/NOIDS-weekly-report-week53-2020.pdf

485 Ibid

486 Inglesby, T. V. et al.(2006) Disease mitigation measures in the control of pandemic influenza. Biosecurity and Bioterrorism : Biodefense Strategy, Practice, and Science, 4(4), 366–375. https://doi.org/10.1089/bsp.2006.4.366

487 World Health Organization Writing Group, Bell, D., et al. (2006). Nonpharmaceutical interventions for pandemic influenza, national and community measures. Emerging infectious diseases, 12(1), 88–94. https://doi.org/10.3201/eid1201.051371

488 Woolhouse, M. (2022). The year the world went mad, p 227. Sandstone Press.

489 Great Barrington Declaration. (2020, October 4). Great Barrington Declaration and Petition. Gbdeclaration.org. https://gbdeclaration.org

490 Ibid

491 Djaparidze, L and Lois, F. (2020, October 23). SARS-CoV-2 waves in Europe: A 2-stratum SEIRS model solution (preprint). medRxiv. https://doi.org/10.1101/2020.10.09.20210146

492 Kulldorff, M. and Bhattacharya, J. (2021, August 2). The smear campaign against the Great Barrington Declaration. Spiked.

https://www.spiked-online.com/2021/08/02/the-smear-campaign-against-the-great-barrington-declaration/

493 Gorksi, D. H. (2020, October 15). Viewpoint: Great Barrington Declaration arguing for herd immunity 'takes page from denialist propaganda playbook'. Genetic Literacy Project. https://geneticliteracyproject.org/2020/10/15/viewpoint-greatbarrington-declaration-arguing-for-herd-immunity-takes-page-from-denialist-propaganda-playbook

494 Kulldorff, M. and Bhattacharya, J. (2021, August 2). The smear campaign against the Great Barrington Declaration. Spiked. https://www.spiked-online.com/2021/08/02/the-smear-campaign-against-the-great-barrington-declaration/

495 Craig, C. (2022, February 23). Show me the difference testing made. (I think there was significant underreporting of respiratory infection outbreaks from care homes in 2019/2020 - pre-covid). Twitter. https://twitter.com/ClareCraigPath/status/1496413712093175811

496 Durrant, T. et al. (2021, January 21). Whitehall Monitor 2021. Institute for Government. https://www.instituteforgovernment.org.uk/sites/default/files/publications/whitehall-monitor-2021_1.pdf

497 Grylls, G. (2022, August 8). Storage of unused PPE costs the taxpayer £700,000 a day. The Times. https://www.thetimes.co.uk/article/storage-of-unused-ppe-costs-thetaxpayer-700-000-a-day-ssmf8hvtw

498 UK Parliament Committees. (2021, October 27). "Muddled, overstated, eyewateringly expensive": PAC damning on Test & Trace that has "failed on main objectives". UK Parliament. https://committees.parliament.uk/committee/127/publicaccounts-committee/news/158262/muddled-overstated-eyewateringly-expensive-pacdamning-on-test-trace-that-has-failed-on-main-objectives

499 Cecil, N. (2021, March 10). "Test and Trace most wasteful public spending programme of all time". Evening Standard. https://www.standard.co.uk/news/politics/test-and-trace-wasteful-public-spending-former-treasury-chief-report-b923249.html

500 Bowler, N. (2022, May 4). As vaccine demand collapses, UK faces £4 billion of waste with 80% of its 650 million dose stockpile unused. The Daily Sceptic. https://dailysceptic.org/2022/05/04/as-vaccine-demand-collapses-u-k-faces-4-billion-of-waste-with-80-of-its-650-million-dose-stockpile-unused

501 Brien, P and Keep, M. (2022, March 29). Public spending during the Covid-19 pandemic. House of Commons Library. https://commonslibrary.parliament.uk/research-briefings/cbp-9309

502 Durrant, T. et al. (2021, January 21). Whitehall Monitor 2021. Institute for Government. https://www.instituteforgovernment.org.uk/sites/default/files/publications/whitehall-monitor-2021_1.pdf

503 Ibid

504 Quoted in Sayers, F. (2022, December 26). Neil Ferguson interview: China changed what was possible. UnHerd. https://unherd.com/thepost/neil-ferguson-interview-china-changed-what-was-possible

505 Ebbs, S. (2021, March 3). Biden calls Texas decision to reopen "Neanderthal thinking". ABC News. https://abcnews.go.com/Politics/biden-calls-texas-decision-reopen-neanderthal-thinking/story?id=76229294

506 Gurdasani, D. et al. (2021, July 7). Mass infection is not an option: We must do more to protect our young. The Lancet. https://www.thelancet.com/journals/lancet/article/PIIS0140-6736(21)01589-0/fulltext

507 Gupta, S. (2022, June 4). Thank you for picking out the perfect summary of my position. Twitter. https://twitter.com/SunetraGupta/status/1533183842336514051

508 Department of Health and Social Care, Office for National Statistics, Government Actuary's Department and Home Office. (2020, April 8). Initial estimates of excess deaths from COVID-19. Gov.uk.

https://assets.publishing.service.gov.uk/government/uploads/system/uploads/a ttachment_data/file/892030/S0120_Initial_estimates_of_Excess_Deaths_fro m_COVID-19.pdf

509 Department of Health and Social Care, Office for National Statistics. (2020, July 15). Direct and indirect impacts of COVID-19 on excess deaths and morbidity: Executive summary. Gov.uk. https://assets.publishing.service.gov.uk/government/uploads/system/uploads/a ttachment_data/file/918738/S0650_Direct_and_Indirect_Impacts_of_COVI D-19_on_Excess_Deaths_and_Morbidity.pdf

510 Roberts, L. (2022, July 26). Lockdown drinking rise could lead to 25,000 extra deaths, warns NHS. The Telegraph. https://www.telegraph.co.uk/news/2022/07/26/lockdown-drinking-rise-could-lead-25000-extra-deaths-warns-nhs

511 BBC Newsnight. (2020, April 2). "We used to admit patients to ICU who were at risk of deterioration, just to observe them… ". Twitter. https://twitter.com/BBCNewsnight/status/1245839392478040065

512 Arbuthnott, G. et al. (2020, October 25). Revealed: how elderly paid price of protecting NHS from Covid-19. The Times. https://www.thetimes.co.uk/article/revealed-how-elderly-paid-price-of-protectingnhs-from-covid-19-7n62kkbtb

513 Lintern, S. (2020, March 25). Coronavirus: U-turn on critical care advice for NHS amid fears disabled people will be denied treatment. The Independent . https://www.independent.co.uk/news/health/coronavirus-nhs-treatment-disabled-autism-nice-covid-19-a9423441.html

514 Arbuthnott, G. et al. (2020, October 25). Revealed: how elderly paid price of protecting NHS from Covid-19. The Times. https://www.thetimes.co.uk/article/revealed-how-elderly-paid-price-of-protecting-nhs-from-covid-19-7n62kkbtb

515 Ibid

516 Ibid

517 The Queen's Nursing Institute. (2020). The experience of care home staff during Covid-19: A survey report by the QNI's international community nursing observatory, p. 16. https://www.qni.org.uk/wp-content/uploads/2020/08/The-Experience-of-Care-Home-Staff-During-Covid-19-2.pdf

518 Ibid

519 Ibid

520 Murdoch, C., Banks, R. and Kanani, N. (2020, April 3). Do not attempt cardiopulmonary resuscitation DNACPR letter 3 April 2020. England.nhs.uk. https://www.england.nhs.uk/coronavirus/documents/do-not-attempt-cardiopulmonary-resuscitation-dnacpr-letter-3-april-2020/

521 May, R., and Powis, S. (2020, April 7). Letter to Chief executives of all NHS trusts and foundation trusts … et al. https://www.england.nhs.uk/coronavirus/wp-content/uploads/sites/52/2020/04/maintaining-standards-quality-of-care-pressurised-circumstances-7-april-2020.pdf

522 Hancock, M. (2020, April 30). Health and Social Care Secretary's statement on coronavirus (COVID-19): 15 April 2020. Gov.uk. https://www.gov.uk/government/speeches/health-and-social-care-secretarys-statement-on-coronavirus-covid-19-15-april-2020

523 Evans, M. (2020, December 20). Hospitals retreat from early covid treatment and return to basics. Wall Street Journal. https://www.wsj.com/articles/hospitals-retreat-from-early-covid-treatment-and-return-to-basics-11608491436

524 Engler, J. (2022, September 12). Were the unprecedented excess deaths curves in Northern Italy in spring 2020 caused by the spread of a novel deadly virus? https://pandauncut.substack.com/p/were-the-unprecedented-excess-deaths

525 University College London. (2020, October 10). Four in 10 extra deaths In Lombardy not linked to Covid-19. UCL News. https://www.ucl.ac.uk/news/2020/oct/four-10-extra-deaths-lombardy-not-linked-covid-19

526 Public Health England. (2020). National Ambulance Syndromic Surveillance System: England. Key messages. Data to: 06 September 2020. https://assets.publishing.service.gov.uk/government/uploads/system/uploads/attachment_data/file/916484/NASS_Bulletin_2020_36.pdf

527 Public Health England. (2021). Emergency Department Syndromic Surveillance System: England. Year 2021: Week 1. Gov.uk. https://assets.publishing.service.gov.uk/government/uploads/system/uploads/attachment_data/file/952272/EDSSSBulletin2021wk01.pdf

528 Pell, R. et al. (2020). Coronial autopsies identify the indirect effects of COVID-19. The Lancet: Public Health: Correspondence, 5(9,E74). https://doi.org/10.1016/S2468-2667(20)30180-8

529 Castoldi, L. et al. (2021). Variations in volume of emergency surgeries and emergency department access at a third level hospital in Milan, Lombardy, during the COVID-19 outbreak. BMC Emergency Medicine, 59(21). https://doi.org/10.1186/s12873-021-00445-z

530 Woolf, S. H. et al. (2020) Excess deaths from COVID-19 and other causes, March-April 2020. Journal of the American Medical Association, 324(5): 510–313. https://doi.org/10.1001/jama.2020.11787

531 The Economic Times: News. (2020, April 2). New York City setting up mobile morgues to deal with growing COVID-19 casualties. The India Times. https://economictimes.indiatimes.com/news/international/world-news/new-york-city-setting-up-mobile-morgues-to-deal-with-growing-covid-19-casualties/articleshow/74955160.cms

532 Dinesh, A. et al.(2021, May 21). Outcomes of COVID-19 admissions in the New York City Public Health System and variations by hospitals and boroughs during the initial pandemic response. Frontiers in Public Health, 9. https://doi.org/10.3389/fpubh.2021.570147

533 Richardson S.et al. (2020). Presenting characteristics, comorbidities, and outcomes among 5700 patients hospitalized with COVID-19 in the New York City area. Journal of the American Medical Association (JAMA), 323(20): 2052–2059. https://doi.org/10.1001/jama.2020.6775

534 COVID-19 Frequently asked questions (FAQs) on Medicare Fee-for-Service (FFS) billing. (2022). Centers for Medicare and Medicaid Services. https://www.cms.gov/files/document/03092020-covid-19-faqs-508.pdf

535 Evans, M. (2020, December 20). Hospitals retreat from early covid treatment and return to basics. Wall Street Journal. https://www.wsj.com/articles/hospitals-retreat-from-early-covid-treatment-and-return-to-basics-11608491436

536 Taylor L. (2021, March 9). Covid-19: Why Peru suffers from one of the highest excess death rates in the world. British Medical Journal 372: n611. https://doi.org/10.1136/bmj.n611

537 Lagadec, P. (2004). Understanding the French 2003 heat wave experience: Beyond the heat, a multi-layered challenge. Journal of Contingencies and Crisis Management, 12(4), 160–169. https://doi.org/10.1111/j.0966-0879.2004.00446.x

538 Poumadere, M. et al. (2006). The 2003 heat wave in France: Dangerous climate change here and now. Risk Analysis 25(6), 1483-94. https://doi.org/10.1111/j.1539-6924.2005.00694.x

539 Vandentorren, S. et al. (2006, October 6). Heat wave in France: Risk factors for death of elderly people living at home, EUROPEAN JOURNAL OF PUBLIC HEALTH, 16(6), 583–591. https://doi.org/10.1093/eurpub/ckl063

540 Lagadec, P. (2004). Understanding the French 2003 heat wave experience: Beyond the heat, a multi-layered challenge. Journal of Contingencies and Crisis Management, 12(4), 160–169. https://doi.org/10.1111/j.0966-0879.2004.00446.x

541 Poumadere, M. et al. (2006). The 2003 heat wave in France: Dangerous climate change here and now. Risk Analysis 25(6), 1489. https://doi.org/10.1111/j.1539-6924.2005.00694.x

542 House of Commons. (2020, June 10). Science and Technology Committee oral evidence: UK science, research and technology capability and influence in global disease outbreaks, HC 136. https://committees.parliament.uk/oralevidence/539/pdf

543 Ward, H. et al. (2021, March 1). REACT-2 Round 5: Increasing prevalence of SARS-CoV-2 antibodies demonstrate impact of the second wave and of vaccine roll- out in England. MedRxiv. https://doi.org/10.1101/2021.02.26.21252512

544 Pastafarian, J. (2022, February 3). Simple life-years argument showing lockdown is a disastrous policy. YouTube. https://www.youtube.com/watch?v=fd8dfbLRGCM

545 Johnson, J. and Rancourt, D. (2022, September 6). Lockdowns did not save lives. Brownstone Institute. https://brownstone.org/articles/lockdowns-did-not-save-lives/

546 Lydall, R. (2022, July 5). "117,000 die on waiting lists for NHS." Evening Standard. https://www.standard.co.uk/news/uk/117-000-die-nhs-waiting-lists-covidbacklog-b1010262.html

547 Woolhouse, M. (2022). The year the world went mad, p. 117. Sandstone Press.

548 Lynn, R. M. et al. (2020, June 25). Delayed access to care and late presentations in children during the COVID-19 pandemic: A snapshot survey of 4075 paediatricians in the UK and Ireland. Archives of Disease in Childhood, 106: e8. http://dx.doi.org/10.1136/archdischild-2020-319848

549 Cortés-Albornoz M. C. et al. (2022) Effects of remote learning during the COVID-19 lockdown on children's visual health: A systematic review. British Medical Journal Open, 12: http://dx.doi.org/10.1136/bmjopen-2022-062388

550 Donnelly, L. (2022, March 23). Sir Chris Whitty: Childhood obesity 'significantly' worse as a result of lockdowns. The Telegraph. https://www.telegraph.co.uk/news/2022/03/23/sir-chris-whitty-school-closures-likely-have-caused-substantial

551 UNICEF. (2020, May 13). Coronavirus could increase child deaths by 6,000 a day, UNICEF warns. UNICEF-UK. https://www.unicef.org.uk/pressreleases/coronavirus-could-increase-child-deaths-by-6000-a-day-unicef-warns

552 Brookings Institution. (2022, March 23). Ensuring young mothers return, remain, and learn in school. Brookings.edu. https://www.brookings.edu/events/ensuring-young-mothers-return-remain-and-learn-in-school

553 Quoted in Turner, C. (2021, February 13). "Explosion" of children with tics and Tourette's from lockdown. https://www.telegraph.co.uk/news/2021/02/13/explosion-children-tics-tourettes-lockdown

554 Zea Vera, A. (2022, February 25). The phenomenology of tics and tic-like behavior in TikTok. Pediatric Neurology, 130, 14–20. https://doi.org/10.1016/j.pediatrneurol.2022.02.003

555 Godar, S. C et al.(2014). Animal models of tic disorders: A translational perspective. Journal of Neuroscience Methods, 238, 54–69. https://doi.org/10.1016/j.jneumeth.2014.09.008

556 Brueck, H. (2022, May 12). 8 teen girls are proof TikTok is not to blame for the rise in social tics, CDC says. Insider. https://www.insider.com/tiktok-tics-what-actually-happened-blaming-social-media-is-misleading-2022-5

557 Prestigiacomo, A. (2020, October 22). New CDC numbers show lockdown's deadly toll on young people. The Daily Wire. https://www.dailywire.com/news/new-cdc-numbers-show-lockdowns-deadly-toll-on-young-people

558 Statistics Canada. (2021, March 10). Provisional death counts and excess mortality, January to December 2020. https://www150.statcan.gc.ca https://www150.statcan.gc.ca/n1/daily-quotidien/210310/dq210310c-eng.htm

559 WFP Staff. (2022, June 21). WFP urges G7: "Act now or record hunger will continue to rise and millions more will face starvation". World Food Programme. https://www.wfp.org/stories/wfp-urges-g7-act-now-or-record-hunger-will-continue-rise-and-millions-more-will-face

560 Office of US Foreign Disaster Assistance. (n.d.). Disaster case report. Ethiopa – drought/famine. Agency for International Development. Retrieved January 16, 2023, from https://pdf.usaid.gov/pdf_docs/PBAAH005.pdf

561 Wikipedia. (n.d.). 1983–1985 Famine in Ethiopia. Wikipedia. Retrieved January 16, 2023, from https://en.wikipedia.org/wiki/1983%E2%80%931985_famine_in_Ethiopia

562 Mathur, C. (2020). COVID-19 and India's trail of tears. Dialectical Anthropology, 44, 239–242. https://doi.org/10.1007/s10624-020-09611-4

563 Green, T. and Bhattacharya, J. (2021, July 22). Lockdowns are killers In the global south. UnHerd. https://unherd.com/thepost/lockdowns-are-killers-in-the-global-south

564 Region of Peel. (n.d.). COVID-19 What to do if your child is dismissed from school, day camp or child care. Region of Peel. Retrieved January 16, 2023, from https://www.peelregion.ca/coronavirus/_media/child-dismissed-protocol-en.pdf

565 Education Reporter. (2020, May 20). "I am wrong sometimes": Teachers' union leader who was overheard making offensive remarks about children admits she is 'blunt too often'. Mail Online. https://www.dailymail.co.uk/news/article-8341959/Teachers unionleader-caught-making-offensive-remarks-children-admits-blunt-often.html

566 Haynes, C. and Mistry, P. (2020, October 1). Warning over student halls fire exit breaches. BBC News. https://www.bbc.co.uk/news/uk-england-leeds-54376899

567 BBC News. (2020, October 15). Covid: York self-isolating students told "wait behind in fire." BBC. https://www.bbc.co.uk/news/uk-england-york-north-yorkshire-54551789

568 BBC News. (2022, February 4). Call for fire safety talks over classroom door plans. BBC. https://www.bbc.co.uk/news/uk-scotland-60261400

569 Ehley, B. (2020, August 13). CDC: One quarter of young adults contemplated suicide during pandemic. Politico. https://www.politico.com/news/2020/08/13/cdc-mental-health-pandemic-394832

570 Suárez-González, A. (2021, July 31). The effect of COVID-19 isolation measures on the cognition and mental health of people living with dementia: A rapid systematic review of one year of quantitative evidence. eClinicalMedicine, 39, 101047. https://doi.org/10.1016/j.eclinm.2021.101047

571 Apex World. (2020, October 5). Milton Keynes, UK: Heartbreaking moment when a son moved closer to his mother at his father's funeral to comfort her ... Twitter. https://twitter.com/apexworldnews/status/1313189351858339843

572 Herby, J., Jonung, L. and Hanke, S. H. (2022). A literature review and metaanalysis of the effects of lockdowns on COVID-19 mortality. Studies in Applied Economics 200. https://sites.krieger.jhu.edu/iae/files/2022/06/A-Systematic-Review-and-Meta-Analysis-of-the-Effects-of-Lockdowns-of-COVID-19-Mortality-II.pdf

573 Franklin, B. (1963) [1755, November 11]. Reply to the Governor, in Labaree, L. W. (ed.), The Papers of Benjamin Franklin, vol 6. p. 242. Yale University Press.

574 Public Health England. (2019, September). PHE Strategy 2020-25, p. 5. https://assets.publishing.service.gov.uk/government/uploads/system/uploads/attachment_data/file/831562/PHE_Strategy_2020-25.pdf

575 F. Furedi cited in Swart, N. (2021, July 27). Is the trade off between freedom and safety an illusion? - Professor Frank Furedi on lockdowns. BizNews.com. https://www.biznews.com/business-unusual/2021/07/27/lockdowns-furedi

576 Keating, G. C. (2018). Principles of risk imposition and the priority of avoiding harm. Revus, 36, 7–39. Retrieved January 31, 2023, from https://journals.openedition.org/revus/4406

577 Lazarus, E. (1883). 'The New Collossus' (poem).

578 Health and Safety Laboratory. (2008). Evaluating the protection afforded by surgical masks against influenza bioaerosols Gross protection of surgical masks compared to filtering facepiece respirators. Health and Safety Executive. https://www.hse.gov.uk/research/rrpdf/rr619.pdf

579 Sullivan, R. (2020, March 4). Don't wear face masks in response to coronavirus, says Chief Medical Officer. The Independent. https://www.independent.co.uk/news/uk/home-news/coronavirus-uk-news-professor-chris-whitty-no-masks-advice-a9374086.html

580 UK Prime Minister. (2020, March 11). PM DCMO Q&A On Coronavirus. Dr Jenny Harries, Deputy Chief Medical Officer, came into Downing Street to answer some of the most commonly asked questions on Coronavirus. Facebook. https://www.facebook.com/watch/?v=486465015562901

581 Reuters Staff. (2020, October 8). Fact check: Outdated video of Fauci saying "there's no reason to be walking around with a mask". Reuters.com. https://www.reuters.com/article/uk-factcheck-fauci-outdated-video-masksidUSKBN26T2TR

582 Jefferson, T. et al.(2020). Physical interventions to interrupt or reduce the spread of respiratory viruses. The Cochrane Database of Systematic Reviews, 11(11), CD006207. 461 https://doi.org/10.1002/14651858.CD006207.pub5

583 Cochrane. (2020, November 20). Do physical measures such as hand-washing or wearing masks stop or slow down the spread of respiratory viruses? Cochrane.org . https://www.cochrane.org/CD006207/ARI_do-physical-measures-such-handwashing-or-wearing-masks-stop-or-slow-down-spread-respiratory-viruses

584 DELVE group. (2020, May 4). DELVE Group publishes evidence paper on the use of face masks in tackling coronavirus (covid-19) pandemic. Royal Society. https://royalsociety.org/news/2020/05/delve-group-publishes-evidence-paper-on-use-of-face-masks

585 Howard, J. (2021, January 11). An evidence review of face masks against COVID19. Perspective: Biological Sciences 118(4): e2014564118. https://doi.org/10.1073/pnas.2014564118

586 Gregory, A. (2020, May 4). Scientists divided over report recommending widespread use of face masks. The Independent. https://www.independent.co.uk/news/health/coronavirus-face-masks-effective-work-study-royal-society-research-uk-a9498796.html

587 Stoneman, J. (2021, July 17). Cloth face masks are "comfort blankets" that do little to curb Covid spread, scientist warns. The Telegraph. https://www.telegraph.co.uk/news/2021/07/17/cloth-face-masks-comfort-blankets-do-little-curb-covid-spread

588 Kelland, K. (2020, June 5). Wear masks in public says WHO, in update of COVID-19 advice. Reuters.com. https://www.reuters.com/article/us-healthcoronavirus-who-masks-idUSKBN23C27Y

589 Miller, I. (2022). Unmasked: The global failure of COVID mask mandates. Post Hill Press.

590 Johnson, G. R. et al. (2011). Modality of human expired aerosol size distributions. Journal of Aerosol Science , 42(12), 839–851. https://doi.org/10.1016/j.jaerosci.2011.07.009

591 Fennelly K. P. (2020). Particle sizes of infectious aerosols: Implications for infection control. The Lancet. Respiratory Medicine, 8(9), 914–924. https://doi.org/10.1016/S2213-2600(20)30323-4

592 Sznitman, J. (2013). Respiratory microflows in the pulmonary acinus. Journal of Biomechanics, 46(2), 284–298. https://doi.org/10.1016/j.jbiomech.2012.10.028

593 Witten, J., Samad, T. and Ribbeck, K. (2018). Selective permeability of mucus barriers. Current Opinion in Biotechnology, 52, 124–133. https://doi.org/10.1016/j.copbio.2018.03.010

594 Wallace, L. E. et al. (2021). Respiratory mucus as a virus-host range determinant. Trends in Microbiology, 29(11), 983–992. https://doi.org/10.1016/j.tim.2021.03.014

595 Sosnowski, T. S. (2021, August). Inhaled aerosols: Their role in COVID-19 transmission, including biophysical interactions in the lungs. Current Opinion in Colloid and Interface Science, 54, 101451. https://doi.org/10.1016/j.cocis.2021.101451

596 Ribbeck K. (2009). Do viruses use vectors to penetrate mucus barriers? Bioscience hypotheses, 2(6), 329–362. https://doi.org/10.1016/j.bihy.2009.07.004

597 Wilkes, A. R. et al. (2001). The bacterial and viral filtration performance of breathing system filters. Anaesthesia, 55(5). https://doi.org/10.1046/j.1365-2044.2000.01327.x

598 Health Advisory and Recovery Team. (2021, July 15). Confusion among mask advocates. https://www.hartgroup.org/confusion-among-mask-advocates

599 NYC Health. (2021, February 18). Mask to the max – Health Department updates mask guidance. NYC.gov. https://web.archive.org/web/20210301230935/https://www1.nyc.gov/site/doh/about/press/pr2021/mask-to-the-max-guidance.page

600 Greenhalgh, T. (2021, January 18). Exactly what I do … @trishgreenhalghh, Twitter. https://twitter.com/trishgreenhalgh/status/1351174932231819264

601 Greenhalgh, T. (2021, July 11). Two tricks to improve the fit of a medical mask. @trishgreenhalghh, Twitter. https://twitter.com/trishgreenhalgh/status/1414300431698780165

602 Greenhalgh, T. (2021, December 24). Canada officially junks the "aerosol generating medical procedure…". @trishgreenhalghh, Twitter. https://twitter.com/trishgreenhalgh/status/1474291559004786688

603 Greenhalgh, T. (2022, January 8). Narrator: we could a) wear better masks (FFP2/N95 minimum)… @trishgreenhalghh, Twitter. https://twitter.com/trishgreenhalgh/status/1479722255462670337

604 Greenhalgh, T. (2021, December 25). Dear @elonmusk …. @trishgreenhalghh, Twitter. https://twitter.com/trishgreenhalgh/status/1474638741830615046

605 The Post Millennial. (2021, December 28). Cloth masks are "not appropriate for this pandemic" says CNN medical expert Leana Wen. @TPostMillenial, Twitter. https://twitter.com/TPostMillennial/status/1475835255454150660

606 Egan, J. and Przybyla, H. (2022, January 19). White House says it will distribute 400 million free N95 masks to protect against Omicron. NBC News. https://www.nbcnews.com/politics/white-house/white-house-says-it-will-distribute-400-million-free-n95-n1287672

607 Rengasamy S, Miller A, Eimer BC, Shaffer RE. Filtration Performance of FDACleared Surgical Masks. J Int Soc Respir Prot. 2009 Spring-Summer;26(3):54-70. https://www.ncbi.nlm.nih.gov/pmc/articles/PMC7357397/

608 American Chemical Society. "A replaceable, more efficient filter for N95 masks." ScienceDaily. ScienceDaily, 21 May 2020. www.sciencedaily.com/releases/2020/05/200521124646.htm

609 Miller, I. (2022b, January 31). As many experts move the goalposts from cloth masks to N95's, it's important to point out that the states in Germany with N95 mandates are actually doing WORSE .... @ianmSC,Twitter. https://twitter.com/ianmSC/status/1488214517372755974

610 Loeb, M. et al. (2022). Medical masks versus N95 respirators for preventing 463 COVID-19 among health care workers: A randomized trial. Annals of Internal Medicine, 175(12), 1629–1638. https://doi.org/10.7326/M22-1966

611 Loh, M. M. et al. (2023) Measurement of SARS-CoV-2 in air and on surfaces in Scottish hospitals, Journal of Hospital Infection (133, 1-7) https://doi.org/10.1016/j.jhin.2022.11.019

612 Dawood A. A. (2021). Transmission of SARS CoV-2 virus through the ocular mucosa worth taking precautions Vacunas (English Edition), 22(1), 56–57. https://doi.org/10.1016/j.vacune.2021.01.007

613 Bundgaard, H. et al. (2021, March). Effectiveness of adding a mask recommendation to other public health measures to prevent SARS-CoV-2 infection in Danish mask wearers: A randomized controlled trial. Annals of Internal Medicine. 174(3), 335–343. https://doi.org/10.7326/M20-6817

614 Silver, A. and Cyranoski, D. (2020, April 15). China is tightening its grip on Coronavirus research. Nature, News. https://www.nature.com/articles/d41586-020-01108-y

615 Quoted in ibid.

616 Eikenberry, S. E. (2020). To mask or not to mask: Modeling the potential for face mask use by the general public to curtail the COVID-19 pandemic. Infectious Disease Modelling, 5, 293–308. https://doi.org/10.1016/j.idm.2020.04.001

617 Adjodah, D. et al. (2020, October 23). Decrease in hospitalizations for COVID-19 after mask mandates in 1083 U.S. Counties. MedRxiv (Preprint). https://doi.org/10.1101/2020.10.21.20208728

618 Adjodah, D. et al. (2020, November 4). Decrease in hospitalizations for COVID19 after mask mandates in 1083 U.S. Counties. MedRxiv (Withdrawn). https://www.medrxiv.org/content/10.1101/2020.10.21.20208728v2

619 Guerra, D. D. and Guerra, D. J. (2021). Mask mandate and use efficacy in statelevel COVID-19 containment. MedXriv. https://doi.org/10.1101/2021.05.18.21257385

620 Drummond, H. (2022). The Face Mask Cult. CantusHead Books.

621 Milton, D. K. et al. (2013). Influenza virus aerosols in human exhaled breath: Particle size, culturability, and effect of surgical masks. PLOS Pathogens, 9(3), e1003205. https://doi.org/10.1371/journal.ppat.1003205

622 Scientific Advisory Group for Emergencies. (2022, May 13). SPI-B, SPI-M and EMG: Considerations for potential impact of plan b measures, 13 October 2021. Gov.uk. https://www.gov.uk/government/publications/spi-b-spi-m-and-emg-considerations-for-potential-impact-of-plan-b-measures-13-october-2021

623 New and Emerging Respiratory Virus Threats Advisory Group (NERVTAG) and the Environmental and Modelling group (EMG). (2021, October 13 updated November 26). EMG-Nervtag update on transmission and environmental and behavioural mitigation strategies, including in the context of Delta. Gov.uk. Retrieved January 31, 2023, from

https://assets.publishing.service.gov.uk/government/uploads/system/uploads/attachment_data/file/1036475/S1395_EMG-Nervtag_Update_on_Transmission_and_Environmental_and_Behavioural_Mitigation_Strategies__including_in_the_context_of_Delta.pdf

624 Mills, M., Rahal, C. and Akimova, E. prepared for The Royal Society SET-C Group. (2020, June 26). Face masks and coverings for the general public: Behavioural knowledge, effectiveness of cloth coverings and public messaging. https://royalsociety.org/-/media/policy/projects/set-c/set-c-facemasks.pdf

625 DELVE group. (2020, May 4). DELVE Group publishes evidence paper on the use of face masks in tackling coronavirus (covid-19) pandemic. Royal Society. https://royalsociety.org/news/2020/05/delve-group-publishes-evidence-paper-on-use-of-face-masks

626 Quoted in Davis, N. (2020, May 4). Report on face masks' effectiveness for covid19 divides scientists. The Guardian. https://www.theguardian.com/world/2020/may/04/scientists-disagree-over-face-masks-effect-on-covid-19

627 Hendrix M.J., et al. (2020, July 17) Absence of apparent transmission of SARSCoV-2 from two stylists after exposure at a hair salon with a universal face covering policy – Springfield, Missouri, May 2020. Morbidity and Mortality Weekly Report 2020, 69: 930-932. http://dx.doi.org/10.15585/mmwr.mm6928e2

628 Frankel, T. C. (2020. June 17). The outbreak that didn't happen: Masks credited with preventing coronavirus spread inside Missouri hair salon. The Washington Post. https://www.washingtonpost.com/business/2020/06/17/masks-salons-missouri

629 Spocchia, G. (2020, May 4). 900 Missouri residents who "snitched" on lockdown rule-breakers fear retaliation after details leaked online. The Independent. https://www.independent.co.uk/news/world/americas/us-politics/coronavirusmissouri-lockdown-breakers-details-online-st-louis-county-a9497271.html

630 Anderson, J. H. (2021, August 11). Do masks work? City Journal. https://www.city-journal.org/do-masks-work-a-review-of-the-evidence

631 Springfield Greene County Health (n.d.). https://www.springfieldmo.gov/5147/Recovery-Dashboard

632 Department of Education. (2022). Evidence summary: Coronavirus (COVID-19) and the use of face coverings in education settings. Gov.uk. https://assets.publishing.service.gov.uk/government/uploads/system/uploads/a ttachment_data/file/1055639/Evidence_summary_-_face_coverings.pdf

633 10 Downing Street. (2020, August 28). PM Boris Johnson and Deputy Chief Medical Officer Dr Jenny Harries host a back to school Q&A. YouTube. https://www.youtube.com/watch?v=ymTwS4XhQMQ

634 Sunday Morning. (2022, July 3). 'Sunday Morning' - 03/07/2022. BBC iPlayer. https://www.bbc.co.uk/iplayer/episode/m00190q6/sunday-morning-03072022

635 Jefferson, T. et al. (2023). Physical interventions to interrupt or reduce the spread of respiratory viruses. Cochrane Database of Systematic Reviews. Cochrane.library.com. https://doi.org/10.1002/14651858.CD006207.pub6

636 Heneghan, C., & Jefferson, T. (2023, February 3). Do Mask Mandates Work? A Look at the Evidence. The Spectator. https://www.spectator.co.uk/article/do-mask-mandates-work

637 Statement on 'Physical interventions to interrupt or reduce the spread of respiratory viruses' review https://www.cochrane.org/news/statement-physical-interventions-interrupt-or-reduce-spread-respiratory-viruses-review

638 Desami, Maryanne (2023, March 15) Did Cochrane sacrifice its researchers to appease critics? https://maryannedemasi.substack.com/p/breaking-did-cochrane-sacrifice-its

639 Marchiori, M. (2020). COVID-19 and the social distancing paradox: Dangers and solutions. Arxiv: Populations and Evolution. https://arxiv.org/pdf/2005.12446.pdf

640 Quote from Devlin, K. (2020, August 26). Boris Johnson calls face masks In classrooms 'nonsensical' hours after latest u-turn. The Independent. https://www.independent.co.uk/news/uk/politics/boris-johnson-face-masks-school-classroom-children-covid-19-a9689741.html

641 Hancock, M.(2022). Pandemic diaries: The inside story of Britain's battle against Covid. Biteback.

642 Glover, J. (2022, December 4). Hancock diaries show Nicola Sturgeon WAS leaking Cobra decisions. Scottish Daily Express. https://www.scottishdailyexpress.co.uk/news/uk-news/hancock-diaries-reveal-nicola-sturgeon-28650902

643 UK Parliament. (2020, July 14). Matt Hancock: "face coverings increase confidence in people to shop." Parliament.uk. https://www.parliament.uk/business/news/2020/july/coronavirus-statement-14-july

644 Dodsworth, L. and Greenhalgh, T. (2021, July 5). Should we ditch face masks after July 19? Two opponents go head to head. The Sun. https://www.thesun.co.uk/news/15501769/ditch-face-masks-july-19-debate/

645 Dodsworth, L. (2021). A state of fear: How the UK government weaponised fear during the Covid-19 pandemic, p. 117. Pinter and Martin.

646 Sidley, G. (2021, June 25). Continued mask wearing won't help us return to normal. The Critic Magazine. https://thecritic.co.uk/continued-mask-wearing-wont-help-us-return-to-normal

647 Smith-Schoenwalder, C. (2020, September 16). CDC's Redfield suggests masks may offer better coronavirus protection than a vaccine. USA News. https://www.usnews.com/news/health-news/articles/2020-09-16/cdcs-redfield-suggests-masks-may-offer-better-coronavirus-protection-than-a-vaccine

648 A politicians logic. (n.d.). YouTube. Retrieved January 17, 2023, from https://www.youtube.com/watch?v=vidzkYnaf6Y

649 Greenhalgh, Trish. (2021, January 20). Joggers and cyclists should wear masks if they can't maintain a physical distance from pedestrians. The Conversation. https://theconversation.com/joggers-and-cyclists-should-wear-masks-if-they-cant-maintain-a-physical-distance-from-pedestrians-153110

650 Greenhalgh, T. et al. (2020, April 9). Face masks for the public during the covid19 crisis. British Medical Journal, 368, m1435. https://doi.org/10.1136/bmj.m1435

651 Bundgaard, H. et al. (2021, March). Effectiveness of adding a mask recommendation to other public health measures to prevent SARS-CoV-2 infection in Danish mask wearers: A randomized controlled trial. Annals of Internal Medicine. 174(3), 335–343. https://doi.org/10.7326/M20-6817

652 Wikipedia. (2023, January 18). Robert Conquest. Wikipedia. https://en.wikipedia.org/wiki/Robert_Conquest

653 Stajduhar, A. et al. (2022, February 7). Face masks disrupt holistic processing and face perception in school-age children. Cognitive Research: Principles and Implications 7, 9 (2022). https://doi.org/10.1186/s41235-022-00360-2

654 Lane, J. et al. (2018) Impacts of impaired face perception on social interactions and quality of life in age-related macular degeneration: A qualitative study and new community resources. PLOS ONE 13(12): e0209218. https://doi.org/10.1371/journal.pone.0209218

655 Stajduhar, A. et al. (2022, February 7). Face masks disrupt holistic processing and face perception in school-age children. Cognitive Research: Principles and Implications 7, 9 (2022). https://doi.org/10.1186/s41235-022-00360-2

656 Deoni, S. C. L. (2021). Impact of the COVID-19 pandemic on early child cognitive development: Initial findings in a longitudinal observational study of child health. MedRxiv (preprint) https://doi.org/10.1101/2021.08.10.21261846

657 Quoted in, Dyer, O. (2021, August 16). Covid-19: Children born during the pandemic score lower on cognitive tests, study finds. The British Medical Journal (News), 374: n2031. https://doi.org/10.1136/bmj.n2031

658 Murray, J. (2020, October 19). Charities fear impact of masks on deaf children in UK schools. The Guardian. https://www.theguardian.com/world/2020/oct/19/charities-fear-impact-of-masks-on-deaf-children-in-uk-schools

659 Royal National Institute for Deaf People. (n.d.). Facts and figures. Rnid.Org. Retrieved January 17, 2023, from https://rnid.org.uk/about-us/research-and-policy/facts-and-figures

660 Gutentag, A. (2021, November 22). What they did to the kids. Tablet Magazine. https://www.tabletmag.com/sections/news/articles/school-closures-covid-alex-gutentag

661 @Neucky (2022, February 12). After A Parent-teacher Conference Over @Zoom, My First Grade Daughter Asked Me, Twitter. https://twitter.com/Neucky/status/1492366114608848896

662 #Masks4All. (n.d.). Cloth masks can help stop the spread of COVID-19, save lives and restore jobs. Masks4all.Co. Retrieved January 17, 2023, from https://masks4all.co/masks4all-why-cloth-masks-must-be-required-in-public-to-help-contain-covid-19

663 de Quetteville, H. (2021, November 30). Why face masks became the symbol of a divided Britain. The Telegraph. https://www.telegraph.co.uk/news/2021/11/30/face-masks-became-symbol-divided-britain/

664 UNICEF-UK. (n.d.). Children in lockdown: What coronavirus means for UK children. https://downloads.unicef.org.uk/wp-content/uploads/2020/04/Unicef-UK-Children-In-Lockdown-Coronavirus-Impacts-Snapshot.pdf

665 Barrett, B. (2021, January 30). Politician on @BBCRadio4 saying children are resilient … @BarbaraMBarrett, Twitter. https://twitter.com/BarbaraMBarrett/status/1355555698990592000

666 Demkowicz, O. (2020, May 21). Whose responsibility is it anyway? Resilience in children and young people's mental health. Policy@Manchester Blogs, Manchester University. https://blog.policy.manchester.ac.uk/posts/2020/05/whose-responsibility-is-it-anyway-resilience-in-child-and-young-person-mental-health

667 UsForThem (n.d.) https://usforthem.co.uk

668 Cole, L. and Kingsley, M. (2022). The children's inquiry: How the state and society failed the young during the Covid-19 pandemic. Pinter and Martin.

669 1,500 members of the Royal College of Paediatrics and Child Health. (2020, June 17). Open letter from UK paediatricians about the return of children to schools. Royal College of Paediatrics and Child Health. https://www.rcpch.ac.uk/sites/default/files/2020-06/open_letter_re_schools_reopening_2020-06-17.pdf

670 Uncommon Knowledge with Peter Robinson. (2021, October 21). What happened: Dr. Jay Bhattacharya on 19 Months of COVID. Hoover Institution. https://www.hoover.org/research/what-happened-dr-jay-bhattacharya-19-months-covid-1

671 Quoted in Education Endowment Foundation. (2021). New study finds "significantly lower achievement", with a "large and concerning gap" for disadvantaged pupils. https://educationendowmentfoundation.org.uk

672 Interview in Cole, L. and Kingsley, M. (2022). The children's inquiry: How the state and society failed the young during the Covid-19 pandemic. Pinter and Martin.

673 Children's Commissioner for England. (2020, April). We're all in this together? Local area profiles of child vulnerability. Https://www.childrenscommissioner.gov.uk. Retrieved January 18, 2023, from https://www.childrenscommissioner.gov.uk/wp-content/uploads/2020/04/cco-were-all-in-this-together.pdf

674 Ibid

675 Children's Commissioner for England. (2020, April 30). Statement by the Children's Commissioner for England. https://www.childrenscommissioner.gov.uk/2020/04/30/statement-on-changes-to-regulations-affecting-childrens-social-care

676 Kingstone, T. and Dikomitis, L. (2021, January 1). The pandemic transformed how social work was delivered – and these changes could be here to stay. The Conversation. https://theconversation.com/the-pandemic-transformed-how-social-work-was-delivered-and-these-changes-could-be-here-to-stay-165993

677 Brewster, B. (2021, February). Policing county lines: Impact of Covid-19. University of Nottingham Rights Lab. https://www.nottingham.ac.uk/research/beacons-of-excellence/rights-lab/resources/reports-and-briefings/2021/february/briefing-policing-county-lines-during-covid-19.pdf

678 National Youth Agency. (2021, March). Gangs and exploitation: County lines. NYA. https://www.nya.org.uk/wp-content/uploads/Between-the-lines-amended-2022.pdf

679 Question Everything. (2022, February 15). Do lockdown measures equate to child abuse? Dr. Zenobia Storah. YouTube. https://www.youtube.com/watch?v=b01xpSAB7lw

680 Government's education catch-up and mental health recovery programmes. (2022, February 3) Hansard 708,

https://hansard.parliament.uk/Commons/2022-02-03/debates/405F870D-812D-4A03-A38C-BDD2BC6D0E3F/Government'SEducationCatch-UpAndMentalHealthRecoveryProgrammes?highlight=separation%20anxiety#contribution-F104B0CB-3A0A-4AF2-8733-E85D989E113C

681 Terry, K. (2022, April 4). Covid lockdowns delayed childrens development, Ofsted report warns. Mail Online. https://www.dailymail.co.uk/news/article-10683471/Covid-lockdowns-delayed-childrens-readiness-primary-school-new-Ofsted-report-warns.html

682 Townsend, E. (2020, June). Child and adolescent mental health in a postlockdown world: a ticking time bomb? University of Nottingham. https://www.nottingham.ac.uk/vision/vision-child-adolescent-mental-health-post-lockdown

683 Cole, L. and Kingsley, M. (2022). The children's inquiry: How the state and society failed the young during the Covid-19 pandemic. Pinter and Martin.

684 UNICEF. (2021, December 9). COVID-19 'biggest global crisis for children in our 75-year history'– UNICEF. UNICEF-Chad. https://www.unicef.org/chad/press-releases/covid-19-biggest-global-crisis-children-our-75-year-history-unicef

685 Committee on the Rights of the Children, United Nations. (2013, May 29). Convention on the Rights of the Child. General comment No. 14 (2013) on the right of the child to have his or her best interests taken as a primary consideration (art. 3, para. 1)*. https://www2.ohchr.org/english/bodies/crc/docs/gc/crc_c_gc_14_eng.pdf

686 Blume, H. (2021, November 17). LAUSD to end weekly COVID tests and spend $5 million on prizes to encourage vaccinations. Los Angeles Times. https://www.latimes.com/california/story/2021-11-17/lausd-to-spend-5-million-for-gift-cards-to-incentivize-students-to-get-covid-19-vaccine

687 Turner, C. (2021, May 2). Schools back mass vaccinations for children as headteachers say 'peer pressure' will boost take up. The Telegraph. https://www.telegraph.co.uk/news/2021/05/02/schools-back-mass-vaccinations-children-headteachers-say-peer

688 The HighWire. (n.d.). Win: ICAN-funded lawsuit halts law allowing 11 yr. olds to get vaccinated without parental consent. The HighWire. Retrieved January 19, 2023, from https://thehighwire.com/videos/win-ican-funded-lawsuit-halts-law-allowing-11-yr-olds to-get-vaccinated without-parental-consent

689 Mencken, H. L. (2004) [1922]. In Defence of Women, p. 29. Dover Publications.

690 Dolan, P. et al. (2010) MINDSPACE: Influencing behaviour through public policy. Cabinet Office and Institute for Government. https://www.instituteforgovernment.org.uk/sites/default/files/publications/MINDSPACE.pdf

691 Government Office for Science. (2020, May 4). List of participants of SAGE and related sub-groups. gov.uk. http://web.archive.org/web/20200504115545/https:/www.gov.uk/government/publications/scientific-advisory-group-for-emergencies-sage-coronavirus-covid-19-response-membership/list-of-participants-of-sage-and-related-sub-groups

692 Kingston University. (2022, February 25). Psychological distress higher in UK than other countries during pandemic, finds new survey involving Kingston University academic. Kingston University, London. https://www.kingston.ac.uk/news/article/2631/25-feb-2022-psychological-distress-higher-in-uk-than-other-countries-during-pandemic-finds-new-survey-involving-kingston

693 Sidley, G. (2021, January 7). The ethics of using covert strategies - A letter to the British Psychological Society (II). Coronababble. https://www.coronababble.com/post/the-ethics-of-using-covert-strategies-a-letter-to-the-british-psychological-society-ii

694 Sidley, G. (2021, September 3). Evasive, disingenuous and wholly unconvincing - the BPS response to ethical concerns about 'nudging.' Coronababble. https://www.coronababble.com/post/evasive-disingenuous-and-wholly-unconvincing-the-bps-response-to-ethical-concerns-about-nudging

695 Dodsworth, L. (2021). A state of fear: How the UK government weaponised fear during the Covid-19 pandemic, p. 94. Pinter and Martin.

696 Dodsworth, L. (2022, March 25). We need to talk about behavioural science – But the powers that be don't want to. Substack. https://lauradodsworth.substack.com/p/we-need-to-talk-about-behavioural

697 UK COVID-19 inquiry: Draft terms of reference. (2022, March 10). Gov.uk. https://www.gov.uk/government/publications/uk-covid-19-inquiry-draft-terms-of-reference

698 Mercer, D. (2020, August 8). Coronavirus: Young people Warned 'don't Kill Granny' as lockdown measures reimposed in Preston. Sky News. https://news.sky.com/story/coronavirus-young-people-warned-dont-kill-granny-aslockdown-imposed-in-preston-12045017

699 BBC News. (2021, January 9). Covid-19: Act like you've got the virus, government urges. BBC.co.uk. https://www.bbc.co.uk/news/uk-55598918

700 Greenhalgh, Trisha, Moral uncertainty: A case study of Covid-19. Patient Education and Counseling. Volume 104, Issue 11, 2021, Pages 2643-2647, ISSN 0738-3991, https://doi.org/10.1016/j.pec.2021.07.022.

701 Asch, S. E. (1956). Studies of independence and conformity: I. A minority of one against a unanimous majority. Psychological Monographs: General and Applied, 70(9), 1-70. https://doi.org/10.1037/h0093718

702 Bennett, R. T. (2021) The light in the heart: Inspirational thoughts for living your best life, p. 106. Roy Bennett.

703 Dollimore, L. (2022, January 23). Compulsory Covid Jabs for NHS staff could be delayed by six months. Mail Online.

https://www.dailymail.co.uk/news/article-10431489/Compulsory-Covid-jabs-NHS-staff-delayed-half-year-Tory-revolt-protests.html

704 Kübler-Ross, E. and Kessler, D. (2014). Life lessons: How our mortality can teach us about life and living. Simon and Schuster.

705 de Foucauld, C. E. (1964). Spiritual Autobiography of Charles of Foucauld (J.-F. Six, Ed.; p. 70). Dimension Books.

706 King James Bible. (n.d.). Psalm 23:4 - The Lord is my shepherd. Bible Hub. Retrieved January 19, 2023, from https://biblehub.com/psalms/23-4.htm

707 ITV News (2021, March 6). Covid: Dalai Lama gets coronavirus vaccine and urges others to come forward too. https://www.itv.com/news/2021-03-06/covid-dalai-lama-gets-coronavirus-vaccine-and-urges-others-to-come-forward-too

708 Newman, J. and Wright, J. (2021, December 22). Jesus would get a vaccine, Archbishop of Canterbury suggests as he says getting the jab is 'not about me and my rights to choose – it's about how I love my neighbour.' Mail Online. https://www.dailymail.co.uk/news/article-10334569/Archbishop-Canterbury-says-unvaccinated-immoral-love-neighbour.html

709 Vatican News. (2021, August 18). Pope Francis urges people to get vaccinated against Covid-19. (2021, August 18). https://www.vaticannews.va/en/pope/news/2021-08/pope-francis-appeal-covid-19-vaccines-act-of-love.html

710 Lee, S. (2021, June 29). Loneliness is strongly linked to depression among older adults, a long-term study suggests. National Institute for Health and Social Care Research. https://evidence.nihr.ac.uk/alert/loneliness-strongly-linked-depression-older-adults

711 United Nations: Human Rights. Office of the High Commissioner. (2020, February 28). United States: Prolonged solitary confinement amounts to psychological torture, says UN expert. https://Www.Ohchr.Org. https://www.ohchr.org/en/press-releases/2020/02/united-states-prolonged-solitary-confinement-amounts-psychological-torture

712 Quoted in Gunia, A. (2020, April 28). Why New Zealand's coronavirus elimination strategy is unlikely to work in most other places. Time. https://time.com/5824042/new-zealand-coronavirus-elimination

713 Covid Action. (n.d.). Frequently asked questions. Covidaction.Uk. Retrieved January 19, 2023, from http://zerocovid.uk/frequently-asked-questions

714 Indie_SAGE. (2020, July 14). Independent SAGE - 14.07.20. YouTube. https://www.youtube.com/watch?v=-QRhwcVMTZI&t=3676s

715 Covid Action. (n.d.). Frequently asked questions. Covidaction.Uk. Retrieved January 19, 2023, from http://zerocovid.uk/frequently-asked-questions

716 Question Everything. (2022, February 1). Prof D. Livermore made a clear statement: we're not going to eradicate covid. @QETalks, Twitter. https://twitter.com/QETalks/status/1488544924089864195

717 Henderson, D. A. and Klepac, P. (2013). Lessons from the eradication of smallpox: An interview with D. A. Henderson. Philosophical transactions of the Royal Society of London. Series B, Biological sciences, 368(1623), 20130113. https://doi.org/10.1098/rstb.2013.0113

718 Martin, D. (2021, February 8). We can be zero! time for a strategic change to combat virus variants? Laboratory News. https://www.labnews.co.uk/article/2031151/we-can-be-zero-time-for-a-strategic-change-to-combat-virus-variants

719 Royal College of Psychiatrists. (2020, May 15). Psychiatrists see alarming rise in patients needing urgent and emergency care and forecast a tsunami of mental illness. www.Rcpsych.ac.uk. https://www.rcpsych.ac.uk/news-and-features/latest-news/detail/2020/05/15/psychiatrists-see-alarming-rise-in-patients-needing-urgent-and-emergency-care

720 Alexanderson, K. et al. (2020, October 14). John Snow memorandum. Johnsnowmemo.com https://www.johnsnowmemo.com/john-snow-memo.html

721 Centre for Policy on Ageing. (2021). How can the care experience of older people in care homes be improved: Findings from five PANICOA studies. Cpa.org.uk. http://www.cpa.org.uk/information/reviews/How-can-the-care-experience-of-older-people-in-care-homes-be-improved.pdf

722 University of Exeter. (2018, February 5). Dementia care improved by just one hour of social interaction each week. https://news-archive.exeter.ac.uk/featurednews/title_637931_en.html

723 The Straits Times. (2022, April 1). China censors Shanghai protest videos as Covid-19 lockdown anger grows. Straitstimes.com. https://www.straitstimes.com/asia/east-asia/china-censors-shanghai-protest-videos-as-covid-19-lockdown-anger-grows

724 Hancock, T. (2022, March 29). China lockdowns cost at least $46 billion a month, academic says. Bloomberg.Com. https://www.bloomberg.com/news/articles/2022-03-29/china-lockdowns-cost-at-least-46-billion-a-month-academic-says

725 The Straits Times. (2022, April 1). China censors Shanghai protest videos as Covid-19 lockdown anger grows. Straitstimes.com. https://www.straitstimes.com/asia/east-asia/china-censors-shanghai-protest-videos-as-covid-19-lockdown-anger-grows

726 Beall Jr., O. and Shyrock, R. (1953). Cotton Mather: First significant figure in American medicine. Proceedings of the American Antiquarian Society, 63(1), 37–274 (172). https://www.americanantiquarian.org/proceedings/44817435.pdf

727 Ministry of Housing, Communities and Local Government. (2021, June 16). Guidance COVID-19: guidance for the safe use of places of worship. Gov.uk. https://www.gov.uk/government/publications/covid-19-guidance-for-the-safe-use-of-places-of-worship-during-the-pandemic-from-4-july/covid-19-guidance-for-the-safe-use-of-places-of-worship-from-2-december

728 Evans, M., Hymas, C. and Stephens, M. (2020, March 30). Police powers: Shopping bags searched and ban on fishing as scope of new laws emerges. The Telegraph. https://www.telegraph.co.uk/news/2020/03/30/police-powers-shopping-bags-searched-ban-fishing-scope-new-laws/

729 BBC News. (2021, April 23). Peterlee woman fined £500 for Covid-breach balloon launch. BBC.co.uk. https://www.bbc.co.uk/news/uk-england-tyne-56863995

730 Ridler, F. (2021, January 28). Two men are each handed £10,000 fines for organising snowball fight. Mail Online. https://www.dailymail.co.uk/news/article-9197711/Two-men-aged-20-23-handed-10-000-fines-organising-mass-SNOWBALL-FIGHT.html

731 The Nottingham Tab. (March 2021). Three Notts students given criminal record after failing to pay £40,000 Covid fine. https://thetab.com/uk/nottingham/2021/03/25/three-notts-students-given-criminal-record-after-failing-to-pay-covid-fine-53906

732 The Spectator. (2022, May 13). 'One elderly man got fined for speaking to people at his allotment' @spectator, Twitter. https://twitter.com/spectator/status/1525037612871913476

733 Parker, H. (2022, November 17). Man, 72, jailed after serving mince pies in covid has heart attack 5 days in. Mirror.co.uk. https://www.mirror.co.uk/news/uk-news/gravely-ill-man-72-jailed-28512895

734 Bowcott, O. and Agency. (2020, May 12). Judge questions coronavirus case against 'homeless' London man. The Guardian. https://www.theguardian.com/uk-news/2020/may/12/judge-questions-coronavirus-case-against-homeless-london-man

735 McFadden, B. (2022, April 14). Pub landlord fined £4,000 for lockdown rule breach says it is 'ridiculous' PM was fined £50. Inews.co.uk. https://inews.co.uk/news/pub-landlord-fined-breaking-lockdown-rules-borisjohnson-pay-1574185

736 Greenhalgh, Trish (2021, 5 December) 'I think the key behavioural change …' @trishgreenhalgh, Twitter (subsequently deleted).

737 Cited in Shapiro, F. R. (2006). The Yale book of quotations, p. 498 Yale University Press.

738 Lord Sumption. (2020, October 27). 'Government by decree - Covid-19 and the constitution'. Cambridge Law Faculty. YouTube. https://www.youtube.com/watch?v=amDv2gk8aa0

739 Hoar, F. (2020, April 21 updated). A disproportionate interference with right and freedoms the coronavirus regulations and the European Convention on Human Rights. Fieldcourt.co.uk. https://fieldcourt.co.uk/wp-content/uploads/Francis-Hoar-Coronavirus-article-on-ECHR-compatibility-20.4.2020-2.pdf

740 Robert Craig. (2020, April 6). Lockdown: A response to Professor King – Robert Craig. UK Human Rights Blog. https://ukhumanrightsblog.com/2020/04/06/lockdown-a-response-to-professor-king-robert-craig

741 Hoar, F. (2020, April 21 updated). A disproportionate interference with right and freedoms the coronavirus regulations and the European Convention on Human Rights. Fieldcourt.co.uk. https://fieldcourt.co.uk/wp-content/uploads/Francis-Hoar-Coronavirus-article-on-ECHR-compatibility-20.4.2020-2.pdf

742 McMillan, G. (2022, November 25). Covid lockdown was 'COMPLETE FAILURE of government' and 'radical experiment' says Lord Sumption. GBnews.uk. https://www.gbnews.uk/health/covid-lockdown-was-complete-failure-of-government-and-radical-experiment-says-lord-sumption/396763

743 Quoted in Gyngell, K. (2022, June 19). PANDA's Nick Hudson on covid and the death of logic – part 3. The Conservative Woman. https://www.conservativewoman.co.uk/pandas-nick-hudson-on-covid-and-the-death-of-logic-part-3

744 Misesmedia. (2022, July 25). An evening with Clifton Duncan. YouTube. https://www.youtube.com/watch?v=hCH_8KtLeVQ

745 BBC News. (2021, December 1). Howard Springs: Australia police arrest quarantine escapees. BBC.co.uk. https://www.bbc.co.uk/news/world-australia-59486285

746 Wikipedia (n.d.). Edmund Burke: False quotations. https://en.wikipedia.org/wiki/Edmund_Burke

747 Medley, G. (2021, January 1). Replying to. @FraserNelson, @doug_no1 and 2 others. Twitter. https://twitter.com/GrahamMedley/status/1472243230213394434

748 Quoted in Yorke, H. (2022, January 13). Downing Street's controversial 'Nudge Unit' accused of exploiting scare tactics during Covid crisis. The Telegraph. https://www.telegraph.co.uk/politics/2022/01/13/downing-streets-controversial-nudge-unit-accused-exploiting

749 Pandemic Flu Response Review Team Cabinet Office. (2010). The 2009 influenza pandemic: An independent review of the UK response to the 2009 influenza pandemic. Pandemic Flu Response Review Team. https://assets.publishing.service.gov.uk/government/uploads/system/uploads/attachment_data/file/61252/the2009influenzapandemic-review.pdf

750 Health Advisory and Recovery Team. (2022, July 20). The SARS-CoV-2 and influenza see-saw. https://www.hartgroup.org/the-sars-cov-2-and-influenza-see-saw

751 Spofforth, Bernie. (2022, November 28). UK - The media want you to forget their disgusting coverage of 'freedom fighters' during UK lockdown protests. Twitter. https://twitter.com/BernieSpofforth/status/1597172271604715522

752 Our World in Data (2023). https://ourworldindata.org/coronavirus

INDEX

Printed in Great Britain
by Amazon

38627849R00283